Particles on Surfaces 1

Detection, Adhesion, and Removal

Particles on Surfaces 1

Detection, Adhesion, and Removal

Edited by
K. L. Mittal
IBM US Technical Education
Thornwood, New York

PLENUM PRESS • NEW YORK AND LONDON

Library of Congress Cataloging in Publication Data

Particles on surfaces 1: detection, adhesion, and removal / edited by K. L. Mittal.
    p.   cm.
    Proceedings of a symposium held in conjunction with the Seventeenth Annual Meeting of the Fine Particle Society, held July 28–August 2, 1986, in San Francisco, California.
    Bibliography: p.
    Includes index.
    ISBN-13: 978-1-4615-9533-5   e-ISBN-13: 978-1-4615-9531-1
    DOI: 10.1007/ 978-1-4615-9531-1
    1. Particles—Congresses. 2. Surfaces (Technology)—Congresses. I. Mittal, K. L. 1945–   . II. Fine Particle Society. Meeting (17th: 1986: San Francisco, Calif.)
TA418.78.P37   1988   88-28841
620′.43—dc19   CIP

Proceedings of a symposium on Particles on Surfaces: Detection
Adhesion, and Removal, held in conjunction with the Seventeenth Annual
Meeting of the Fine Particle Society, held July 28–August 2, 1986,
in San Francisco, California

© 1988 Plenum Press, New York
Softcover reprint of the hardcover 1st edition 1988
A Division of Plenum Publishing Corporation
233 Spring Street, New York, N.Y. 10013

PREFACE

This volume chronicles the proceedings of the Symposium on Particles on Surfaces: Detection, Adhesion and Removal held under the auspices of the Fine Particle Society in San Francisco, July 28-August 2, 1986.

The study of particles on surfaces is extremely important in many areas of human endeavor (ranging from microelectronics to optics to biomedical). A complete catalog of modern precision and sophisticated technologies where particles on surfaces are of cardinal importance will be prohibitively long, but the following eclectic examples should underscore the concern about particles on a variety of surfaces. In the semiconductor world of shrinking dimensions, particles which, a few years ago, were cosmetically undesirable but functionally innocuous can potentially be killer defects now. As the device sizes get smaller, there will be more and more concern about smaller and smaller particles. In the information storage technology, the gap between the head and the disk is very narrow, and if a particle is trapped in the gap that can have very grave consequences. The implications of particulate contamination on sensitive optical surfaces is all too manifest. So the particulate contamination on surfaces is undesirable from functional, yield and reliability points of view.

This symposium was organized with the following objectives in mind: to bring together active practitioners in this field; to provide a forum for discussion of the latest research and development activities in this area; to provide opportunity for cross-pollination of ideas; and to highlight topics which needed intensified effort. The response to this Symposium was extremely gratifying and the general consensus was that a comprehensive symposium on this topic was both timely and needed. Concomitantly, the final technical program contained 51 papers covering various ramifications of particles on surfaces. Apropos, the program comprised both invited overviews and original research contributions. It should be recorded that there were enlightening (not exothermic) discussions both formally and informally throughout the duration of the Symposium.

As for this proceedings volume, it contains a total of 28 papers divided into four sections as follows: General Papers; Particle-Substrate Interaction and Particle Adhesion; Particle Detection, Analysis and Characterization; and Particle Removal. The topics covered include: sources of particles and deposition of particles on surfaces; particle-substrate interaction and factors which influence such interaction; particle adhesion measurement; various ways to detect, analyze and characterize particles on surfaces; various ways to remove particles from a variety of surfaces; particle prevention and implications of particulate contamination on surfaces. It should be recorded here that all papers were peer reviewed and suitably modified before acceptance and inclusion in this proceedings volume, as the peer review is a desideratum to maintain the standard of publications.

May I add here that this is the Premier Volume on this topic and I certainly hope it will be valuable to both the neophyte interested in learning about particles on surfaces and to the veteran researcher who wishes to know the latest developments. Apropos, this volume is christened Volume 1 as the proceedings of the Second Symposium will be labelled Volume 2. As a matter of fact, based on the interest and tempo of activity in the world of particles on surfaces, we have decided to hold symposia on this topic on a biennial basis.

Acknowledgements: First, it is my great pleasure to mention that this Symposium was jointly organized by yours truly and Dr. M.B. (Arun) Ranade and my sincere thanks are extended to him. Also my thanks go to the Fine Particle Society for sponsoring this event. Next I am thankful to the appropriate management of IBM Corporation, particularly A. Hermann, for allowing me to organize this symposium and to edit this volume. Special thanks are due to Lisa Honski of Plenum Publishing Corp. for her continued interest in this project. The time and effort of the reviewers is earnestly appreciated for making valuable comments. On a personal note, my thanks are extended to my wife, Usha, for helping me in many ways during the editing of this volume. Last, but not least, the co-operation, contribution and patience of the contributors is gratefully acknowledged without which this book would not have seen the light of day.

K.L. Mittal
IBM U.S. Technical Education
500 Columbus Ave.
Thornwood, NY 10594

CONTENTS

PART I. GENERAL PAPERS

FINE PARTICLES ON SEMICONDUCTOR SURFACES: SOURCES, REMOVAL AND

IMPACT ON THE SEMICONDUCTOR INDUSTRY

Stuart A. Hoenig

Department of Electrical and Computer Engineering
University of Arizona
Tucson, Arizona 85721

The impact of fine particles and organic contamination
on device yield is very serious. We have investigated several
technologies in this area. They include: 1) Application of
thermophoresis for the prevention of surface contamination.
2) The use of electrets for collection of particles that might
otherwise settle on surfaces. 3) The use of dry ice snow as a
cleaning medium for the removal of particulates and organic
contamination.

INTRODUCTION

Fine (less than 5 micrometers in diameter) particles are almost
everywhere in the atmosphere but, in general, they do not significantly
affect industrial processes. Semiconductor manufacturing operations
involve a very special environment where particles and organic vapors can
present serious problems in terms of loss of yield.

The general use of High Efficiency Particle Absolute (HEPA) filters
removes almost all of the particulates that come in with the "make up"
air. The resultant clean room particle spectrum is wholly dependent upon
the operation of the clean room and the personnel involved.

Exposure of collection plates in operational clean rooms has
indicated that the major contaminant is lint from employee garments
followed by hair, cosmetics and clusters of skin cells[1,2]. All of these
entities are "employee related" and can be greatly reduced by proper
gowning and attention to "clean room rules." This has been accomplished
rather effectively in Japan, but many US facilities leave much to be
desired.

In this connection, it is interesting to note that hair, lint and
clusters of skin cells will "not" be observed by the usual laser airborne
particle counter and must be detected by optical microscopy or SEM
scanning of witness plates. The problem here is associated with size
(over 50 μm) or the long, slender, shape of the lint fragments. A study
of particles observed on witness plates in clean rooms is presented in
reference 3. A true measure of the contamination in a clean room will

not be achieved unless particle counting is supplemented by witness plates and particle characterization.

Another category of particulates is associated with the wafer production process itself. The author's experience, after examining many witness plates from operational semiconductor facilities, is that silica fragments and dry photoresist are the next most common category of contaminants. The number of particles tends to increase as particle size decreases in agreement with the results of other investigators[4].

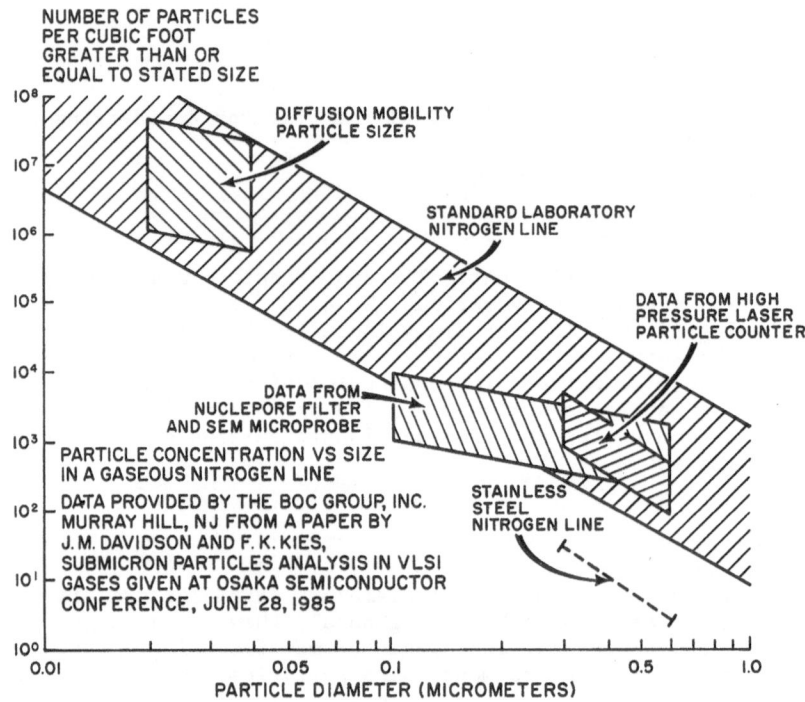

Figure 1. Particle Concentration vs Size in a Gaseous Nitrogen Line.

The problem with process gases may be even worse as shown in Figure 1. The particle count increases as the size decreases all the way to 0.01 μm.

If there are corona discharge ion generators in the area the number of submicron (e.g., 0.1 micrometer) particles may be very large. This seems to be associated with the condensation of organic vapors from the environment on the negative corona points[5]. It has been suggested that these corona points may be generating showers of submicron particulates that cannot be detected without a condensation nuclei counter (CNC).

Ultraviolet light in the 1800 A range has been identified as a mechanism for the conversion of organic vapors into small particles[6]. The problem here is that no source of this radiation is normally available in clean rooms. We are investigating effects of this type; it may well be that some organic vapors are converted to particulates by the short wave radiation leaking from fluorescent light fixtures.

There are some questions about the origins of organic vapors in the clean room. At present, suspicion has been leveled at sealants and at vinyl curtains or polymeric flooring materials[7]. Vinyl tile is some 50% plasticizer, usually dioctyl phthalate (DOP). DOP and other plasticizers have a very significant vapor pressure as shown in Figure 2. Condensation of organic vapors and the solidification of the condensate into droplets on wafers has been reported[8].

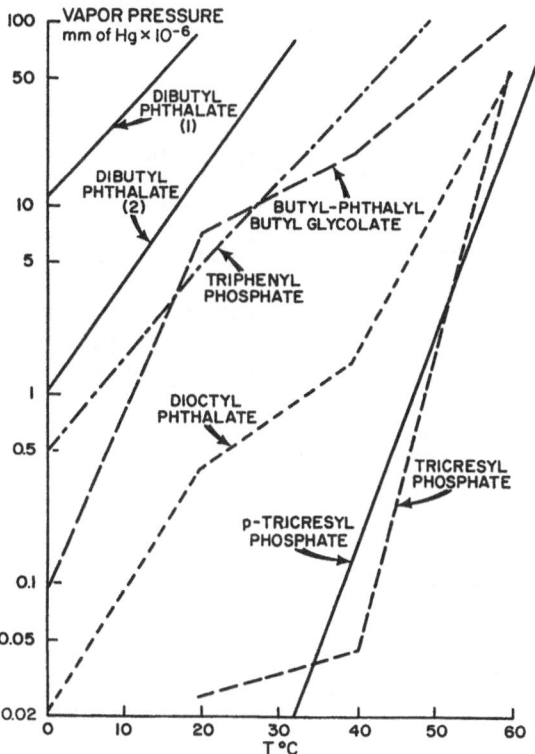

Figure 2. Vapor Pressure Data for Several Plasticizers, from L. Holland, <u>Vacuum Deposition of Thin Films</u>, p. 50, Chapman and Hall, London (1966).

In any real contamination problem, we must recognize that the deposit is not just particles but particles that are held more or less strongly to the surface by a layer of condensed water vapor and organics*. The fact that particle adhesion increases with relative humidity[9] is evidence of this effect. This suggests that we must consider not only the particle problem but the organic vapor situation as well.

- - - - - - - - - -

* There do not seem to have been any reports of the direct measurement of the effects of deposited organics on particle adhesion. However, the experience with water vapor suggests that adhesion will increase if organic deposits are present on a surface.

IMPACT OF PARTICULATE CONTAMINATION ON THE SEMICONDUCTOR INDUSTRY

The direct impact of any form of contamination is a reduction in yield. However, yield statistics are precisely the information that very few semiconductor companies are willing to release. The data in Figure 3 and 4 represent what little information is available to the author for publication. It is generally accepted folklore in the industry that 50% of the yield loss is associated with errors and omissions while the other 50% can be ascribed to contamination. We noted earlier that the Japanese seem to have been more successful in employee training and motivation. This may be the controlling factor in the "yield and cost per device" data presented in Figure 5. We, in the USA, cannot allow this situation to continue indefinitely.

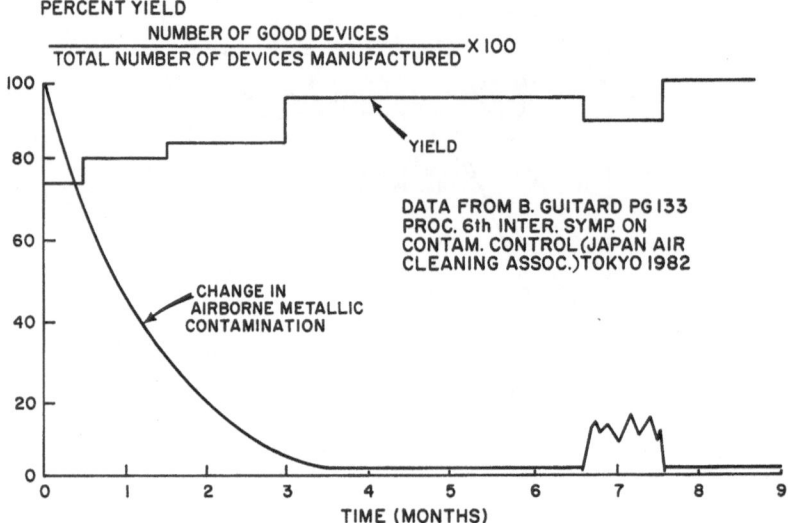

Figure 3. Percent Yield of Good Devices Compared with the Change in Airborne Metallic Contamination.

PREVENTING CONTAMINATION

In a sense contamination is like disease, far easier to prevent than to treat. We have been investigating some innovative technologies in this area with the idea that the wafer environment will never be totally free of particulates. We must, therefore, try to protect the wafer by reducing the opportunities for deposition or by capturing the particles on some other surface.

Thermophoresis offers the opportunity to protect surfaces against the deposition of small (under 5 μm) particles. Experimental results by Davies[10] indicate that if there is a 50 degree Kelvin per centimeter temperature gradient near a heated surface, the upward velocity of a one micrometer particle, of unit density, will be 0.15 cm/sec. In contrast, the gravitational settling velocity of a particle of this type will be 0.003 cm/sec. Clearly the particle will "never" arrive at the surface.

There have been some suggestions that the velocity of 0.15 cm/sec mentioned above should be compared with the velocity from kinetic theory or electrostatic attraction. The diffusion effect will be associated with Brownian motion and we have noted that will be "away" from the heated substrate. Electric field forces may be very large but cannot be taken

into account unless some assumption is made about the electrostatic
fields involved.

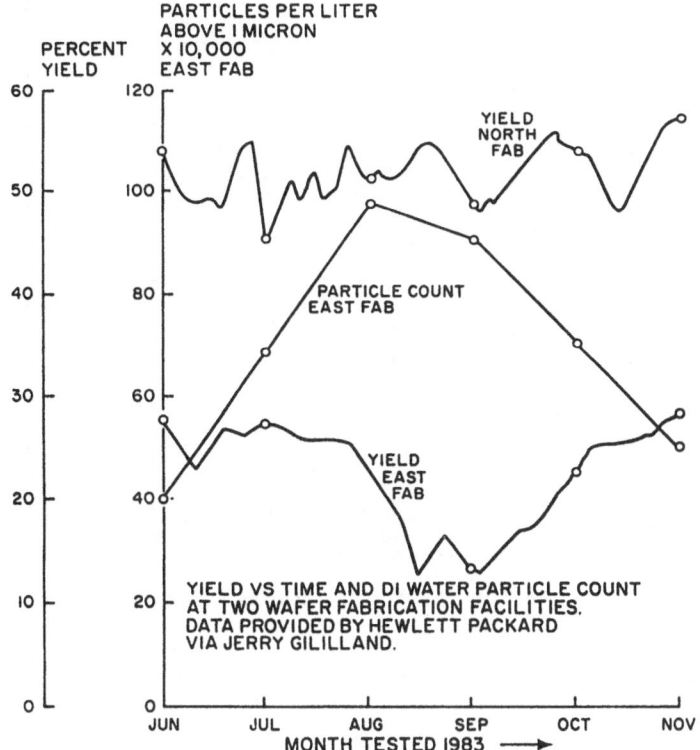

Figure 4.   Yield vs Time and DI Water Particle Count at Two Wafer
            Fabrication Facilities.

Thermophoretic protection techniques have the added advantage that the
effect "increases" as particle size decreases. This offers the
opportunity to protect wafers against submicron particulates and even
organic vapors that might otherwise become surface contaminants.

In Figure 6A, B we show some photographs of a 100 mm wafer in an
environment of 0.1 µm ammonium chloride particles. In Figure 6A, the
wafer is at ambient temperature and the particles deposit on the surface.
In Figure 6B, the wafer is some 16°C above ambient temperature and
deposition of the particulates was greatly reduced. Very similar effects
can be observed with organic vapors and, in Figure 7, we show the results
of exposing a series of 75 mm wafers at various temperatures to an
environment contaminated with glycerol vapors. At about 50°C (25°C above
ambient) deposition of the vapor is greatly reduced. A program to
investigate the applications of this technology to the semiconductor
industry is underway at the University.

There is some question here about the effect (s) of thermophoresis on
organic vapors. Certainly the heated substrate will raise the local vapor
pressure of any organic that does deposit on the surface thereby inducing
desorption. However, we must recognize that organic vapors are composed
of large molecules that will be subject to thermophoretic forces. Under
the circumstances we should expect that they will not be deposited on
heated surfaces.

7

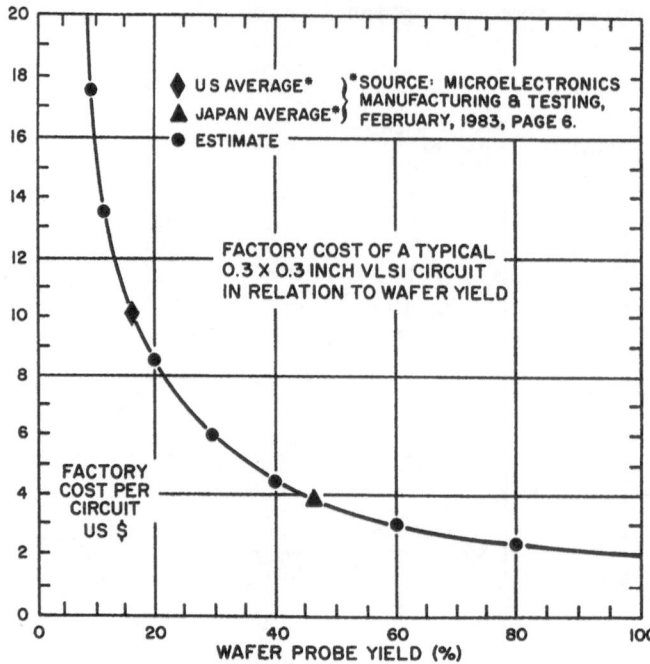

Figure 5. Yield vs Factory Cost of a Typical 0.3x0.3 Inch VLSI Circuit.

Electrets offer an effective mechanism for capture of particulates and organic vapors (provided that they have a permanent dipole moment). Electrets are manufactured by heating a plastic in an electrostatic field. The resultant material will hold a surface voltage of some 1000 V for periods of four years or longer in a 100% relative humidity environment.

A recent paper by Yost and Steinman[11] discusses the effects associated with charged surfaces and charged or uncharged particulates. For a surface with a charge of 1000 V the electrostatic attraction on a 1 μm particulate will be larger than the gravitational force if the particle is within 1 cm of the surface. For (+) charged particles near a (-) charged surface (or vice versa) the electrostatic forces are "always" larger than the gravitational forces.

Experimental studies of the removal of particulates from air by electret filters have been reported[12]. The electret materials are significantly more effective than the usual uncharged media. Another example of the effects of electrostatic forces on particle collection is shown in Figure 8. Here a + charged surface was used to collect particles that had been charged to a negative potential by an ion generator. When the ionizer and collector were "off" dust removal was very slow; when the ionizer and collector were "on" dust removal was very rapid.

This suggests that electrets might well be used for dust collection in semiconductor manufacturing equipment and at least one company actually has a program of this type "under way." We might note that some potential users have expressed concern that the electret generated electrostatic field gradient might damage ESD sensitive wafers. While we cannot rule out this effect <u>a priori</u>, we should note that Teflon[TM] electrets have a very high surface resistance so that it is impossible to draw any significant current from an electret.

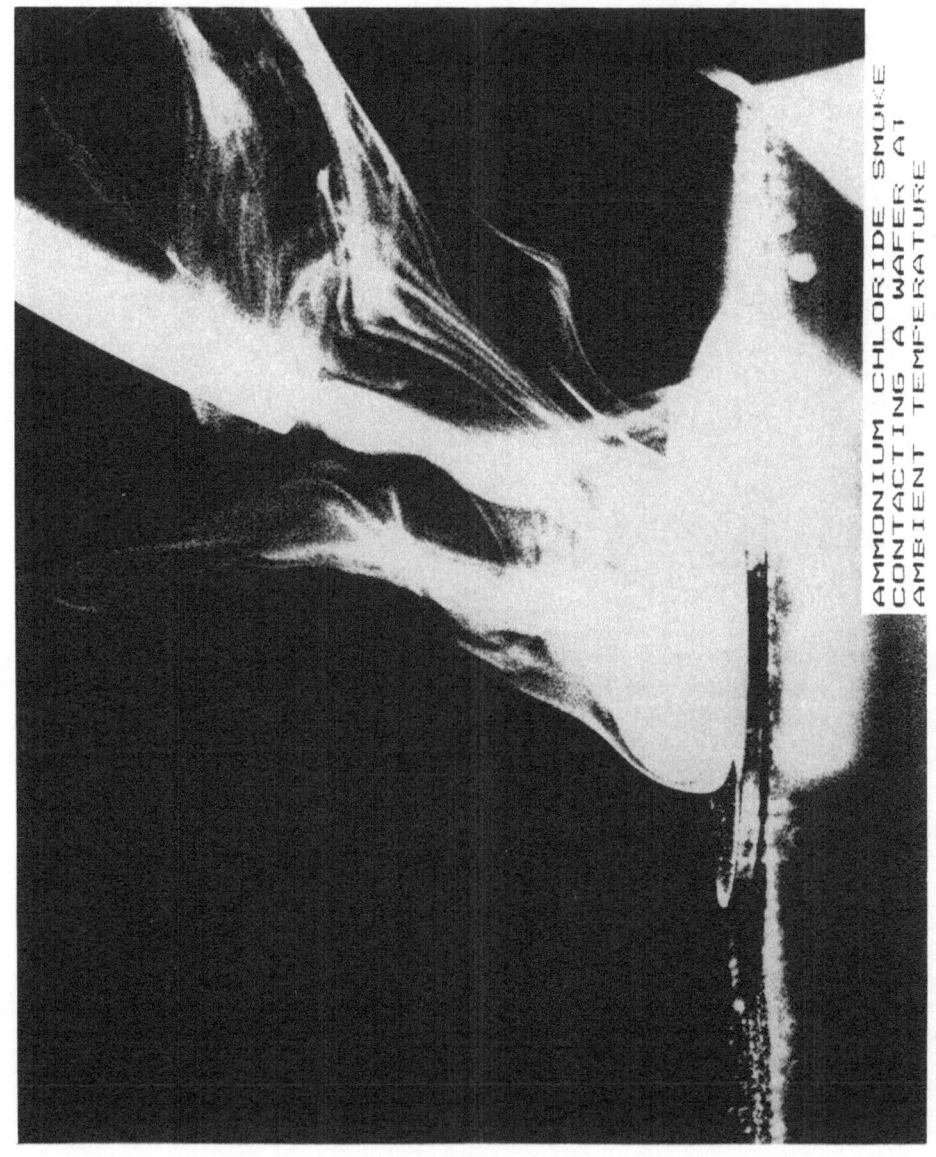

AMMONIUM CHLORIDE SMOKE CONTACTING A WAFER AT AMBIENT TEMPERATURE

Figure 6A. Ammonium Chloride Smoke Contacting a Wafer at Ambient Temperature.

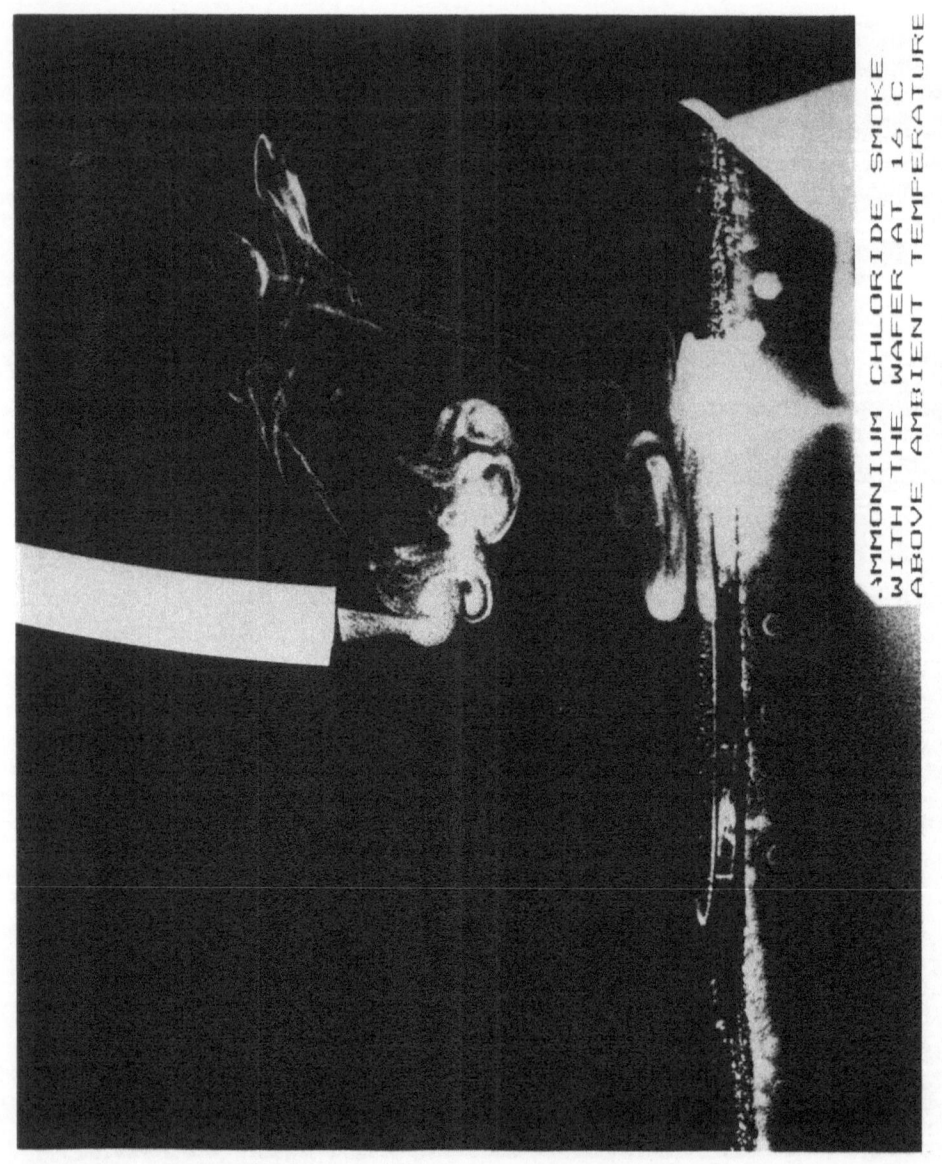

Figure 6B. Ammonium Chloride Smoke with the Wafer 16°C above Ambient Temperature.

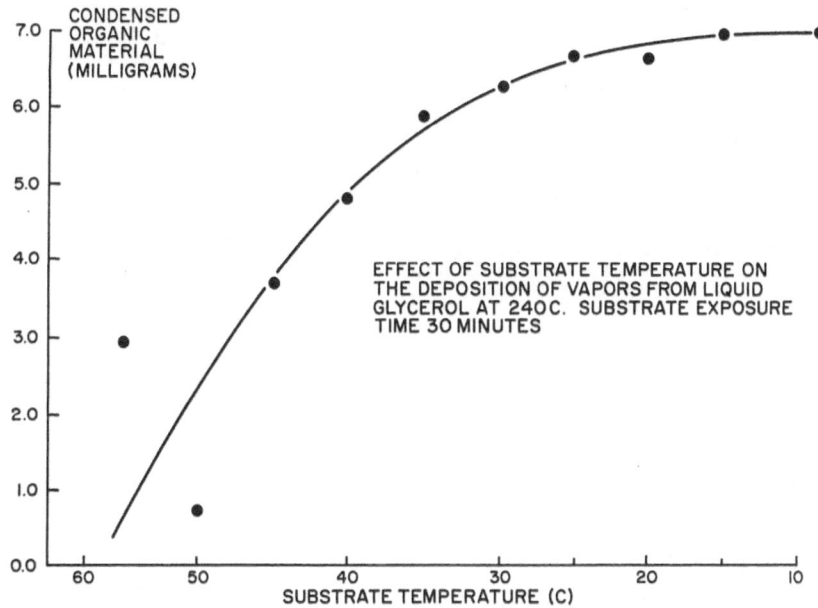

Figure 7. Effect of Substrate Temperature on the Deposition of Vapors from Liquid Glycerol at 240°C.

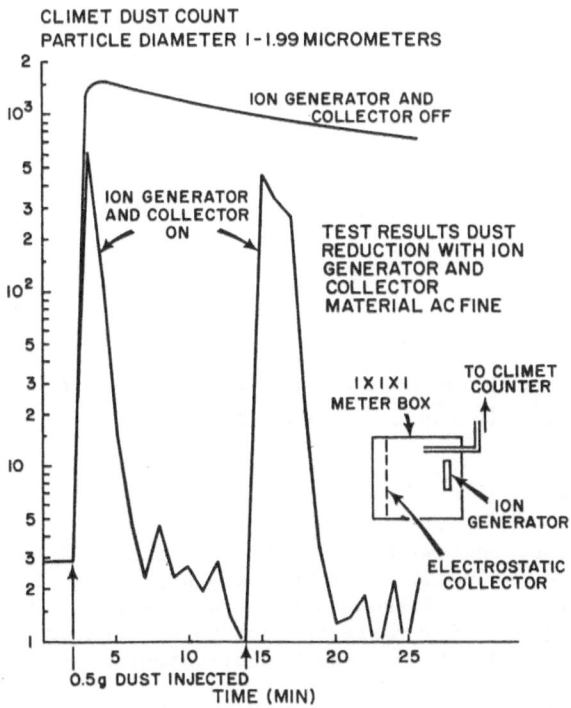

Figure 8. Climet Dust Count - Ion Generator and Collector On vs Ion Generator and Collector Off.

We expect to set up a small test program whereby ESD sensitive wafers would be stored in an electret lined box for, say, 48 hours. If no damage is observed we can expect that the hazard is very limited.

The ability of an electret to draw organic vapors from the air before they can deposit on a wafer is of some interest. If the vapor molecules have a permanent dipole moment they will tend to move in an electrostatic field gradient. Most of the organic plasticizers (e.g., DOP) are long chain molecules that might be expected to have a dipole moment. Experiments under way in our laboratory have indicated that the deposition of an organic material (e.g., glycerol) is affected by the local electrostatic field. Studies with other organics are planned for the future.

Table I. Efficiency of Removing Particles from Optical Surfaces.

		REMOVAL EFFICIENCY FOR PARTICLES > 5 μM	LOWEST ACHIEVED CONCENTRATION/CM^2 > 5 μM
1. SCRUBBING AND DRAGGING WITH LENS TISSUE USING ETHANOL AND ACETONE†		99.6 - 99.98%	2 - 40
2. SPRAYING WITH LIQUID SOLVENT, 5-30s DURATION TRICHLOROTRIFLUOROETHANE (FREON TF*)	345 KPA (50 PSI)	97% 3% FOR > 1 μM PARTICLES	1500
" " "	6.9 MPA (1000 PSI)	99.7 - 99.9% 81% FOR > 1 μM PARTICLES	10 - 35
WATER	17 MPA (2500 PSI)	98 - 99.5%	2 - 60
3. ADHESIVE STRIPPABLE COATING SCOTCH 2253†		95 - 98%	500
4. ULTRASONIC AGITATION OF FREON TF, 1-2 MINUTE DURATION*		24 - 92% 1% FOR > 1 μM PARTICLES	9000 - 70000
5. SEQUENTIAL CLEANING OPERATION ULTRASONIC AGITATED TWD-602*/ FOLLOWED BY FREON TF LOW PRESSURE SPRAY/VAPOR DEGREASE/IMMERSION IN BOILING SOLVENT/ULTRASONIC AGITATION IN RINSE TANK		92%	4000
6. COMPRESSED GAS JET FOR 10 s DURATION MICRO-DUSTER** 690 KPA (100 PSI)		50 - 61%	5000 - 5800
7. VAPOR DEGREASING IN FREON TF*		11 - 28%	65000 - 80000

* E.I. DUPONT DE NEMOURS & CO. INC., WILMINGTON, DE
† 3M CO., ST. PAUL, MN
** TEXWIPE CO., INC., HILLSDALE, NJ

REMOVAL OF CONTAMINATION

Cleaning techniques that might be applied to semiconductor wafers have been under study and evaluation from the very beginning of the industry. The efficiencies of some well known systems are shown in Table I. The results for particles under 5 μm leave much to be desired. At present, most manufacturing organizations make use of deionized (DI) water and mechanical scrubbing systems to remove particulates. DI water and brush cleaning are effective but the wafers must be taken to special cleaning machine and carefully dried to avoid contamination. We suggest that a need exists for an in situ cleaning system that does not involve a liquid like water.

We have been investigating the use of dry ice snow as a cleaning technology. The system is shown in Figure 9; clean liquid CO_2 is drawn from a tank and allowed to expand to form dry ice snow. If the snow is blown across a dusty surface the dry ice snow slides over the surface and "pushes away" the dust.

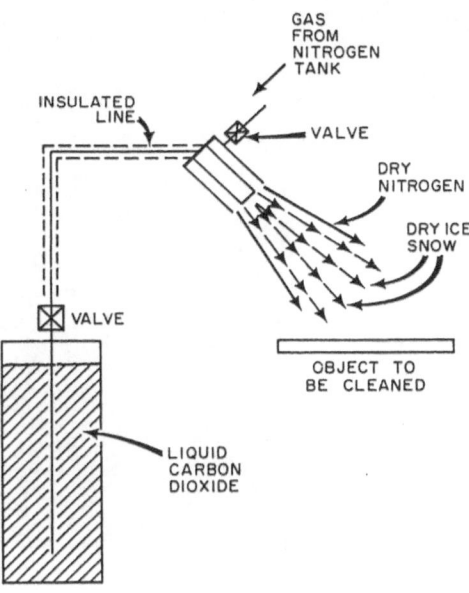

Figure 9.   Schematic Drawing of Dry Ice Snow Cleaning Technology.

Experimental studies with soft materials (e.g. germanium and lithium niobiate) have indicated that the dry ice snow does not scratch these materials. In Figure 10 we show photographs of a germanium mirror surface before and after cleaning. The same pattern of random scratches is visible indicating that the dry ice cleaning did not damage the surface.

Other results with 0.5 μm particles on silicon wafers are shown in Figure 11. Removal seems to have been quite effective particularly when it is appreciated that the suspension medium for these particles contains a detergent that increases particle adhesion when it dries. This suggests that polystyrene particles present a severe challenge to any cleaning technology.

There have been some concerns about the possibility of ESD damage or condensation of organics associated with dry ice cleaning processes. Evaluation of the potential for ESD was done with the help of a large semiconductor manufacturing company. A series of ESD sensitive wafers were cleaned with dry ice snow and then evaluated for ESD problems. The study indicated that no ESD occurred and we suggest that this is not a problem with the current dry ice cleaning system.

The problems of organic condensation are not as easily dismissed. The flow of dry ice snow across the wafer does induce some cooling (5 to 10°C) that can induce condensation of organics and/ or water vapor. If condensation occurs, the deposit can be very difficult to remove[7]. We have approached this problem from two directions:

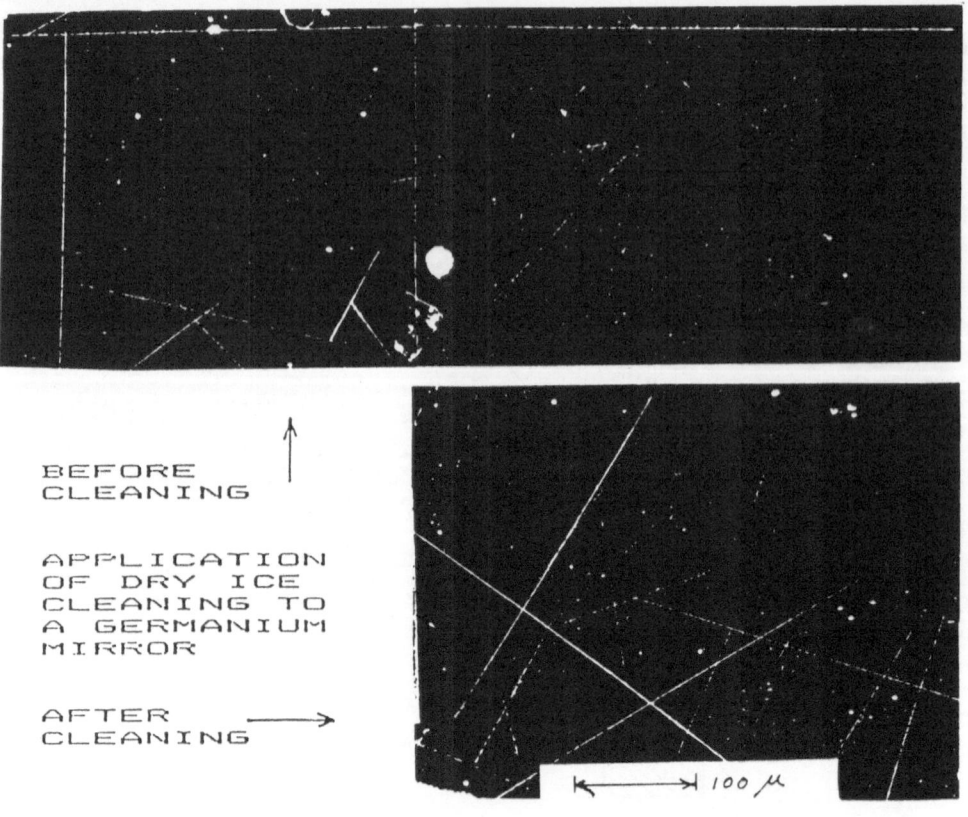

BEFORE
CLEANING

APPLICATION
OF DRY ICE
CLEANING TO
A GERMANIUM
MIRROR

AFTER
CLEANING

Figure 10.   Application of Dry Ice Cleaning to a Germanium Mirror.

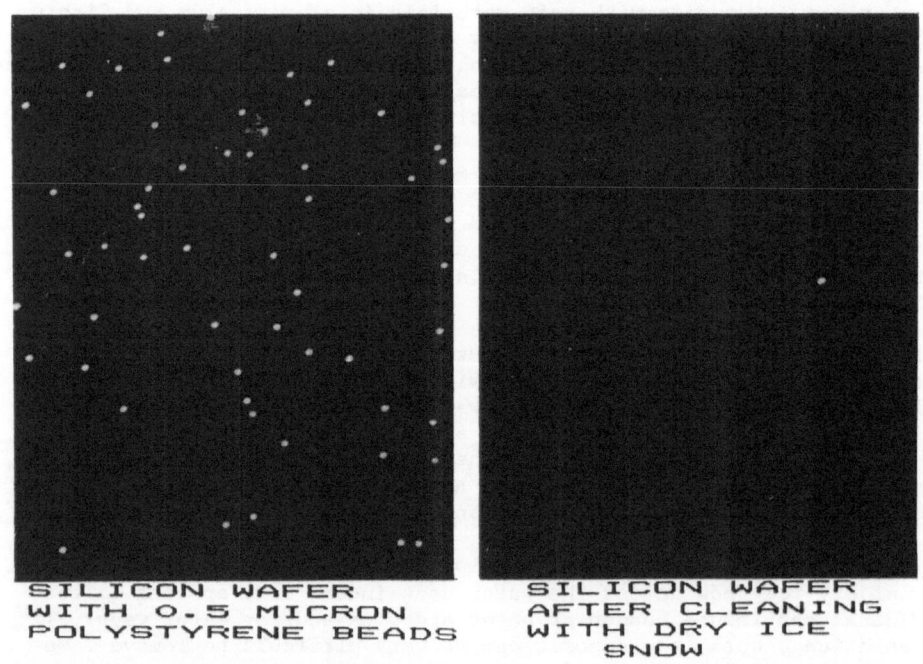

SILICON WAFER
WITH 0.5 MICRON
POLYSTYRENE BEADS

SILICON WAFER
AFTER CLEANING
WITH DRY ICE
SNOW

Figure 11.   Before and After Photographs of a Silicon Wafer Cleaned with
Dry Ice Snow.

1. Providing an environment of dry nitrogen using the system shown in Figure 9. This is 100% effective provided that if tank nitrogen is used it is carefully filtered to the 0.2 μm level at the point of use. Failure to filter tank nitrogen will result in a heavy deposit of submicron particulates that will appear as background haze in a laser surface scanning system.

2. Holding the wafer a few degrees above ambient with a hot chuck or an infrared lamp. We have found this to be a more effective technology in that the heated wafer is protected against contamination "after cleaning" by the thermophoretic effects discussed above.

3. Making certain that the liquid CO_2 is free of particulates. Recent experience in our laboratory indicates that there may be a significant variation. in particle content from tank to tank even with the cleanest (class 4) material.

In this connection we might note that the Airco Central Research Laboratory in Murray Hill, NJ has developed a new technology for generating very clean liquid carbon dioxide. Tests with this material are currently under way. The dry ice cleaning system is being adapted to tanks, bottles and pipes. The efficiency of the process is being evaluated with ultraviolet light induced fluorescence[12].

CONCLUSIONS

We have reviewed some of the contamination problems that will confront the semiconductor industry in the future. Some of the technologies that might be employed to detect or remove contamination have been discussed as well as prospects for what might be called the "next generation."

ACKNOWLEDGEMENTS

Other personnel that assisted with this program include Ervin Smith, Richard Gimmi and Ilya Glinsky. Financial support was provided by the National Science Foundation. All of this support is gratefully acknowledged.

REFERENCES

1. E.W. Moore, Microcontamination 3, No. 9, 65 (1985).
2. Q.T. Philips, W.D. Auser, J.M. Baldwin and G.J. Washington, J. Environmental Sci. XXVII, No. 5, 27 (1983).
3. K. Madden and J. Ramsey, Test Measurement World, 4, No.2, 54 (1984).
4. R.P. Donovan, B.R. Locke and D.S. Ensor, Solid State Technol. 28, No.9, 139 (1985).
5. B.Y.H. Liu, D.Y.H. Pui, W.O. Kinstley, and W.G. Fisher, "Aerosol Charging and Neutralization and Electrostatic Discharge in Clean Rooms", University of Minnesota. Particle Technology Lab. Publication No. 589, June 1986.
6. R. Zalabsky and S. Twomey, J. Res. Atmos. 13, No. 2, 147 (1979).
7. J.N. Ramsey, Appl. Surface Sci. 20, 413 (1985).
8. L.H. Fergason, Microcontamination 4, No. 4, 33 (1986).
9. W.J. Whitfield, in "Surface Contamination: Genesis, Detection and Control" K.L. Mittal, Ed, Vol. 1, pp. 73-81, Plenum Press, New York (1979).

10. C.N. Davies, "Aerosol Science", Chap. 6 and 7, Academic Press, New York (1966).
11. M. Yost and A. Steinman, Microcontamination $\underline{4}$, No. 6, 18 (1986).
12. C. Kanaoka, in "Aerosols," B.Y.H. Liu, D.Y.H. Pui and H.J. Fissan, editors, p. 613, Elsevier, New York (1984).

CLEANING SEMICONDUCTOR SURFACES: FACTS AND FOIBLES

Arvind Khilnani

Institute for the Future
2740 Sand Hill Road
Menlo Park, CA 94025

This paper is intended as a brief description of the
fundamentals of cleaning and monitoring surfaces in the
semiconductor industry. First a brief review of why
cleaning is important in semiconductor manufacturing is
presented. Next, a brief description of the contaminants
and their sources is presented. Finally, how the wafers
are cleaned is discussed. Anomalies in cleaning methods,
pertinent to the semiconductor industry, are highlighted
along the way.

INTRODUCTION

It is expected that as integrated circuits become increasingly more
complex, contaminants will come to play a more dramatic role and the
demand for cleaner manufacturing methods will increase. Because
semiconductor phenomenology is inherently a surface mechanism, it is only
natural that practitioners of the art have focused dominantly on surface
cleanliness and its converse--surface contamination.

There are five essential elements of defect generation which must come
into play if a contaminant is to be of substantive concern to the
manufacturer. These are shown in Table I. In order for a defect to come
into being, there must first be some source of contamination. This is
generally dust, dirt, or foreign matter. Operator particulates such as
sputum, epidermal flakes, hair and clothing compose an appreciable
percentage of such foreign material. Heavy metals from equipment used in
manufacturing also contribute a substantial amount.

Next, these contaminants must somehow migrate to the product. Airborne
and liquid-borne migration have been the classical transport mechanisms
towards which clean room control has been directed. More recently, the
actual materials constituents themselves have come under sharper focus.

Having migrated to the wafer, a contaminant must then somehow become
trapped or otherwise interact with the wafer. Not all trapped particles
create defects. Consequently, some of the issues in monitoring deal with
how to differentiate a non-killer defect from a killer defect. Killer
defects at the surface generally create pinholes, material voids, cracks,
or material bridges at critical points.

The final stage consists of detection of the defect. It is the latter stage that has created a great deal of confusion in the industry, for most papers in the literature focus upon the phenomenology of detection, with little regard to the seriousness of the interactions themselves, or even whether an interaction has actually occurred and defect been generated. Consequently, it is this issue which is just now coming to the attention of the industry.

The semiconductor industry utilizes a variety of equipment to maintain wafer surface cleanliness. This equipment is termed 'cleaning systems' for the purpose of this study. Cleaning systems have to do with the removal of such contaminants before they have had a chance to interact with the surface and create a defect. By and large, this means removing the contaminant at the point of interaction--on the surface of the wafer, or at the mask. However, wafer cleanliness is so critical to operations that contamination prevention is also an important effort. Prevention also implies an effort upstream to eliminate the transport mechanism and even the source itself. Consequently, the act of cleaning pervades manufacturing--from filtration methods for clean air and clean chemicals, to materials purification methods. This report refers to cleaning systems in the wider context commonly encountered in the semiconductor industry.

HOW CLEAN IS CLEAN?

This section will briefly examine all of those activities which lead up to cleaning and monitoring. It will set the stage for the subsequent section, which addresses the methodologies and applications.

The obvious place to start in this chapter is to ask the question: 'Just what is a clean surface, and how clean is clean?' Mittal, in introducing his two volumes on surface contamination, points out that there is no universally acceptable defintion of a 'clean' surface. He then goes on to say that, generally speaking, a clean surface is one which is free from contaminants or any unwanted matter or energy (see Mittal[1]).

However for the solid state scientist, a more descriptive definition can be had by imagining that some pure crystal, such as a diamond, or a silicon ingot, is suddenly cleaved. The two open faces that result from such cleavage are 'clean' by almost any definition conceivable. Nevertheless, dirtiness soon sets in. Regardless of the surrounding environment, the new faces will begin to adsorb material from their surroundings. Some of this will be considered to be normal or even desirable, such as the buildup of an oxide film on the new surface as it seeks a minimum energy state with its surroundings.

What is happening at this hypothetical clean surface is that it is being bombarded by the gas molecules around it. At room temperature and ambient pressure, roughly 1.0×10^{23} collisions per second will occur on each square centimeter of surface. Silicon contains 5×10^{22} atoms per cc. This is roughly equivalent to 2×10^{15} atoms per square centimeter on its surface. Consequently, each surface atom will be bombarded about 50 million times per second. Some of these bombarding atoms will stick; some will penetrate the lattice to become contaminants just below the surface; some will knock off electrons and charge the surface so that it attracts other particles borne by the passing air. Soon the newly cleaved surface will be dirty and methods of cleaning must then be considered. Clearly, the choice of the cleaning method to be used will depend upon the recent history of the surface and upon its surroundings. It will also depend on the nature of the surface and the nature of the contaminant. Understanding the nature of the contaminant is facilitated by knowing its source.

SOURCES OF CONTAMINANTS AND THEIR SIZE DISTRIBUTIONS

As mentioned above, once contamination occurs, time works against its removal. The destructive elements of contamination set in. Contaminants can come from many sources. In a semiconductor facility, contamination is potentially derived from five key sources:

  *Air-borne particulates

  *Liquid-borne impurities and particulates

  *Human 'dust'

  *Manufacturing processes, facilities and equipment

  *Materials impurities

Each of these sources contributes uniquely to the contamination problems in modern semiconductor facilities. They will be discussed momentarily. Meanwhile, in order to provide a more encompassing view of contaminants, our purposes might be better served by first examining particle sizes. This is important, because--as will be shown--almost all sources of contaminants ultimately are derived from some 'particle' that has been carried along to its point of interaction by some transport mechanism.

Particles of immediate interest in the clean room generally range in size from about 100 microns in diameter down to about 0.5 microns. These are the sizes which interact directly to cause either lithographic or film defects. Particles of emerging interest are those smaller than 0.5 microns. The advent of VLSI has focused a great deal of attention on these.

Upon arriving at a wafer surface--whether by way of an air carrier or via a liquid carrier--some of the particles will be chemisorbed and will become contaminants. These will interact atomically and will essentially become impurities. Current purity levels in the industry range from about the 'five nines' level down into the 'seven nines' level. On an absolute basis, this is equivalent to a range from 10 parts per million to about 100 parts per billion.

Figure 1 delineates these impurity concentration ranges in somewhat more detail and contrasts them with current-day specification levels. To better envision this, recall that the atomic density of silicon is 4.96×10^{22} atoms per cc. Typical dopant impurities range in concentration from about 1.0×10^{14} to around 1.0×10^{19} atoms per cc. This is equivalent to a range from 10 to 100,000 ppb. The range overlaps the purity specifications of most liquids. For example, the SEMI standard for most semiconductor grade chemicals calls for an impurity level of gold, iron, sodium, and chlorine of roughly one ppm. Such ranges of values are equivalent to the same range of routine impurity doping densities used within silicon. Serious difficulties would occur if this level of contaminant worked its way into silicon. By and large, this does not happen. Still, there is enough concern about impurity levels to cause an industry-wide reappraisal of the allowed impurity levels of high grade materials.

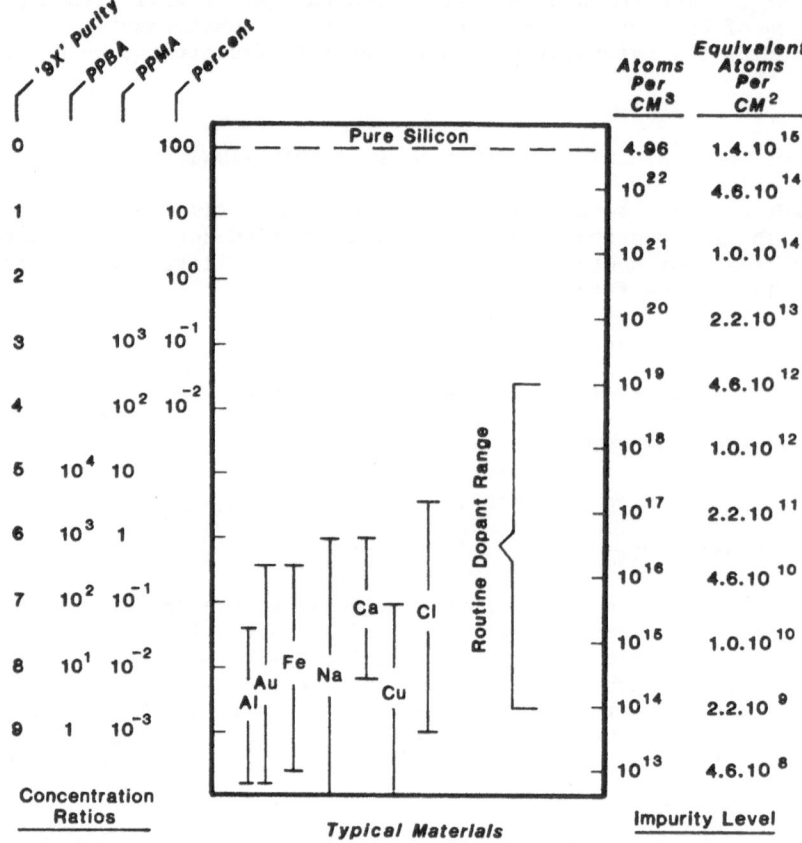

Figure 1. Comparison of various impurity levels and concentration ratios. Sample material range shown as listed by specification-level (upper bar) and minimum detectability (lower bar).

Historically, people were the greatest contributers to particle contamination. Improvements have reduced such contribution in recent years to the point that people now only contribute to contamination at a rate that is about equivalent to that contributed by equipment. Few published figures are available however. Jim Harper at Veeco Integrated Automation has made some studies which indicate people now contribute slightly less than equipment. On the other hand, Millipore, in a commercial brochure, shows people contributing slightly more. Millipore's documents[2] show the contribution as in Table I.

Table I. Relative Contribution of Sources of Contamination.

People:	34%
Equipment:	25%
Process:	25%
Materials:	8%
Air:	7%

These values appear to be reasonably consistent with most of those reported in the literature.

Solutions to the people contamination issue subsequently brought equipment-caused contamination to the forefront. Ion implanter end stations were focused upon quite heavily in the late seventies. They broke wafers. The broken wafers created silicon dust that would adhere to subsequently processed wafers. Then it was found that other equipment was equally as dirty. Robotics came to be used as a method for reducing equipment contamination until it was discovered that robots are almost as dirty as people, then the issue turned to that of making clean robots. Jim Harper, of Veeco Integrated Automation, offered the following comments in private communication. "Robots can be modified to improve their cleanliness. Our measured results show about 750 particles-per-part improvement:"

Table II. Particles Generated by Robot Motion.
Before and After Modifications to Improve Cleanliness.

| Robot Axis | Particles Generated Per Minute | |
	As Received	Following Modifications
Waist	2,100	0
Shoulder	2,900	15
Elbow	100	3
Wrist 1	8,700	0
Wrist 2	12,000	18
Wrist 3	1,200	0
TOTAL	27,000	36

He then adds, 'The data suggest to me that robots might be very much like people--some good (clean) and some bad (dirty). I will even suggest that it is possible to rehabilitate a bad one with proper training and make it a good productive citizen. Perhaps I have not seen a clean robot, but I have seen a cleaner robot.'

Liquid-borne particulates have come under close scrutiny in recent years. Much of the early work was done by Tolliver at Motorola. Tolliver[3] provides some of the most definitive studies of particles in semi-conductor grade fluids. He gives detailed results on particle counts and distributions. Much of Tolliver's work indicates that liquid particle sizes and types are similar to those air particulates identified by Willeke and Whitby[22]. For example, particle count of foreign matter in fluids also increases with decreasing size. The typical quantity of particles at two micron diameter is about 570 particles per cc, and it is about 0.50 particles per cc at five micron diameter (both data points are taken on prefilterd D.I. water).

Particles in fluids are generally defined as undissolved materials in suspension. They may be either organic or inorganic in nature. Inorganic materials generally consist of silt, iron, glass, metals, airborne particles, fiberglass and submicron particles such as colloids. Organic materials consist of bacteria, algae, resin, pollen and polymers.

Virtually all liquids used in clean rooms have been pre-filtered in some manner. Most liquid filters range in size down to about 0.2 micron. Consequently, much of the material found in liquids can be suspected of having been generated at or near the semiconductor factory itself. Weiss[4] states that much of this contamination is generated through the transportation, handling and storing of fluids. This seems to be generally true. Consequently, most liquid filtration schemes are directed at point-of-use solutions.

Two immediate conclusions can be drawn from the data. First, it is observed that upstream pre-filtering of the D.I. (Tolliver's analysis) water in the facility system itself works very well for particles larger than 5 micron. Virtually none of the particles in this size range arrived at the point-of-use filter. Second, the downstream liquid is still not particulate free, even after point-of-use filtration. Consequently, it can be suspected that some of the difficulties alluded to as being caused by the tools themselves may be nothing more than ineffective filtration.

Tolliver used the D.I. water taken from the above-mentioned point-of-use filter and mixed it with both hydrofluoric and sulfuric acid. He then remeasured the particle count. The data clearly indicate that the acids contained more particulates than did the D.I. water.

For example, the 3.0 micron particle count was 48,205 particles per liter before mixing, and while not using point-of-source filtration. Afterwards, it was 75,640. Tolliver does not give the mix ratio; however, most HF acids are diluted from 5 to 10 times, or more. If we assume that 10:1 mix was used, then the HF particle count exceeded 320,000 particles per liter. Sulfuric acid appears worse still, with over twice the contamination of hydrofluoric acid in photoresists.

Tolliver's results clearly point to the advantages of using point-of-use filtration to eliminate particulates, even though that was not the purpose of his paper.

Dillenbeck[5] reports similar results for bottled liquids. Particle counts range from low values of a few thousand to upwards of one million particles per 50 milliliters.

Some of this particulate material is derived from the source material itself. However, much of it comes from the container and the piping. Przybytek[6] has performed studies on the size distribution and contaminant types for various bottles:

"What is the composition of various particles [in bottles]? Glass, rust and other metallic oxides, salt, bacteria, diatoms, metal, fibers, carbon, viscous goo, and plastics have all been identified in various samples. The source of contamination is sometimes obvious. For example, the simple operation of removing a screw cap from a bottle can introduce particulate contamination into the system."

In Przybytek's study, each cap was removed and retightened 20 times to exaggerate the generation of particulates by cap abrasion. He was somewhat surprised to discover that the more flexible polypropylene caps generated an order of magnitude more particulates than did the less flexible Bakelite phenolic caps.

Przybytek's results dramatically point out that clean bottles are not clean, just as Tolliver's results did for acids and piped D.I. water. The effect is even more striking when these results are compared with class 100,000 cleanroom air. The results suggest that liquids used in VLSI cleanrooms are not even as clean as is class 100,000 air.

DEPOSITION AND FILTRATION MECHANISMS

The previous data make clear the potential for contaminants appearing at a wafer's surface. It also clarifies the transport vehicles and implies that the first line of defense is removal of the contaminant via proper filtration at the point-of-use. Following that, the second line of defense is mitigation of the inherent deposition mechanisms. Deposition mechanisms consist of diffusion, sedimentation, impaction, turbulent deposition, and electrostatic attraction. These same mechanisms occur in both gases and liquids.

Sedimentation is dominantly brought about through gravitational attraction. Diffusion is limited to very small particles of the order of 0.1 micron or less. Impaction is a function of the carrier velocity and its vector force propelling a particulate towards the wafer surface. Turbulent deposition occurs in nonlaminar flow situations. Electrostatic attraction acts in the same manner as sedimentation, but on charged particles.

A great deal of work has been carried out on these effects at the Particle Technology Laboratory at the University of Minnesota. Liu, Pui, Rubow and Szymanski[7] recently presented results of measurements on all these forces as a function of particle sizes. Data are given for particles ranging from 0.01 to 10 microns in size.

For particles of 0.1 microns in size--where diffusion would normally be expected to play a dominant role--diffusion alone contributed just 10% of the total deposition while electrostatic effects contributed from 10% to 90%, depending upon the method used. Clearly, electrostatic effects dominate for small particles.

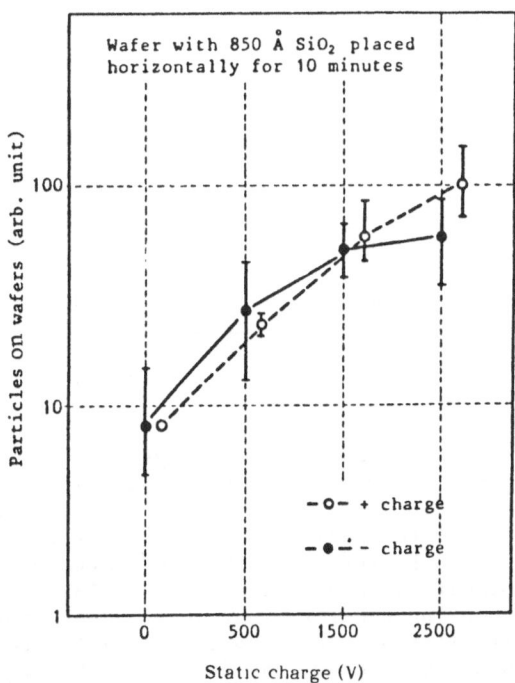

Figure 2. Relationship between particle adhesion and static charge. (From Ref. 8)

These results are corroborated by studies of Japanese factories. Nakamura[8] presented figures of electrostatic charge buildup on actual wafers at Toshiba. His results are reproduced in Figure 2.

In a cleanroom, electrostatic charge buildup is strongly correlated to particle adhesion. Figure 3 points out the adverse role which equipment can play in contributing to electrostatic deposition.

While these deposition mechanisms play an adverse role when acting to deposit particles, they can and are also used quite effectively as well in providing filtration. IBM uses pre-charged wafers as getters. Such wafers act as sacrificial lambs, preceding other wafers through lithographic tools to clean the mask sets by collecting dust as they go (see Fredericks et al. [9]). The inverse of electrostatic charging is, of course, charge removal. This effectively creates an electrostatic filter. Charge removal can be accomplished by providing local plasmas above the wafer surface. Typically, ultraviolet light is used to create ozone, just as in routine bacteriological instruments. Zafonte and Chiu[10] describe a method similar to this for removing resist residue. Pak and Verkuil[11] also describe a similar method in use at IBM for maintaining clean wafers.

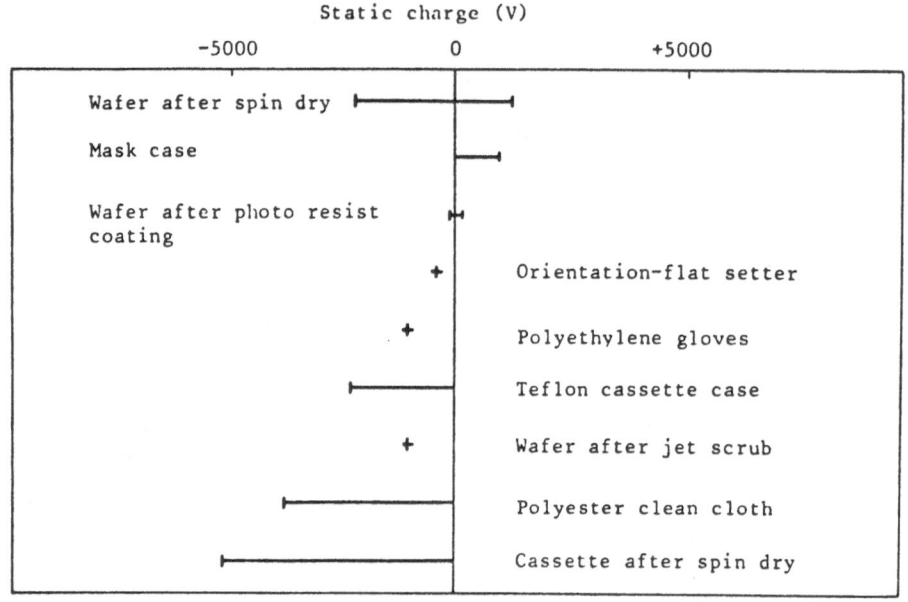

Figure 3. Static charge on equipment in the clean room. (From Ref. 8)

The use of deposition mechanisms also plays a crucial role in filtration. For example, many--perhaps even most--individuals in the semiconductor industry appear to believe that HEPA Filters are ineffective in the submicron region and do not filter submicron particles well. Consequently, ULPA Filters have appeared. However, it has been known for years that diffusion phenomena cause HEPA Filters to become very efficient at removing submicron particles (see Donovan et al. [13]). But it is not at all clear how this knowledge has failed to permeate the semiconductor industry. As recently as November 1985 at the Microcontamination Conference in San Jose, California, panel members were specifying 1990 particulate count specifications down in the 0.1 micron range.

To take the issue one step further, in March 1985, Rubow [12] presented new results reconfirming such efficient submicron filtration. However, lack of knowledge on this very subject appears to have generated an ongoing disagreement that is being carried in letters appearing in Microcontamination Magazine. (See those articles in Microcontamination under Liu, April-May, 1985; Blitshteyn, August, 1985; and letters to the editor October & November 1985).

Regardless of these erroneous beliefs, however, it has recently been shown that in modern VLSI cleanrooms, submicron particle contamination does, in fact, decrease in accordance with the very filtration efficiency described in the literature, Donovan, Locke, Osburn and Caviness at the Research Triangle Institute and MCNC [13], did the work. They studied five state-of-the-art VLSI cleanrooms and found that volumetric size distribution deviated substantially from Fed. Std. 209B. They showed that the quantity of particles failed to increase below 0.1 micron in size, exactly as that large body of knowledge about HEPA Filters suggested it should. Their results are reproduced in Figure 4.

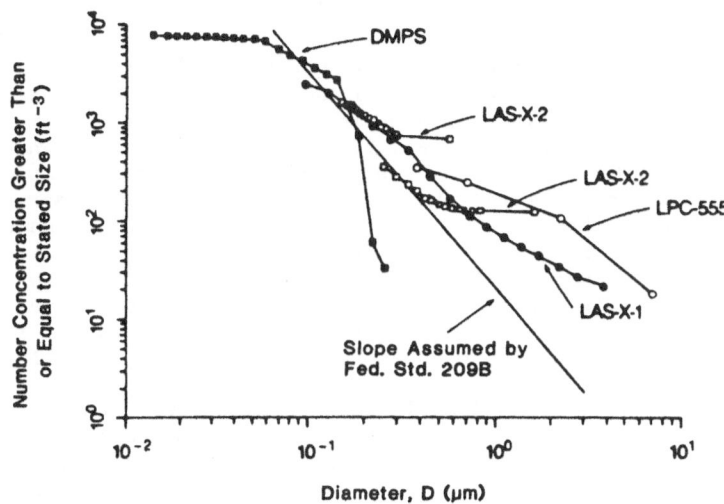

Figure 4. Average cumulative size distribution of particles in a university laboratory (From Ref. 13).

ADHESION MECHANISMS

Upon arriving at a wafer's surface, a contaminant must somehow stick to that surface, otherwise it will not remain and will not contribute to defect generation. Unfortunately, this is not difficult. Moreover, the smaller the particle, the stronger the relative adhesion forces.

Two categories of phenomena set in. When the particulate first attaches itself to the surface, the adhesion forces are primarily physical. However, adsorption begins to take place almost immediately at points of contact--thus time and temperature begin to take their toll. Eventually, some portion of the particulate is adsorbed through chemisorption. Once that begins, the process is irreversible and the contaminant can no longer be removed in its entirety through cleaning. Some form of mild surface etching must also be used. This is why it is so important to remove contaminants quickly.

Turning first to the physical phenomena, these are found to be related more to particulates, than to film-type contaminants. But having once passed through a high temperature process, the residue of such particulates exists as an adsorbed film. So if the wafer is cleaned prior to a high temperature operation, the cleaning process can be allowed to be dominantly of a physical nature.

The four primary physical forces of adhesion that bind particles to a surface are 1) capillary, 2) van der Waals, 3) electrostatic image, and 4) electrical double-layer. These are well described in the literature. Two references of particular value which discuss these forces from a semiconductor vantage point are Bhattacharya and Mittal [14] and Bowling [15,23]. The former authors approach the issues from a more generic viewpoint. Bowling looks at some very specific applications.

Capillary action can result in a strong adhesion force. Condensation of water vapor or other process liquids collects in the gaps between the particle and the wafer surface, thus forming a meniscus. The surface tension of this meniscus draws the surfaces together, resulting in capillary attraction.

There are several components to van der Waals forces. The one of greatest concern here is the so-called van der Waals-London force, attributed to both van der Waals and London. The van der Waals-London attraction is an intermolecular adhesion force. Atoms that compose a particle are instantaneous dipoles. The dispersion interaction between these dipoles as well as the dipoles induced in neighboring atoms produces the attractive force. Detailed studies of the van der Waals-London forces show that van der Waals-London attraction decreases with the first power of the particle diameter (see Bowling [23]). But the mechanical stress necessary to remove particles decreases with the third power of a particle's diameter. Thus, the smaller the particle, the greater the difficulty in overcoming the van der Waals-London force. The entire cleaning process also becomes more difficult as a result.

Electrostatic image forces apply to particles larger than 5 micron. In conductors, this type of charge is neutralized by contact charge flow and thus electrostatic attraction becomes of minimal interest. In nonconductors, however, electrostatic attraction can be significant.

Electrical double-layer force is associated with particles below 5 micron. A surface contact potential is created between two different materials based upon each material's respective local energy state. The resulting surface charge buildup needed to preserve charge neutrality sets up a double-layer charge region, which creates the electrostatic attraction.

Some orders of magnitude are helpful here. In an air environment, the attractive forces between a one micron glass particle and a wafer surface are as follows: capillary - 0.045 dynes; van der Waals-London - 0.014 dynes; electrostatic image - 0.001 dynes; and electrical double-layer - 0.003 dynes. In contrast, the gravitational force on such particles is four orders of magnitude less than is the electrostatic image force.

A common misconception about cleaning is that solubility of the particulate, per se, is a determining criterion for a good cleaning solvent. While high solubility is important, it is not particularly helpful for insoluble materials such as heavy metals. Where these metals are being held on the surface largely by capillary forces, a good cleaning solvent is one that will cause the elimination of the capillary forces through total immersion and will thereby counteract this major force of attraction. A good solvent also provides charge neutralization, thus also reducing electrostatic forces. Nevertheless, even with these mitigating factors, the remaining physical forces of adhesion can still far exceed the forces that can be applied to remove the particle.

26

Bhattacharya and Mittal[14] have shown that the force of removal is a function of angle of the applied force. This can be seen with aid of the diagram shown in Figure 5.

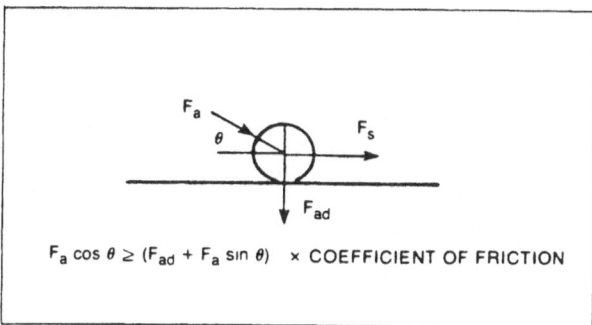

$$F_a \cos \theta \geq (F_{ad} + F_a \sin \theta) \times \text{COEFFICIENT OF FRICTION}$$

Figure 5.   The relationship between an applied surface cleaning force, its angle of incidence, and the coefficient of friction involved. (From Bhattacharya and Mittal [14]).

Here F_a is the force needed to remove the particulate, Fad is the force of adhesion and theta is the angle of the applied force. As can be seen, forces tangential to the wafer will be the most effective, and forces normal to the surface will be least effective. At equilibrium, the force necessary to shear off the particle is thus:

$$F_a = \frac{kF_{ad}}{cos\theta - ksin\theta}$$

Where k is the static coefficient of friction.

The magnitude of these forces of adhesion are as follows:

Capillary:	$z\pi\gamma d$
Electrostatic image force:	$Q^2/(\varepsilon d^2)$
Electrical double layer force :	$\pi\varepsilon_0(\Delta\phi)^2 d/(2z)$
van der Waals force:	$h\omega d/(16\pi d^2)$

The variables and constants in these formulas are:

d	=	Effective particulate diameter, in cm.
z	=	Effective separation of particulate from the surface, ~ 4 A
γ	=	Water surface tension ~ 73 dynes per cm.
$\Delta\phi$	=	Work function difference ~ 1V
$h\omega$	=	van der Waals' constant ~ 7.2 eV
ε	=	Relative dielectric constant
ε_0	=	Permittivity of free space = 8.9 pF/m.

CLEANING METHODS

The cleaning process involves either the application of an external force to the particle in order to overcome the adhesion forces, or the application of a brief 'polish' or 'etch' to remove chemisorbed materials.

For physical removal, an air jet may be used to shear or roll the particles off the surface. For dry, uncharged particles, only the van der Waals-London and the electrical double-layer forces are present. Zimon [16] has shown that air or nitrogen blow-off guns are effective in removing particles larger than 10 micron from surfaces but are ineffective in removing particles ranging from 1 to 10 micron in size, In more recent experiments conducted by Bhattacharya and Mittal [14], it was shown that air-jet effectiveness could be extended to half micron particles by taking into account the rolling action of the particle from the surface. However, the experiments also demonstrated that particles below 0.5 micron could not be blown off.

The use of a liquid as the surrounding medium, rather than air or vacuum, helps reduce adhesion forces. Liquid cleaning methods, which employ spraying or agitation, tend to lower the van der Waals-London forces and to generate double-layer electrostatic repulsion forces that further facilitate particle removal. In fluid dynamic studies, force transfer to a particle that is separated from a surface by a liquid film has been shown to increase if the viscosity of the liquid is lowered. Brandreth and Johnson [17] examined particle removal from surfaces through the use of common solvents such as alcohol mixtures with fluorocarbons. Different types of Freon and acetone proved highly effective in removing particles at relatively low accelerations. The relative success of acetone, Freon and other organic cleaners compared with water has been attributed to the lower surface tension of organic cleaners.

Particles bound with capillary adhesion are difficult to remove despite the use of such liquids. Capillary adhesion, which binds particles deposited from a liquid suspension, is virtually impossible to overcome by chemical means. Even baking such liquid-deposited particles at 180°C (a temperature higher than the liquid's boiling point) does not facilitate removal with an air jet. Subsequent attempts with ultrasonic and megasonic methods are not successful either. According to Bhattacharya and Mittal [14], the particles can only be removed through mechanical scrubbing. The implication for semiconductor manufacturers is clear: if the chemical cleaning fluid used reposits particles, then mechanical means will be necessary to clean the wafer.

Ultrasonic and megasonic agitational equipment is commonly used in chip manufacturing to clean semiconductor surfaces. Various types of ultrasonic devices, operating at 20 kHz, have been effective in dispersing particles on surfaces immersed in liquids. This dispersive action is thought to occur through the collapse of cavitation bubbles at the agglomerate interface. The growth and collapse of bubbles are caused by pressure. These fluctuations occur above and below ambient pressure. They are created with high-intensity sound waves that are, in turn, caused by mechanical disturbances. Bubble collapse conditions must be optimized in order to maximize the strength of dispersion effects. Megasonic cleaning systems operate at frequencies of 850-900 kHz. The higher frequencies of these devices have been found to produce a cleaning action more efficient than that of ultrasonic devices. In this method, particle removal is executed through a high-pressure wave mechanism rather than through cavitation. Megasonic systems have proved effective in the removal of 0.3 micron particles using a hydrogen peroxide solution, see Shwartzman et al. [18].

Although ultrasonic and megasonic cleaning systems have worked in practice, the cleaning theory is not fully understood. Consequently, each new cleaning situation must be treated separately, and the cleaning potential of the devices can only be determined on a trial-and-error basis.

Electrostatic techniques have often been used in contamination prevention of semiconductor equipment. However, the application of such techniques to particulate removal is new and experimental. Researchers Cooper et al. [25]. and Hays [26] have tried a variety of experiments only to conclude that electric fields much greater than the breakdown field for air are needed for removal. Research today clearly indicates that complete particle layer detachment by an electric field is not practical.

Coming full circle then, these studies collectively tend to indicate that the application of mechanical force is the best answer to particulate removal--a point which proponents of scrubbers have made time and again. Stowers [19] has made a study of the effectiveness of all types of cleaners and concluded the same. He was the first to show that nothing beats a simple wipe down with a lens tissue. He then went on to show the effectiveness of a Freon spray (see Stowers and Patton [24]). Stowers' work was done on laser optical lenses for the SHIVA project. His results are also appropriate for wafer cleaning because lenses have oxide coatings similar to those used in semiconductors.

While these methods are effective, by now it should be clear that the chemicals used in the cleaning play an equally important role. Among these, the so-called 'RCA clean' is the most widely known and has become the industry standard. It is based upon an entire sequence of cleaning steps. These first remove organic contaminants by oxidation, then trace metals and chemisorbed ions are removed by solubilization. This is followed by an optional HF etch to remove a few layers of the oxide. Eventually, this sequence of steps was further augmented with the RCA megasonic clean to combine the removal of particulates with the desorption taking place.

The RCA clean became of such widespread use that its inventors' original paper was the fourth most cited among papers ever published in the Science Citation Index, see Kern et al. [20]. According to authors:

"Extensive analytical studies and device reliability and life testing by many independent researchers have confirmed the process, now widely known as "RCA Standard Clean," to be the most effective cleaning method known for attaining the degree of purity that is imperative in the fabrication of sensitive silicon semiconductor devices. Furthermore, the process is safe and relatively simple, has attractive economic and ecological advantages, uses readily available high-purity solid-free and volatile reagents, and was accepted by the American Society for Testing and Materials as a standard procedure. Actually, the process is so widely employed that most authors refer to it without citing our original work, apparently assuming it to be common knowledge...."

CLEANING PRACTICES

A typical semiconductor process will employ a large number of cleaning operations. These include lithography (coat, expose, develop and etch), low temperature depositions (ion implant and films), and high temperature operations (doping, diffusion and oxidation). Modern process engineers regard many, if not most, of these steps as critical with regard to wafer cleanliness. In its survey of the semiconductor industry VLSI Research [21] identified common pre-and post-operational cleaning practices.

Wafers are not generally cleaned upon being received at a lithography step. At lithography, the process activity generally relies upon the wafers having been received clean. Only 10% of lines employ pre-cleaning immediately prior to resist application. An even lesser percentage-7.5%-preclean incoming wafers from the vendor. However, post cleaning or resist stripping of photoresist is performed for every mask level. Stripping occurs after etching or ion implanting, when photoresist is used to delineate implanted areas.

For low temperature film deposition processes, end users commonly employ wafer precleans to remove contaminants. Ion implant is of equal concern, but wafer precleans are prohibited here because of the prevalent use of photoresist to shield non-implanted areas. Those implant steps using oxides or nitrides to mask off non-implanted areas generally use wafers 'cleaned' at prior operations or use wafers from operations employing an operational post clean.

At low temperature film deposition wafer cleanliness concerns include chemical vapor depositions of oxides, nitrides, and polysilicon as well as any deposition of metals. Post-operational cleans are relatively frequent here. They are emphasized at 30% of deposition-related process steps.

High temperature operations, whether for oxidation, dopant pre-deposition or drive-in, are of considerable concern with regard to metallic and sodium ion contamination. Ninety-percent of all high temperature oxidations involve some form of wafer precleaning. Those oxidations not receiving wafer precleans are generally transferred from furnace boat to furnace boat in the same module. Diffusion or drive-in wafer precleans are used in 80% of these operations. This clean is oriented towards the removal of accrued particulates arriving with the wafers or towards removal of highly doped glass films resulting from unwanted furnace dopant depositions. Post-operational cleans for dopant depositions and drive-ins (performed by 30% and 10%, of the industry, respectively) are likewise intended to remove doped glass.

With few exceptions, practitioners report that particulate contamination is critical for all mask levels of photolithography (see Ref. 21). Respondents in this study were further polled as to the criticality of ionic contamination at deposition, diffusion and implant process steps. Respondents were nearly unanimous in citing deposition as the critical process. It is interesting to note that chemical vapor deposition (CVD) in general and specific depositions of poly-silicon and nitride, commonly deposited by CVD methods, were cited by nearly 75% of the industry.

Only 12.6% of the surveyed process lines express satisfaction with today's cleaning methods. The balance report a variety of problems ranging from 'bottlenecks' to 'nightmares'.

Equipment cleaning is a leading concern among 25% of the product lines. Deposition equipment is the most frequently cited. Problematic wafer clean operations include photoresist removal after ion implantation (9.7% of respondents) and wafer precleans at deposition and diffusion (5.6% of respondents).

Cleaning problems are further aggravated by the nature of the contaminant. Many contaminants are a result of equipment and/or indirect material residues. Major residues cited during this study include equipment redeposits- 28.6% of respondents, incomplete removal of photoresist- 27.2%, and residues from chemicals applied in the process- 26% of respondents.

The monitoring of wafer and material cleanliness has likewise changed little in recent years. Optical inspection of product or test wafers for particulates has been used as a direct measure of wafer cleanliness and as an indirect measure of the cleanliness of materials employed prior to inspection. Equipment contribution to particulate contamination has been evaluated in the same fashion.

Wafers subjected to precleans and subsequent metal and other film depositions are commonly examined under high intensity light sources with the naked eye. Highlighted particles or haze may be further examined by optical microscopes, often with Nomarsky or dark field observations.

An additional method employed for dielectric films is the use of test wafers with deposited metal "dots" to establish capacitors using the deposited film as a dielectric. Dielectric breakdown voltages are measured. The incidence and breakdown voltage values are related to the incidence of particulate contamination.

With the exception of monitoring house gases and de-ionized water at the plant "pad" or centralized facilities level, materials are rarely, if ever, directly monitored within the fabrication are. D.I. water resistivity is one exception. Other indirect material purities, such as specialty gases and chemicals, are either monitored at incoming inspection or the product line relies on vendor assays, certificates of compliance and/or reputation. Forty percent of the surveyed product lines rely on vendor testing of received materials, and only 10% perform additional tests at incoming inspection.

There is considerable commonality across fabs in the staging and logistics of cleaning. While variations exist, the predominant practice is described in the flow chart of Figure 6.

Resist removal is conventionally located in the proximity of etch stations. With few exceptions, resist removal is not mask-level dependent. A centralized resist strip station(s) is predominant. The station(s) serve both post-implant and post-etch resist removal.

Wafers are subsequently returned to lot boxes and manually transported to and staged near the subsequent furnace tube or deposition module. All furnace tube deposition, diffusion and oxidation, pre-and post-cleans take place at stations located as near as possible to the process equipment involved.

Normally, one clean station serves a "bank" of furnace tubes. Furnace tube "banks" are typically allocated to doping/drive-in, oxidations, highly critical oxidations, chemical vapor depositions, and a utility bank. The latter would include alloy and reflow type operations.

Product lines make every effort to minimize storage between cleans and the execution of the proess step. The preference is to clean, dry and load directly to process boats. Matching the clean station output to specific process batch loads is difficult to achieve. Interim post-clean storage takes place under laminar flow or in dry boxes, if extended delays are expected.

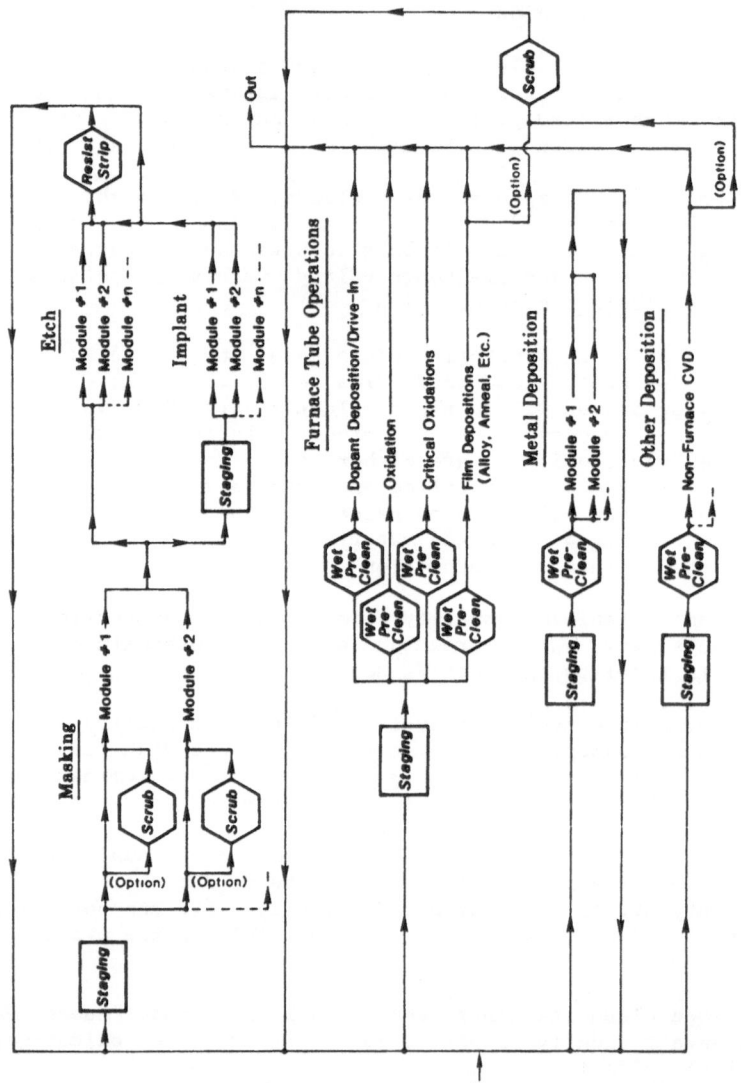

Figure 6. Principal wafer clean stations and staging sites.

32

Post-cleans may be followed by operations in the same area or by transport to other operational modules. Wafers are not normally returned to lot boxes when the subsequent operation takes place in the same, or in an adjacent tube bank. Here, the preferred procedure is to off-load the wafer from the process hardware to cleaning hardware and/or to directly reload the subsequent operation. When prohibited, the post-clean is delayed and/or treated as the previously described preclean.

Some product lines now employ scrub methods such as a post-clean for certain film depositions such as polysilicon, silicon nitride and some CVD oxides. The scrub station or stations are centralized and located near the deposition area but removed from other pre-and post-clean stations.

With these exceptions, most process engineers believe wafers outputted from furnace tube operations and most other depositions are as clean as can be achieved. The wafers are placed in lot boxes, manually transported, and staged at masking or the appropriate module.

Wafers arriving at masking from these sources are not normally precleaned. Notwithstanding, a limited number of product lines do employ or are considering in-line wafer scrubbing with resist dispense. As proponents currently represent less than 10% of existing product lines, a masking preclean is shown as an option in the flow chart.

Metal deposition and non-furnace tube depositions use process equipment that is not as easily modularized as is a bank of furnace tubes. Notwithstanding, clean stations are provided as near as possible to the respective equipment load stations. Every attempt is made to load directly from clean boats onto equipment fixtures. Typically, a clean station is provided for every two deposition systems.

Minimum delay is desired between precleans and the subsequent operation. Sequential operations in the same furnace bank module are usually affected by boat transfers as opposed to returning wafers to lot boxes. With these exceptions, wafer lots are stored, transported and staged in lot boxes. The common major storage areas are input staging to masking, implant, furnace tube operations, metal deposition and non-tube deposition areas.

CONCLUSIONS

To conclude then, it appears that the semiconductor industry continues to lag behind state-of-the-art in cleaning science. Liquid chemicals are far dirtier than they should be. They are only just starting to get the attention they deserve. Electrostatic forces are double-edged swords. Electrostatic induced mobility is a serious contributor to contamination. Yet, the same force is a powerful cleaning tool. Particles in the 0.3-0.5 micron range are the biggest problem to be dealt with. Yet discussions on filtering and cleaning have moved into the 0.1 micron range. Combinations of cleaning mechanisms are shown to be necessary. Yet most cleaning equipment have a single mechanism focus. In view of these foibles, the question becomes: Can and will the semiconductor industry 'catch up'?

VLSI Research put that question to the industry. Those surveyed were asked the question, "Are you working on new cleaning or monitoring methods?" The response is disappointing. Only one-third of the industry is working on new cleaning and monitoring methods. The remainder are content in their ways. Of those working on new methods, scrubbing techniques and chemical formulation dominate. Table III shows the industry's activities vis-a-vis new cleaning and monitoring methods.

The message is clear: One-third of the firms are working to catch up. Furthermore, they are on the right tracks. As for the rest, it may be too late to clean up their act.

Table III. New Methods For Cleaning And Monitoring.
Are You Working On New Cleaning Or Monitoring Methods?
End User's Response:

No	55%	
Yes	34%	
Will Discuss	17%	
Won't Discuss	17%	
Decline	6%	
Don't Know	5%	
TOTAL	100%	65 Respondents

What Methods?   (Based On 11 Respondents Willing To Discuss)

Scrubbing	36%
Chemical Cleaning Formulations	27%
Spray	10%
Material Cleaning	10%
Monitoring	9%
Other	8%
Total	100%

Source: VLSI Research, Inc.

ACKNOWLEDGEMENT

The author wishes to express his appreciation to Dr. K.L. Mittal for valuable suggestions.

REFERENCES

1. K.L. Mittal, editor, "Surface Contamination: Genesis, Detection, and Control," Vol. 1 and 2, Plenum Press, New York, 1979.
2. "Commitment to Zero Defects- The Millipore Solution: Contamination Control of Semiconductor Process Fluids," Millipore Cat. No. PB881, March 1985.
3. D.L. Tolliver, "Changing Factors of Process Chemicals," Semiconductor International: Transition to One Micron Technology, P.A. Cahners Special Publication, 1985.

4. A. Weiss, Semiconductor International, 5, No. 7, 55-64
    (July,1982).
5. K. Dillenbeck, Microcontamination, 57-62 (December 1984).
6. J.T. Przybytek, Microcontamination, 3, No. 6, 51-54 (June 1985).
7. B.Y.H. Liu, D.Y.H. Pui, K.L. Rubow and W.W. Szymanski, Electrostatic
    effects in aerosol sampling and filtration, presented at the British
    Occupational Health Society Annual Conference, York, England, April
    3-6, 1984; Particle Technology Laboratory Publication No. 525,
    University of Minnesota, Revised November 1984.
8. M. Nakamura, S. Kanro and Y. Sumitomo, Technical Proceedings
    from the SEMICON/WEST'85, May 21-23, 1985, pp. 134-138, San Mateo,
    California.
9. E.C. Fredericks, S.M. Katz, R.P. Mueller and G.G. Via, IBM Tech.
    Disclosure Bull., 26, No. 12,6625 (May 1984)
10. L. Zafonte and R. Chiu, Proc. SPIE. Soc. Opt. Eng. (USA), 470, 164-
    175 (1984).
11. M.S. Pak and R.L. Verkuil, IBM Tech. Disclosure Bull., 26, No. 9,
    4832 (February 1984).
12. K.L. Rubow, Microcontamination, 3, No. 3, 39-43 (March 1985).
13. R.P. Donovan, B.R. Locke, C.M. Osburn and A.L. Caviness, J.
    Electrochem. Soc., 132, 2730-2738 (November 1985).
14. S. Bhattacharya and K.L. Mittal, Surface Technol., 7, 413-425 (1978).
15. R.A. Bowling, Technical Abstracts of the SEMICON/Southwest '85 Trade
    Show, p. 16, SEMI, Mt. View, CA, October 1985.
16. A.D. Zimon, "Adhesion of Dust and Powder," Plenum Press, New York,
    1969.
17. D.A. Brandreth and R.E. Johnson, Jr., in "Surface Contamination:
    Genesis, Detection and Control," K.L. Mittal (ed), Vol. 1, p. 83,
    Plenum Press, New York, 1979.
18. S. Shwartzman, A. Mayer and W. Kern, RCA Review, 46, 80-105
    (March 1985).
19. I.F. Stowers, J. Vac. Sci. Technol., 15, No. 2, 751-754 (1978).
20. W. Kern, S. Shwartzman and A. Mayer, RCA Engineer, 28, No. 4, 99-105
    (July/August 1983).
21. "Cleaning and Contaminant Monitoring Systems for the Semiconductor
    Industry, A Business Opportunity Analysis," VLSI Research, Inc., San
    Jose, CA, April 1986.
22. K. Willeke and K.T. Whitby, J. Air Pollution Control Assoc., 25,
    529-534 (1973).
23. R.A. Bowling, these proceedings.
24. I.F. Stowers and H.G. Patton, in "Surface Contamination: Genesis,
    Detection and Control," K.L. Mittal, editor, Vol. 1, pp. 341-349,
    Plenum Press, New York, 1979.
25. D.W. Cooper, H.L. Wolfe and R.J. Miller, these proceedings.
26. D.A. Hays, these proceedings.

EFFECT OF CHEMICAL CLEANING SEQUENCING ON PARTICLE ADDITION/REDUCTION ON SILICON WAFERS

Charlie A. Peterson

FSI International, Inc
322 Lake Hazeltine Drive
Chaska, MN 55318-1096

The purpose of this study was to determine what effects, if any, differing chemical sequences had on the particle levels of silicon wafers. The chemical solutions studied are: sulfuric acid, hydrogen peroxide mix (SPM); ammonium hydroxide, hydrogen peroxide, DI water mix (APM); hydrochloric acid, hydrogen peroxide, DI water mix (HPM); dilute hydrofluoric acid (DHF).

The findings show that some changes in chemical sequencing can significantly alter particle levels (e.g., APM and DHF ordering; the use of SPM), while others show no effect (e.g., HPM and APM ordering).

INTRODUCTION

The thorough elimination of contamination is a major concern in the production of semiconductor devices. Ions, metals, organics and particles are all important contaminants that must be removed from the wafer surface. The various chemical mixtures and sequences used in the semiconductor industry each have strengths and weaknesses in removing the different types of contaminants. The negative impacts of ions and heavy metals have been documented.[1-5] Several studies on the metal removal ability of various chemical mixtures and sequences[6-12] and the efficiency of various solutions to remove organic contamination[6,13-15] have also been reported. In the case of particles, a recent review has detailed many of the problems associated with their presence.[16] While much effort has been expended on trying to remove the particles from the wafer environment (the air, chemicals, water and gases that come into contact with the wafer), only a few studies report particle counts on the wafer as a function of the chemical sequences used to clean the wafer. To add to this data base, this study was undertaken to further explore and document the effects on particle levels of changing sequences with four common cleaning solutions. These solutions, and their abbreviations, are listed in Table I.

Table I. Cleaning Solutions

SPM, $\underline{S}$ulfuric acid hydrogen $\underline{P}$eroxide $\underline{M}$ix

- $4H_2SO_4 : 1H_2O_2$

APM, $\underline{A}$mmonium hydroxide, hydrogen $\underline{P}$eroxide, DI Water $\underline{M}$ix

- $1NH_4OH : 1H_2O_2 : 5H_2O$

HPM, $\underline{H}$ydrochloric acid, hydrogen $\underline{P}$eroxide, DI water $\underline{M}$ix

- $1HCl : 1H_2O_2 : 5H_2O$

DHF, $\underline{Di}$lute $\underline{H}$ydro$\underline{f}$luoric acid

- $1HF : 100H_2O$

EXPERIMENTAL

Particle counts were obtained using an Aeronca WIS-150 Particle Scanner, which was calibrated with a surface contamination standard obtained from VLSI Standards, San Jose. The particle scanner was run with an edge exclusion of 4 mm. Data are reported as changes in particle levels (i.e., post-processing particle levels minus pre-processing particle levels) for $<0.5\ \mu m$ and $>0.5\ \mu m$ size ranges.

All wafers were 100 mm <100> silicon. Two separate groups of tests were run approximately one month apart. All comparisons are made within one group. There were three wafers per run in wafer carrier slots 2, 13 and 24 (i.e., second position from the bottom, middle of the carrier and second position from the top). This was done to insure that any effects seen were not position dependent, but were representative of the entire carrier. Three runs were made for each chemical sequence. All runs were made in an FSI ZEUS Acid Processing System. The chemicals were semiconductor low mobile ion grade from Allied Chemical. The seven different chemical sequences tested are listed in Table II. The initial particle counts and the changes in particle counts, which occurred with processing, are shown for both sets of data in Table III. The values given are the average and standard deviation for the nine wafers used in that chemical sequence; i.e., three wafers per run, three runs per sequence. All wafers, unless otherwise indicated, were first processed using the FSI "B" Clean (see Table II), so that they would have a common starting basis.

Table II. Cleaning Solution Sequences

FSI "A" Clean - SPM + APM + DHF + HPM

FSI "B" Clean - SPM + DHF + APM + HPM

FSI "C" Clean - DHF + APM + HPM

APM + HPM

HPM + APM

SPM + DHF + APM

APM

Table III. Wafer Particle Levels

SET 1

| CLEAN | STARTING PARTICLE LEVELS | | | | PARTICLE ADDITIONS | | | |
| | < 0 . 5µm | | > 0 . 5µm | | < 0 . 5µm | | > 0 . 5µm | |
	$\bar{X}$	σ	$\bar{X}$	σ	$\bar{X}$	σ	$\bar{X}$	σ
FSI "B" *	11.9	5.5	5.6	3.0	2.9	7.8	-1.3	2.6
FSI "A"	23.6	13.8	8.6	5.6	18.8	10.0	4.2	5.3
FSI "C"	20.2	10.9	11.7	8.2	8.8	13.3	3.8	10.7
APM	11.6	7.4	9.0	6.2	0.2	3.2	-1.6	5.0
HPM − APM	9.0	6.1	3.7	3.2	-0.1	3.5	0.3	3.2
APM − HPM	12.2	7.8	4.3	3.0	3.3	11.0	-0.4	2.2
SPM-HF-APM	9.6	4.4	3.3	2.7	-0.7	6.7	-0.8	3.7

SET 2

| CLEAN | STARTING PARTICLE LEVELS | | | | PARTICLE ADDITIONS | | | |
| | < 0 . 5µm | | > 0 . 5µm | | < 0 . 5µm | | > 0 . 5µm | |
	$\bar{X}$	σ	$\bar{X}$	σ	$\bar{X}$	σ	$\bar{X}$	σ
FSI "C" *	9.8	6.7	12.1	7.7	86.3	74.7	19.6	24.6
FSI "C"	64.9	9.9	25.6	7.7	38.3	23.9	11.3	13.5
FSI "B" *	5.1	2.3	17.4	6.1	36.4	8.0	-4.8	7.5
APM − HPM	37.0	8.5	11.0	4.3	4.9	14.8	1.8	4.1
HPM − APM	46.8	17.0	16.8	8.5	6.2	6.1	0.7	5.6
APM-HF-HPM	55.9	18.5	14.0	7.3	263.2	83.3	84.2	21.9

* = Wafers as received from wafer manufacturer

RESULTS AND DISCUSSION

Effect of APM and DHF Ordering

Previous studies[9,11,12] have shown that caustic peroxide solutions can be very effective in reducing wafer particle counts following an HF process. Table IV again demonstrates that a dilute ammonium hydroxide/hydrogen peroxide solution will reduce the particle count after a dilute HF process. Therefore, from the standpoint of particles, it is highly recommended that the solutions be ordered DHF followed by APM, rather than APM followed by DHF. (If particles are not an issue, the opposite sequence is recommended.[9,11,12]

Effect of SPM

The main function of SPM in many cleaning recipes is to remove any heavy organics which are present on the wafer surface.[6] Table V illustrates the effect on particle counts both when using SPM and when removing it from the process. It can be seen, at least for these virgin wafers, that the use of SPM made a significant difference in particle performance. Therefore, the use of SPM would be recommended from a particle standpoint.

The Effect of HPM and APM Ordering

As seen in Table VI, the use of HPM following an APM step has very little effect on particle performance. With this small sample, there is no significant statistical difference between the two sets of data. If one considers metallic contamination, however, the order is quite significant.[9,11,12]

Table IV. Effect of APM and DHF Ordering on Particles Added

	Particles Added	
(Set 1)	$\leq 0.5\mu m$	$>0.5\mu m$
SPM + APM + DHF + HPM (FSI A)	18.8	4.2
SPM + DHF + APM + HPM (FSI B)	2.9	-1.3
(Set 2)		
APM + DHF + HPM	263.2	84.2
DHF + APM + HPM	38.3	11.3

Table V.   Effect of SPM on Particles Added

	Particles Added	
	$\leq 0.5\mu m$	$>0.5\mu m$
SPM + DHF + APM + HPM (FSI B)	36.4	-4.8
DHF + APM + HPM (FSI C)	86.3	19.6

Table VI.   Effect of HPM on Particle Added

	Particles Added	
(Set 1)	$\leq 0.5\mu m$	$>0.5\mu m$
SPM + DHF + APM + HPM (FSI B)	2.9	-1.3
SPM + DHF + APM	-0.7	-0.8

Table VII. Effect of APM and HPM Ordering on Particles Added

(Set 1)	Particles Added	
	$\leq 0.5\mu m$	$> 0.5\mu m$
APM + HPM	3.3	−0.4
HPM + APM	−0.1	0.3
(Set 2)		
APM + HPM	4.9	1.8
HPM + APM	6.2	0.7

It can also be seen in Table VII that in both cases, the ordering of APM and HPM made no statistically significant difference on particle performance.

CONCLUSION

The data presented illustrates that some changes in chemical sequencing can have a dramatic effect on wafer particle levels. Yet others, at least with this small data set, show no statistically significant difference between the ordering of the chemical sequences.

The reader is reminded again, however, that when chemical sequences are changed, not only particle performance, but ionic contaminants, atomic contaminants, and organic contaminants are affected.

ACKNOWLEDGEMENT

The author wishes to thank Todd Elftmann, FSI, for carrying out the experimental work for this paper.

REFERENCES

1. S.R. Hofstein, IEEE Trans. On Elect. Dev. ED-14, 749 (1967).
2. E.H.Snow, A.S.Grove, B.E.Deal, and C.T.Sah, J. Appl. Phys. 36, 1664 (1965).
3. A.Goetzberger, and W.Shockley, J. Appl. Phys. 31, 1821 (1960).
4. W.M.Bullis, Solid State Elect. 9, 143 (1966).
5. D.C.Burkman, C.A.Peterson, L.A.Zazzera, R.Kopp, and D.S. Becker, FSI Technical Report - TR 317 (1988). Available from author.
6. D.C.Burkman, Semiconductor International, 4(7), 103 (1981).
7. W.Kern, RCA Review, 31, 207 (1970).
8. W.Kern, RCA Review, 31, 234 (1970).
9. W.Kern, and D.A.Puotinen, RCA Review, 31, 187 (1970).
10. R.L.Meek, T.M.Buck, and C.F.Gibbon, J. Electrochem Soc. 120, 1241 (1973).
11. D.S.Becker, W.R.Schmidt, C.A.Peterson, and D.C.Burkman, in "Microelectronics Processing: Inorganic Materials Characterization", L. A. Casper, Ed., ACS Symposium Series No. 295, pp. 366-376, American Chemical Society, Washington, DC 1986.

12. D.C.Burkman, W.R.Schmidt, C.A.Peterson, B.F.Phillips, in "Proc. Semiconductor 83 International", Birmingham, England Sept. 1983. Also available form the authors as FSI TR217.

13. Feder, D. O., and D.E.Koontz, in "Cleaning of Electronic Device Components", ASTM STP No. 246, American Society for Testing and Materials, Philadelphia, 40, 1959.

14. D.A.Peters, and C.A.Decker, J. Electrochem. Soc. <u>126(5)</u>, 883 (1979).

15. J.A.Amick, Solid State Technol., <u>19(11)</u>, 47 (1976).

16. J.R.Monkowski, in "Treatise on Clean Surface Technology", Vol.1, K.L.Mittal, Ed., pp. 123-148, Plenum Press, New York, 1987.

17. K.D.Beyer, and R.H.Kastl, J. Electrochem. Soc. <u>129(5)</u>, 1027 (1982).

18. J.R.Mehta, FSI Technical Report TR270, 1985. Available from the author.

MEASURING AEROSOL PARTICLE CONCENTRATION IN CLEAN ROOMS

AND PARTICLE AREAL DENSITY ON SILICON WAFER SURFACES

R. P. Donovan, B. R. Locke, and D. S. Ensor

Research Triangle Institute
P. O. Box 12194
Research Triangle Park, NC   27709-2194

Particle deposition velocity is the name given the coefficient relating particle buildup on a silicon wafer surface to the aerosol particle concentration adjacent to that surface. This paper reviews methods of measuring: (1) aerosol particle concentration and (2) surface particle areal density. High flow rate (1 cfm) optical particle counters are most suitable for clean room aerosol measurements, and laser scanners can detect surface particles as small as 0.2 to 0.3 μm, although the optical properties of the surface on which the particles rest play an important role in the detection. The aerosol particle measurement can be easily extended to 0.01 to 0.02 μm; no practical instrumentation yet exists for measuring ultrafine (<0.1 μm) surface particles.

INTRODUCTION

Understanding and controlling particle buildup on a silicon wafer (or any other surface) requires knowledge of both the origin of the contaminating particles and the mechanisms and forces whereby particles, generated at a site remote to the wafer, are captured and held by the wafer surface. While high aerosol particle concentration per se does not guarantee low yields, the flux of surface particle buildup (particles/area/time) typically does depend directly on the aerosol particle concentration adjacent to that surface as expressed by the following:

$$\text{surface particle flux} = \begin{pmatrix} \text{deposition} \\ \text{velocity} \end{pmatrix} \begin{pmatrix} \text{aerosol particle} \\ \text{concentration} \end{pmatrix} . \qquad (1)$$

In Equation (1) the coefficient relating surface particle flux to aerosol concentration has the dimensions of velocity and is called the deposition velocity. Under conditions of small deposition velocity, particle flux can remain low even in the presence of significant aerosol concentration. Conversely, high surface particle flux can be experienced during wafer exposure to seemingly modest aerosol particle concentrations when deposition velocity is high. To minimize surface particle flux requires understanding of the forces and variables affecting particle deposition velocity (such as flow velocity and electrical forces). As a prerequisite for measuring particle deposition velocities, this paper discusses the measurement of the two parameters, aerosol particle concentration and particle areal density on a wafer. Both are needed in order to calculate particle deposition velocity via Equation (1) for a specific set of experimental conditions.

AEROSOL PARTICLE MEASUREMENT

A wide variety of instrumentation now exists for measuring aerosol particle concentrations and size distributions. In clean rooms, the most appropriate and commonly used instrument is the optical particle counter (OPC) which has the desirable properties of counting single particles in real time and of classifying particles according to size by analyzing the signal scattered from each particle.[1] Some OPCs sample at flow rates of 500 cm^3/s and can detect aerosol particles as small as 0.2 μm. Lower flow rate instruments can measure even smaller sized particles--to 0.09 μm or even slightly smaller. The smallest sized particle optically detectable,[2] however, is probably on the order of 0.05-0.06 μm.

The small measuring volume required by the smallest particle-detecting instruments means very low sampling flow rates ($\sim$1 cm^3/s) so that operation in a clean room becomes impractical because of the small number of particles counted at such low flow rates--the 500 cm^3/s units are much better suited for measuring aerosol particles at low concentration. At present, however, only one 500 cm^3/s unit claims to detect 0.1 μm particles (Particle Measuring Systems' LPC 110).

Growing concern over the threat posed by particles smaller than the 0.1 to 0.2 μm limits of today's clean room optical particle counters has raised questions regarding aerosol particle concentrations in the size decade between 0.01 μm (100 Å) and 0.1 μm (1000 Å). In typical urban ambient air, shown in Figure 1, aerosol particle concentration varies inversely with particle size to 3rd or 4th power (this empirical data fit is often called the Junge power law). The sub 0.1 μm portion of the aerosol plot is collected using such instruments as electrical mobility analyzers and diffusion batteries, each in conjunction with a condensation nuclei counter. These instruments perform satisfactorily when constructing size distribution plots from the relatively high aerosol particle concentrations found in outdoor ambient air. At the aerosol particle concentrations typical of state-of-the-art clean rooms, however, counting statistics again limit the usefulness of such conventional instruments as is true for the low flow rate optical particle counters.[4] For example, the commercially available version of the differential mobility analyzer (TSI's DMPS Model C3932) produces unstable size distributions[5] when total aerosol particle concentrations decrease below about 1 cm^{-3}. With a custom diffusion battery and two condensation nuclei counters[6]--a nonstandard configuration--crude size distributions can be deduced over the ultrafine size range at total particle concentrations on the order of 0.001-0.01 cm^{-3}. Typical plots of size distributions found in various clean rooms at rest (no personnel in the clean room; no operations active) appear in Figure 2. The flattening of the size distribution below about 0.1 μm reflects the grade efficiency characteristics of today's high quality fibrous filters. Operating on a size distribution such as illustrated in Figure 1 with a typical HEPA filter size efficiency curve (Figure 3) produces curves similar to those experimental curves[8] presented in Figure 2.

Unfortunately, neither of the two instrument combinations used to collect the data on which the two solid curves of Figure 2 are based is easy to use as a routine monitor and, thus, neither is likely to enjoy widespread acceptance in the user community. The same criticism is no

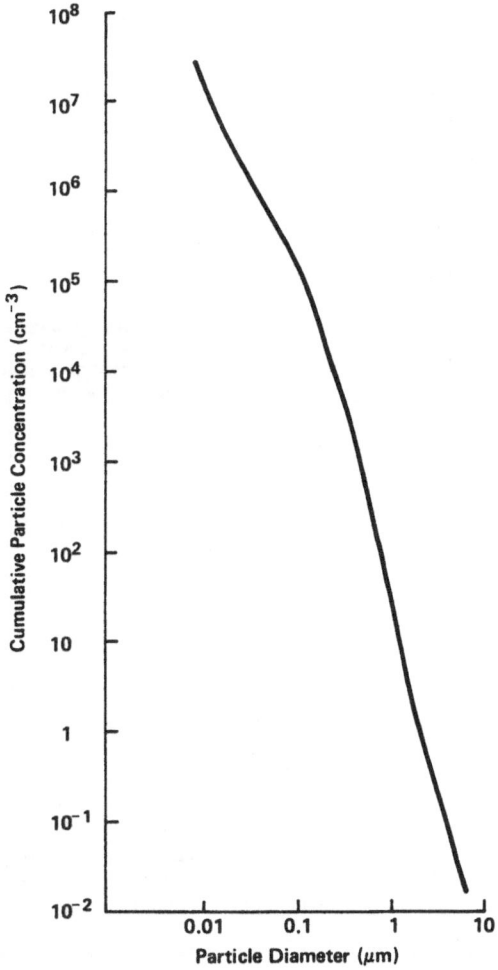

Figure 1. Cumulative particle number distribution, Los Angeles smog.[3]

longer true of the third curve of Figure 2, based on a single ultrafine particle datum read from a condensation nuclei counter (CNC). That single point, plotted in Figure 2 at 0.01 μm and used in conjunction with conventional OPC data, forms the basis for drawing the dashed curve appearing at the bottom of Figure 2. The dashed portion of that curve is drawn to be similar in shape to the top two curves and to terminate on the 0.01 μm axis at the concentration measured by the CNC. While this procedure works adequately for clean rooms at rest in which the background aerosol particle concentration is probably dominated by the filtration of outside ambient air,[9] it is likely to be less satisfactory for operating clean rooms in which internal particle sources dominate the aerosol particle

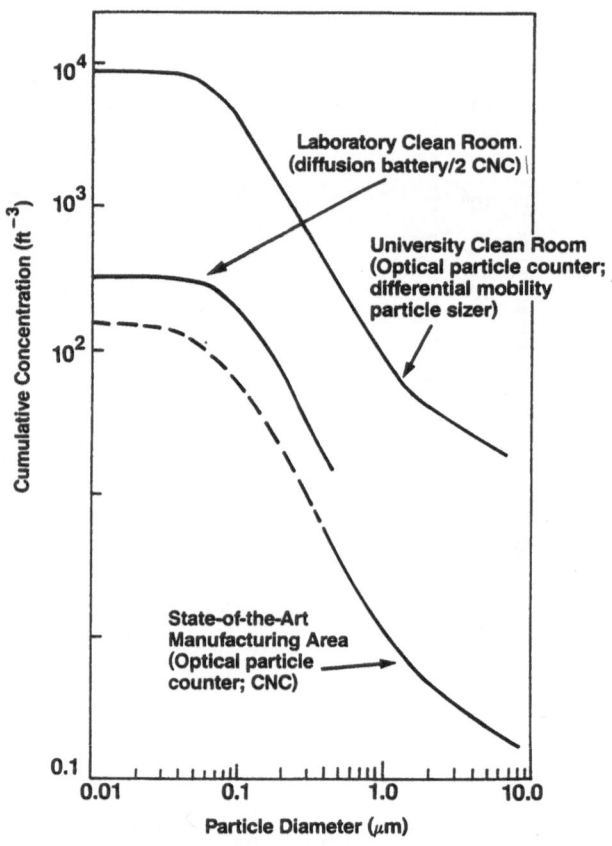

Figure 2. Aerosol particle spectra in clean rooms at rest.

concentration and size distribution. If, however, total particle concentration greater than a certain size is all that is needed (rather than a detailed size distribution), then the CNC covers the ultrafine aerosol particle regime quite adequately. Indeed, when used in conjunction with an OPC, the CNC even lends itself nicely to the measurement of a coarse cumulative size distribution plot, spanning the size range 0.01 to 5 μm. For example, Figure 4 presents a 5-day record of particle concentration in

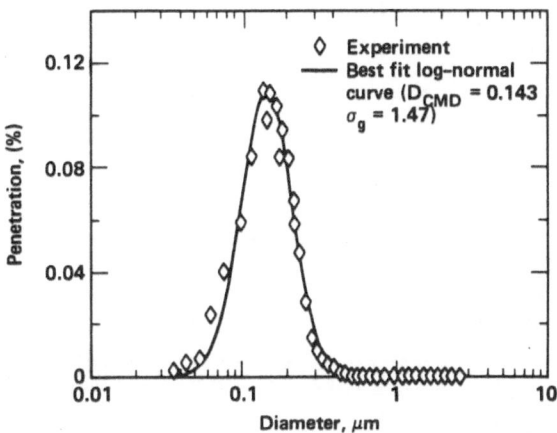

Figure 3.   Particle penetration through a high-efficiency
fibrous filter.[7]

a solvent hood of a semiconductor facility as measured simultaneously by a
CNC (Sensor 1) and an OPC (Sensor 4). This particular OPC (TSI Model
3755) reads out average particle concentration in two channels: concentration of aerosol particles greater than 0.5 μm and concentration of
aerosol particles greater than 5 μm. The CNC (TSI Model 3760) reads out
concentration of particles greater than 0.01 μm. The at-rest size distribution, based on just two data points, might well be drawn as shown in
Figure 5, following the general shape of the at-rest data of Figure 2.
The size distribution of the spike recorded on Tuesday evening, however,
is far less certain. Drawing the curve, shown in dashes, to follow the
at-rest shape is hard to justify and in the absence of intermediate data
points a straight line through the three points is probably all that
should be shown. More detail would depend on understanding the source of
these particles, appearing late on Tuesday afternoon, right after normal
work hours. This Tuesday spike affected a large area as can be seen by
the OPC flow monitor (Sensor 6) adjacent to the hood and also by the CNC
data (not shown) recorded simultaneously in a neighboring hood.

Many other spikes also appear in the CNC record, none of which appear
to be detected by the OPC. Most of these spikes occur during normal working hours and are assumed to be activity related--a hot plate, for example, would be expected to emit small particles that would be detected by
the CNC but not the OPC. The floor OPC (Sensor 6) identifies working
hours quite clearly. No concurrent CNC data for this floor position were
collected, however.

The simplicity and ready availability of the CNC make it the best
choice at present for monitoring ultrafine aerosol particles in clean
rooms. In spite of the fact that the CNC provides no size information, it
can be used to identify ultrafine spikes by comparing its output with
conventional OPC data, as in Figure 4.

Insofar as deposition velocity measurements go, the measurement of
aerosol particle concentration is well ahead of the measurement of surface

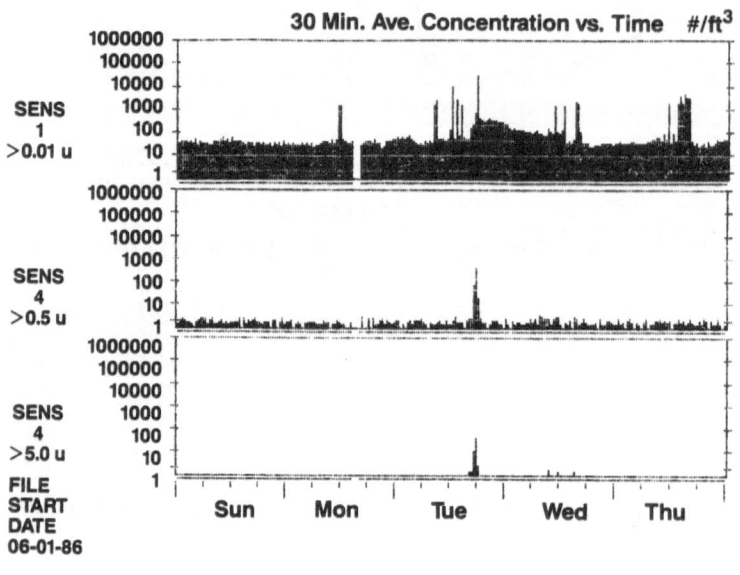

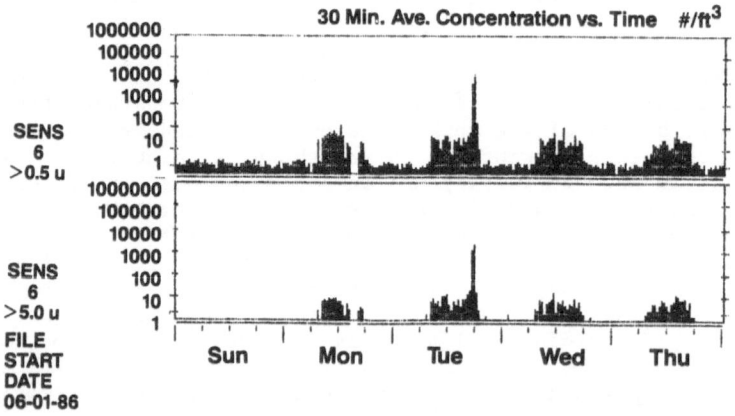

Figure 4. Five-day record of clean room aerosol particle
concentration (30 min averages) as measured by:
    Sensor 1: CNC in a solvent hood
    Sensor 4: OPC in same hood
    Sensor 6: OPC on floor adjacent to hood.

particles which at present are carried out primarily by optical methods,[10] meaning that no method now exists for measuring ultrafine particles on surfaces other than electron microscopy, a notoriously slow and prohibitively expensive method for routine monitoring. Since the measurement of ultrafine particles on a wafer surface is very difficult, the CNC measure of ultrafine aerosol particles may turn out to be the best indication of

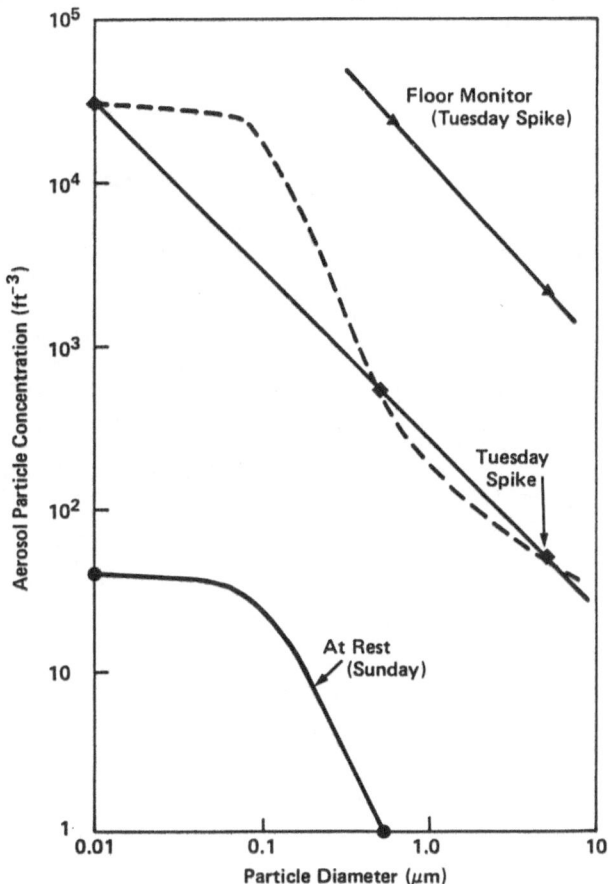

Figure 5. Size distributions from CNC/OPC data of Figure 4.

problems attributable to diffusional deposition of ultrafine particles on a wafer surface. As discussed next, optical detection of larger (>0.2 μm) particles on surfaces even lags optical detection of aerosol particles in both understanding and performance.

MEASUREMENT OF PARTICLE AREAL DENSITY ON AN OPTICALLY FLAT SURFACE

The measurement of particles on a silicon wafer is most commonly performed optically using a laser scanner. A number of such instruments is commercially available. This section discusses the performance of just one specific unit, a wafer inspection system (WIS-150) manufactured by Aeronca Electronics, Inc. (now Estek) of Charlotte, NC (Figure 6). While all data presented have been collected on the WIS-150, many features and properties are common to the general problem of light scattering from a particle on a reflecting surface and so are of broader interest.

Scattering of light from a spherical particle alone, as in an aerosol particle passing through an OPC, is well understood. Solutions to Maxwell's equations for incident radiation with wavelength similar to the diameter of the scatterer have been published by Mie[11] and others. Even though the predicted intensities of scattered light are complex functions of particle size and index of refraction, these intensities can be calculated precisely and have been experimentally confirmed. Thus, this particular optical problem is well understood. Not so for scattering from a sphere resting on a reflecting plane such as a polystyrene latex (PSL) particle on the surface of a bare silicon wafer. Figure 7 illustrates

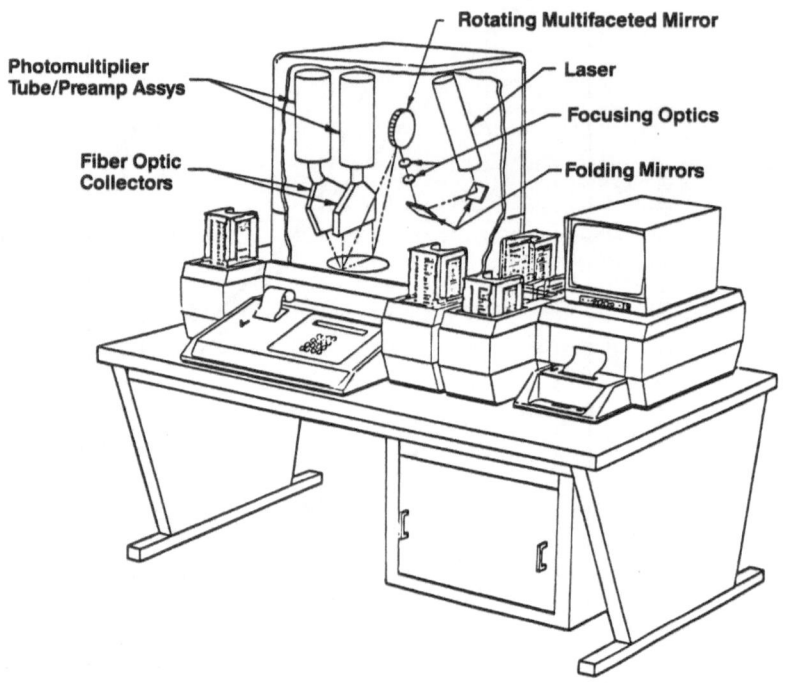

Figure 6. Schematic of the Aeronca WIS-150 surface scanner.

several different paths whereby incident light can reach a detector located directly above the particle. Calculations in the Mie regime[12] in which each path is treated independently of the other predict the two paths involving reflection to produce a larger scattered light intensity than the classical Mie backscattering alone (Figure 8). This result comes about because the path B scatterings are in the forward direction while path A consists of backscattering. Mie solutions show forward scattering intensities to be much greater than backscattering.

As an additional complication to the problem, interference between the reflected beam and the incoming beam, not considered in the independent paths of Figure 7, has been postulated by Knollenberg[2] to alter the form of the wave incident on the particle. A rigorous solution to Maxwell's equations, including the reflecting boundary, would have to include such interference effects. Such rigorous solutions for the scattering of light from a particle resting on a surface are now being reported.[13]

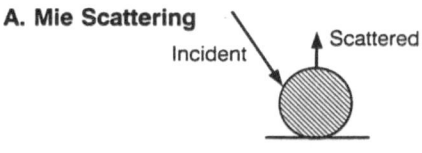

A. Mie Scattering

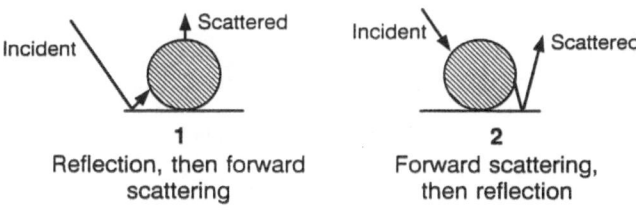

B. One Scattering, One Reflection

Figure 7. Scattering options for particle resting on a reflecting plane.

What we report here is a calibrating series of measurements in which PSL spheres of known size have been deposited on silicon wafers and then subjected to repeated measurements by the WIS-150 at ever decreasing values of threshold voltage settings. The threshold voltage sets the minimum value of photodetector output that is counted. At high threshold voltages few scattered light signals generate a voltage pulse big enough to be counted; at low threshold voltage most scattered light is counted. When the primary source of light scattered from a wafer is an array of deposited monodisperse spheres, the pattern seen in a series of repeated scans with incrementally decreasing values of threshold voltage is: (1) an initial region of little change in counts as the threshold voltage is reduced from its high starting value; (2) followed by a sharp increase in counts as the threshold voltage setting is reduced below the level corresponding to light scattered from the specific monodisperse sphere size deposited; and (3) finally another plateau in count with continued reduced threshold voltages until the noise signal levels are reached. The step rise in count associated with a specific PSL sphere size provides a means of calibrating the laser scanner.[10] An increase in cumulative particle count as a function of decreasing threshold voltage can thus be converted into a plot of cumulative concentration as a function of equivalent PSL sphere size. This procedure has been carried out for PSL spheres deposited on bare polished silicon wafers and also on oxidized silicon wafers of various oxide thicknesses. The results of these calibration runs are summarized next.

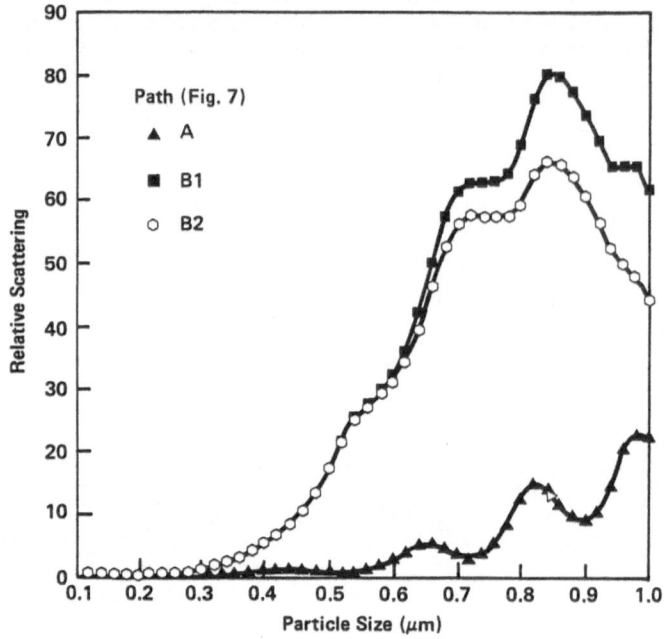

Figure 8.   Calculated relative scattering intensities for optical paths defined in Figure 7.[12]

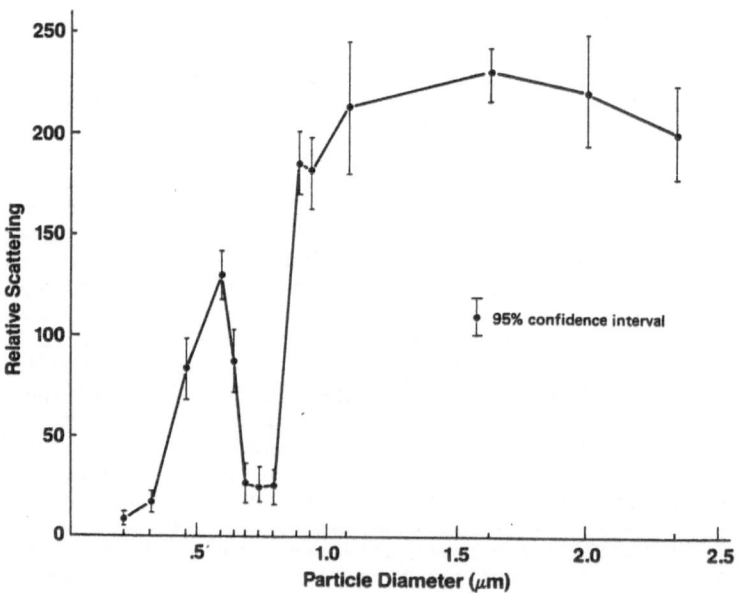

Figure 9.   Summary of relative scattering versus particle diameter for polystyrene latex (PSL) on bare silicon wafers for threshold T-1.

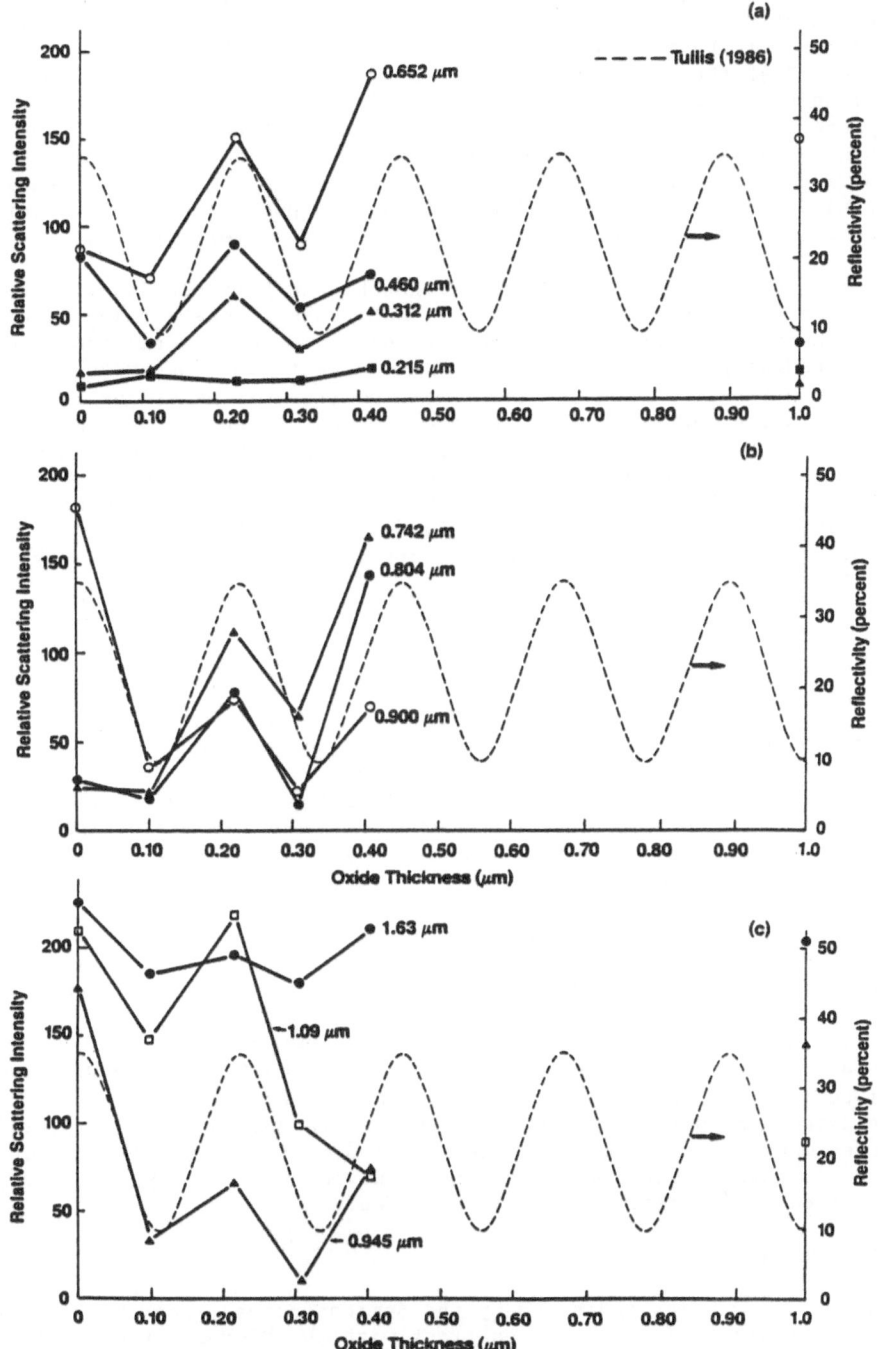

Figure 10. The effect of oxide layer thickness on the WIS-150 response for various polystyrene latex sphere sizes.

Figure 9 shows the relationship between light scattering intensity and PSL sphere size for spheres deposited on bare polished silicon wafers. The striking feature of this plot is the nonmonotonic relationship between particle size and light scattering, especially for particle sizes near the wavelength of the illuminating laser (633 nm). Qualitatively, this feature is part of the Mie solutions for light scattering from an isolated particle. That the reflecting surface also plays an important role in the scattering can be seen by comparing the scattered light intensity from a PSL sphere deposited on bare silicon with that scattered from the same size sphere deposited on an oxidized wafer. These intensities are not the same as illustrated in Figure 10. Furthermore, the scattering intensity varies with oxide thickness, and for at least a portion of the submicrometer particle size range, in phase with the surface reflectively of the oxidized wafer (the dashed reflectivity curves shown in Figure 10 are from Tullis.[14] Shadowing of the incident light by the particle itself also changes the scattered light intensity. And, when a nonabsorbing layer such as an oxide is inserted between the particle and the silicon, the reduced shadowing may explain the increased scattering observed for thicker oxides near the reflectivity maximum (the 0.4 μm vs the 0.2 μm oxides). This effect may be the standoff effect cited by Knollenberg[15]-- oxide layers of optical thickness equal to integral half wave lengths ("absentee" layers) do not alter the "electric field structure" but do displace the particles from the reflecting silicon surface, reducing or eliminating the shadowing that occurs under near vertical illumination. It is evident only in the intermediate particle sizes (0.6 to 0.9 μm) in Figure 10.

SUMMARY

The smallest sized PSL surface particle measurable by laser scanning instruments is in the 0.2 to 0.3 μm size range. This size limit matches that of the 500 cm^3/s OPCs most commonly used for measuring aerosol particle concentration in clean rooms. Thus, deposition velocity measurements for submicrometer particles larger than, say, 0.3 μm can be carried out using present instrumentation. With some refinement, these optical techniques can probably be extended down to 0.1 μm equivalent PSL particle size.

Below 0.1 μm the measurement of surface particles by a laser scanner no longer is practical or possible. Imaging by an electron microscope is a feasible but costly alternative. Such measurements, combined with CNC aerosol particle data, could extend the measurement of deposition velocity to include ultrafine particles. Such measurements have not yet been reported--indeed, the measurements of particle deposition velocity in the 0.1 μm to 1.0 μm range are not yet very common. However, as semiconductor device dimensions push into the submicrometer range, interest in ultrafine particles mounts and further examination of ultrafine particle behavior becomes important. Instrumentation for studying ultrafine particle deposition velocity is barely adequate at present.

ACKNOWLEDGMENT

This research was sponsored by the Semiconductor Research Corporation through a manufacturing science contract with the Microelectronics Center of North Carolina.

REFERENCES

1. R. P. Donovan, B. R. Locke, and D. S. Ensor, Real Time Measurements of Single Submicron Aerosol Particles in Clean Rooms, Solid State Technology, 28, 9, pp. 139-148 (1985).

2. R. G. Knollenberg, The Measurement of Particle Sizes Below 0.1 Micrometers, J. Environ. Sci., 28, 1, pp. 32, 41-47 (1985).

3. K. T. Whitby, R. B. Husar, and B. Y. H. Liu, in"Aerosols and Atmospheric Chemistry," G. M. Hidy (ed.), Academic Press, Inc., New York, pp. 237-264, 1972.

4. R. P. Donovan, B. R. Locke, C. M. Osburn, and A. L. Caviness, Ultrafine Aerosol Particles in Semiconductor Clean Rooms, J. Electrochem. Soc., 132, 11, pp. 2730-2738 (1985).

5. B. R. Locke, R. P. Donovan, D. S. Ensor, and C. M. Osburn, Semiconductor Clean Room Particle Spectra, in "Aerosols," B.Y.H. Liu, D.Y.H. Pui, and H. Fissan (eds.), Elsevier Science Publishing Company, Inc., pp. 669-672, 1984.

6. B. R. Locke, R. P. Donovan, D. S. Ensor, and A. M. Caviness, Assessment of the Diffusion Battery for Determining Low Concentration Submicron Aerosol Distributions in Microelectronics Clean Rooms, J. Environ. Sci., 28, 6, pp. 26-29 (1985).

7. W. Bergman, A. Biermann, W. Kuhl, B. Lum, A. Bogdanoff, H. Hebard, M. Hall, D. Banks, M. Mazumder, and J. Johnson. Electric Air Filtration: Theory, Laboratory Studies, Hardware Development, and Field Evaluations. UCID-19952, Lawrence Livermore National Laboratory, University of California, Livermore, CA, 1984.

8. D. S. Ensor, R. P. Donovan, and B. R. Locke, Particle Size Distributions in Clean Rooms, in Proceedings, 8th International Symposium on Contamination Control, September 1986, A.S.C.C.A., Via Prina, 15, 20154 Milano, Italy.

9. G. J. Sem, A Case for Continuous Multipoint Particle Monitoring in Semiconductor Clean Rooms, in Proceedings of the 32nd Annual Technical Meeting of the Institute of Environmental Sciences, IES, Mount Prospect, IL, pp. 432-438, 1986.

10. B. R. Locke, and R. P. Donovan, Particle Sizing Uncertainties in Laser Scanning of Silicon Wafers Calibration/Evaluation of the Aeronca Wafer Inspection Station 150, Journal of the Electrochemical Society, 134, 7 (1987).

11. H. C. Van de Hulst, "Light Scattering by Small Particles," Dover Publications, Inc., New York, 1981.

12. R. C. Yu, P. C. Reist, and D. Leith, Light Scattering by Particles on Reflective Surfaces, Unpublished note, UNC, Chapel Hill, NC, 1986.

13. G. L. Wojcik, D. K. Vaughan, and L. K. Galbraith, Calculation of Light Scatter from Structures on Silicon Surfaces, SPIE vol. 774 Lasers in Microlithography, pp. 21-31 (1987).

14. B. J. Tullis, Measuring and Specifying Particle Contamination by Process Equipment: Part III, Calibration, Microcontamination $\underline{4}$, 1, pp. 51-55, 86 (1986).

15. R. G. Knollenberg, The Importance of Media Refractive Index in Evaluating Liquid and Surface Microcontamination Measurements. In: Proceedings of the 32nd Annual Technical Meeting of the Institute of Environmental Sciences, IES, Mount Prospect, IL, pp. 501-511, 1986.

PARTICULATE CONTAMINATION ON WAFER SURFACES RESULTING

FROM HEXAMETHYLDISILAZANE/WATER INTERACTIONS

M.A. Logan*, D.L. O'Meara, J.R. Monkowski*, H. Cowles**

J.C. Schumacher Company
1969 Palomar Oaks Way, Carlsbad, CA 92009
*Monkowski-Rhine, Incorporated
9250 Trade Place, San Diego, CA 92126
**Great Western Silicon
11515 East Riggs Road, Chandler, AZ 85249

A major step in the photolithographic processing of
silicon wafers is the application of hexamethyldisilazane
(HMDS) as a photoresist adhesion agent. At this time the
preferred application technique is one known as vapor
priming. In the vapor prime system, the silicon wafers
are heated in a vacuum oven, where they are exposed to
vapors of HMDS which are pulled from a container of the
HMDS liquid. In vapor prime systems as they are presently
configured, there is significant opportunity for contamination by atmospheric moisture. HMDS is known to be
hygroscopic and to react with moisture to form ammonia and
other reaction products. For the most part, the major
result ascribed to this reaction has been a degradation of
the adhesion promotion properties of the HMDS. In this
investigation, however, we have looked at particulate contamination generated by the HMDS/H_2O interaction. A model
is proposed to describe the in-situ generation of particles from an HMDS/H_2O reaction, and considerations for
minimizing contamination in an HMDS vapor prime system are
discussed.

INTRODUCTION

An important step in the photolithographic processing of silicon wafers
is the application of hexamethyldisilazane (HMDS or $((CH_3)_3Si)_2NH$) as a
photoresist adhesion promoter.[1] The preferred technique for application of
HMDS is vapor-priming, in which the silicon wafers are exposed to HMDS
vapors in a vacuum oven.

In vapor prime systems, contamination by atmospheric moisture is common.
Hexamethyldisilazane is hygroscopic and reacts with water to form ammonia
and trimethylsilanol.[2] Previous researchers felt this reaction caused a
degradation of the adhesion promotion properties of the HMDS,[1] and thus

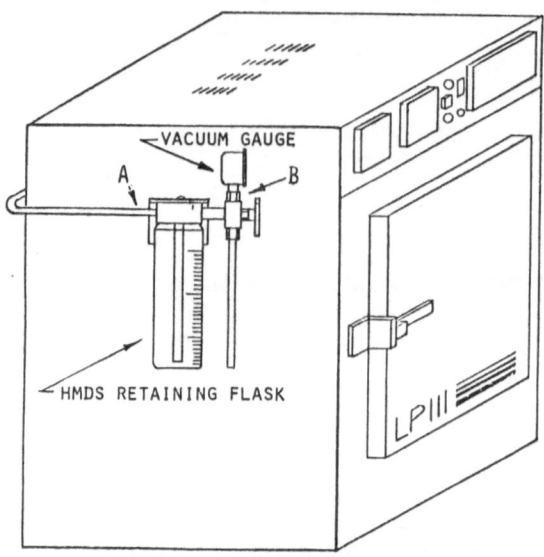

Figure 1.   Conventional HMDS vapor prime system.

caused the loss of HMDS effectiveness. In our experiments, however, we have found that water contamination not only leads to lift-off problems, but also causes wafer particulate contamination. In this paper, we describe our study of the mechanisms responsible for particulate contamination on wafer surfaces resulting from $HMDS/H_2O$ interactions.

MATERIALS AND METHODS

Fig. 1 shows a typical vapor prime system (Yield Engineering Systems Model LP III, supplied by Yield Engineering Systems to the J.C. Schumacher Company). In the actual vapor-priming operation, HMDS vapors are introduced into the heated oven (typically at 150°C) where the wafers are located. Exposure to HMDS vapors lasts approximately five minutes.

In the conventional system, the HMDS retaining flask (shown in Fig. 1) is filled from a larger HMDS container or bottle through the dip-tube extending down from the vacuum gauge (labelled B). The larger container is opened, and positioned so the dip-tube extends into the liquid. By opening the valve below the vacuum gauge, the HMDS is pulled by the pressure drop into the retaining flask. The valve is then closed, the large container is taken away, and the valve is opened momentarily to allow the HMDS remaining in the tube to be drawn into the retaining flask. Once the HMDS is in the flask, vapors can be introduced into the heated oven via the tube labelled "A". In this study, the HMDS introduced in this manner contained less than 100 ppm H_2O in the large container.

Alternatively, an HMDS addition was made from a quartz reservoir containing predried HMDS to minimize H_2O contamination. The experimental set-up is diagrammed in Fig. 2. Ignoring the H_2O container for the moment, this alternative approach involves the use of a hermetically sealed quartz reservoir containing low moisture (<50 ppm H_2O) HMDS supplied by the J.C. Schumacher Company (JCS). Since this quartz container is connected directly into the system, there is little opportunity for atmospheric moisture to enter the system.

The H_2O container shown in Fig. 2 was included to allow the controlled addition of water vapor into the HMDS vapor flow. The liquid nitrogn cold trap retains effluent gases.

Wafers used in these experiments were 100 mm Monsanto test-quality wafers. After priming with HMDS, the wafers were characterized by wafer-surface scanning, scanning electron microscopy (SEM) with x-ray microanalysis, secondary ion mass spectrometry (SIMS), Auger electron spectroscopy (AES), and x-ray photoelectron spectroscopy (XPS or ESCA).

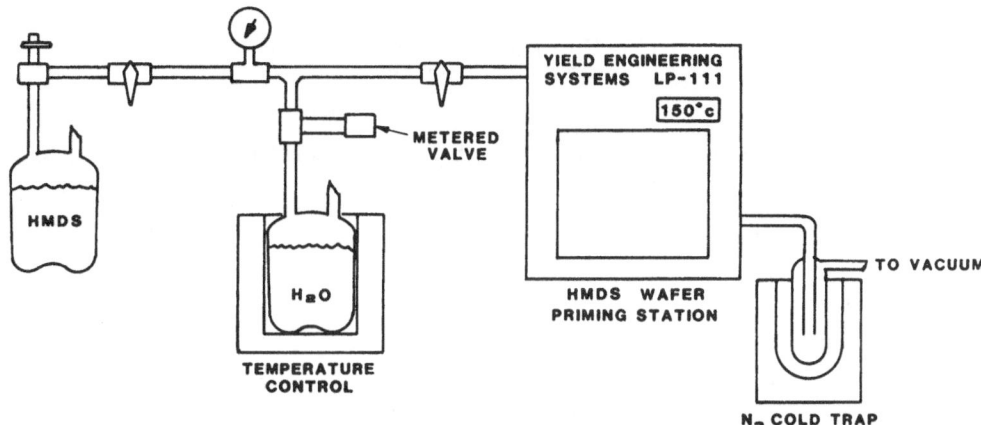

Figure 2. Experimental set-up allowing introduction of controlled amounts of moisture contamination.

Surface scanning of wafers was executed to determine the particle population on the wafer surfaces. A Hamamatsu wafer particle scanner was used, with the particle size range set to 0.5 - 3.0 μm. The scanning electron microscope was an Amray 1000 SEM with Kevex energy-dispersive x-ray microanalysis. The Kevex unit uses an ultrathin window, allowing detection of all elements of atomic number 6 or greater. The instrument was used with 10-20 kV acceleration beams and has a specified resolution of 70 Angstroms.

Auger electron spectroscopy and static SIMS were performed on a PHI 560 system. The chamber was equipped with a small spot (1 μm spot size) electron gun, a differentially pumped argon ion gun, a quadrupole mass spectrometer, a double-pass cylindrical mirror analyzer, and a differen-tially pumped sample introduction system. Angle-resolved XPS was done with a Physical Electronics ESCA Lab system. The ESCA system used a dual anode 15 keV, 400 W source, and a hemispherical analyzer. All data were obtained with a Mg anode (K at 1253.6 eV). The operating pressure was less than 5 X 10^{-9} torr, and the angular range was 15-90°.

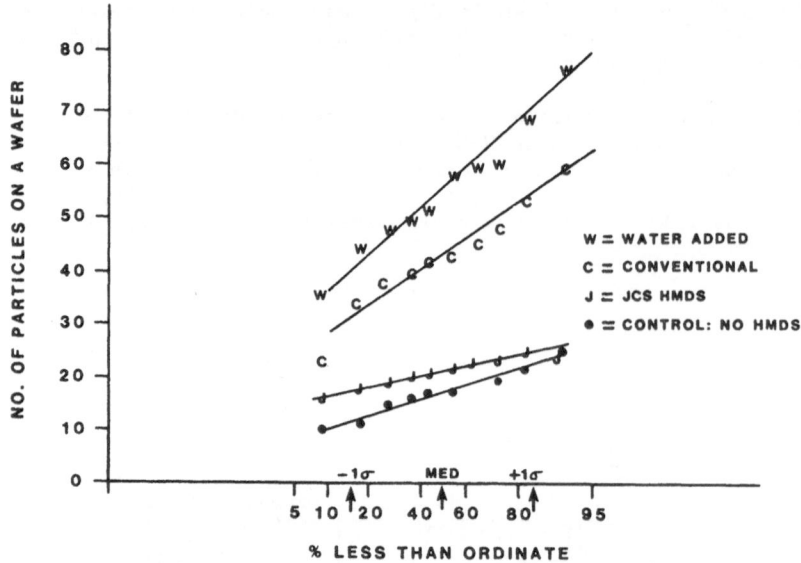

Figure 3. Statistical distribution of wafer particle levels for different HMDS wafer priming conditions.

RESULTS AND DISCUSSION

Wafer Particle Measurements

Fig. 3 shows the results of particle measurements for the different wafer priming operations. The data are plotted on a linear/normal chart, and show the key statistical parameters. The control wafers were taken directly from the wafer box, and did not go through a priming operation; their median particle count is 17 per wafer (in the range of 0.5 - 3.0 μm diameter).

Of the wafers going through the HMDS priming operation, those using the low-moisture HMDS in the hermetically sealed quartz container show the lowest number of additional particles (median = 21). An alternative low-contamination introduction method using low-moisture HMDS in a membrane-sealed glass bottle (data not included in Fig. 3) showed essentially the same results, with a difference in the median particle count between this introduction method and control wafers of 3. In this case, the bottle was sealed with a membrane impervious to water, and was introduced into the system such that the membrane was penetrated without allowing contamination by any atmospheric moisture.

For the conventional HMDS system, the particle counts on the wafers were substantially higher (median = 45). In view of the atmospheric exposure and access to moisture, particularly during the refill process, the particle generation was believed to result from the interaction of HMDS and H_2O. To further substantiate this hypothesis, the low-moisture (JCS) HMDS was run with approximately 100 ppm H_2O added to the system (using the system of Fig. 2). Indeed, the particle counts were elevated (median = 55), and were close to the counts found on wafers run using the conventional system.

Fig. 4a shows an SEM micrograph of a particle found on one of the wafers exposed to the HMDS/H_2O vapor mixture. This particle is one of the largest found. It clearly shows the open, porous structure typical of these particles.

The reaction between HMDS and H_2O to form particles occurs in the liquid phase as well as in the vapor phase. Fig. 4b shows an SEM micrograph of a solid phase formed over a period of 10 days in a room temperature mixture of 1:1 H_2O:HMDS. Note the similar microstructure as seen in the particle of Fig. 4a.

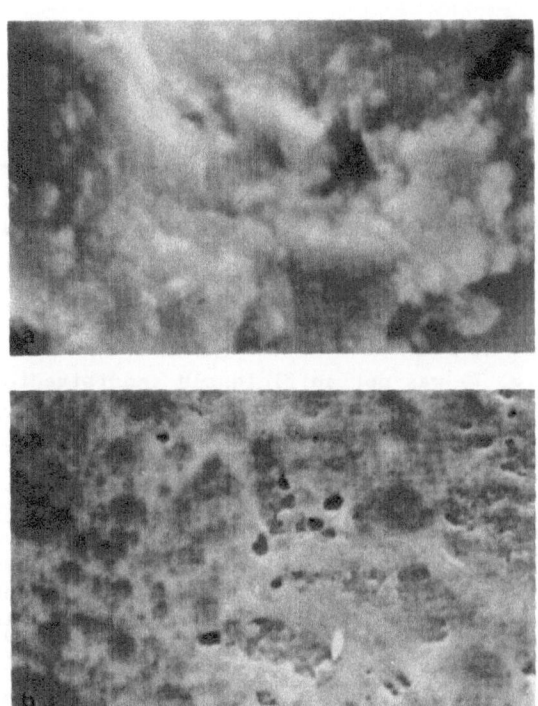

Figure 4. (a) SEM micrograph of particle contamination on a primed wafer surface (1000 ppm H_2O added to HMDS). (b) SEM micrograph of solid reaction product formed in a 1:1 solution of HMDS:H_2O.

The particles were further analyzed by SEM x-ray microanalysis, as shown in Fig. 5. The spectrum, typical for vapor-primed wafers, shows a large amount of silicon and carbon and a small amount of oxygen. Particles found on wafers not primed with HMDS typically exhibit less carbon and more oxygen. Nitrogen was not found in any of the particles.

Surface Chemical Analysis

Fig. 6a and 6b show SIMS spectra of wafer surfaces for an unprimed wafer and a wafer exposed to low-moisture HMDS. This is a large-area technique, so contributions from particles constitute a minor part of the signal. In Fig. 6b, the expected Si-Me (methyl, CH_3) groups are noted. The

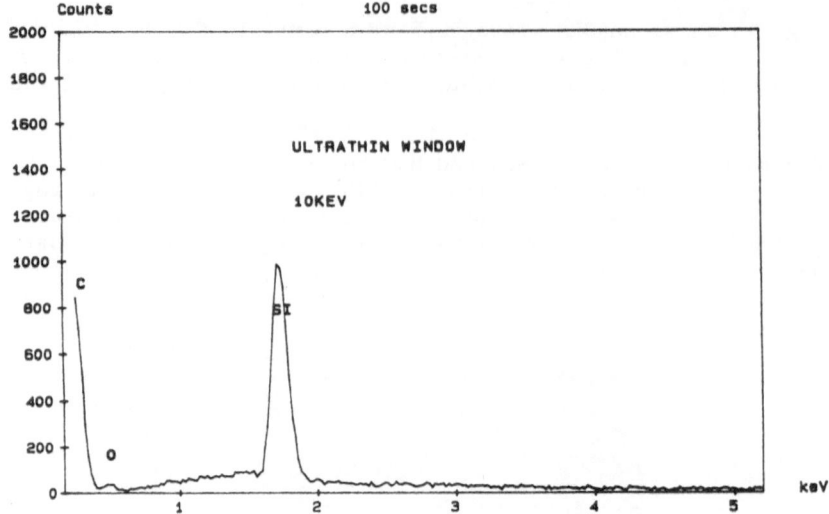

Figure 5. SEM x-ray spectrum of particulate contamination showing the presence of carbon, oxygen, and silicon.

AES technique (Fig. 7) was not sufficiently sensitive to differentiate between the different wafers, but it did substantiate the presence of carbon, silicon, and oxygen in the very near surface region.

Fig. 8 is a typical XPS spectrum, again showing the presence of the silicon and carbon. In this case, however, differences between differently treated wafers could be detected. Fig. 9a shows carbon concentration as a function of reciprocal angle. These data show the higher concentration of carbon in the low-moisture-HMDS primed wafers as compared to either the unprimed wafers or the wafers primed with moisture-contaminated (1000 ppm) HMDS. Deconvolution of this XPS data from angles 15-75° yields the concentration profiles shown in Fig. 9b. The graph shows that carbon is in the near surface region in all cases. Significantly, the wafer exposed to the low-moisture HMDS shows the largest amount of surface carbon; the wafer exposed to an HMDS/H_2O vapor mixture shows no more carbon than the control wafer.

Proposed Model for Particle Formation

The adhesion promotion property of HMDS is believed to occur via the following reaction:[1]

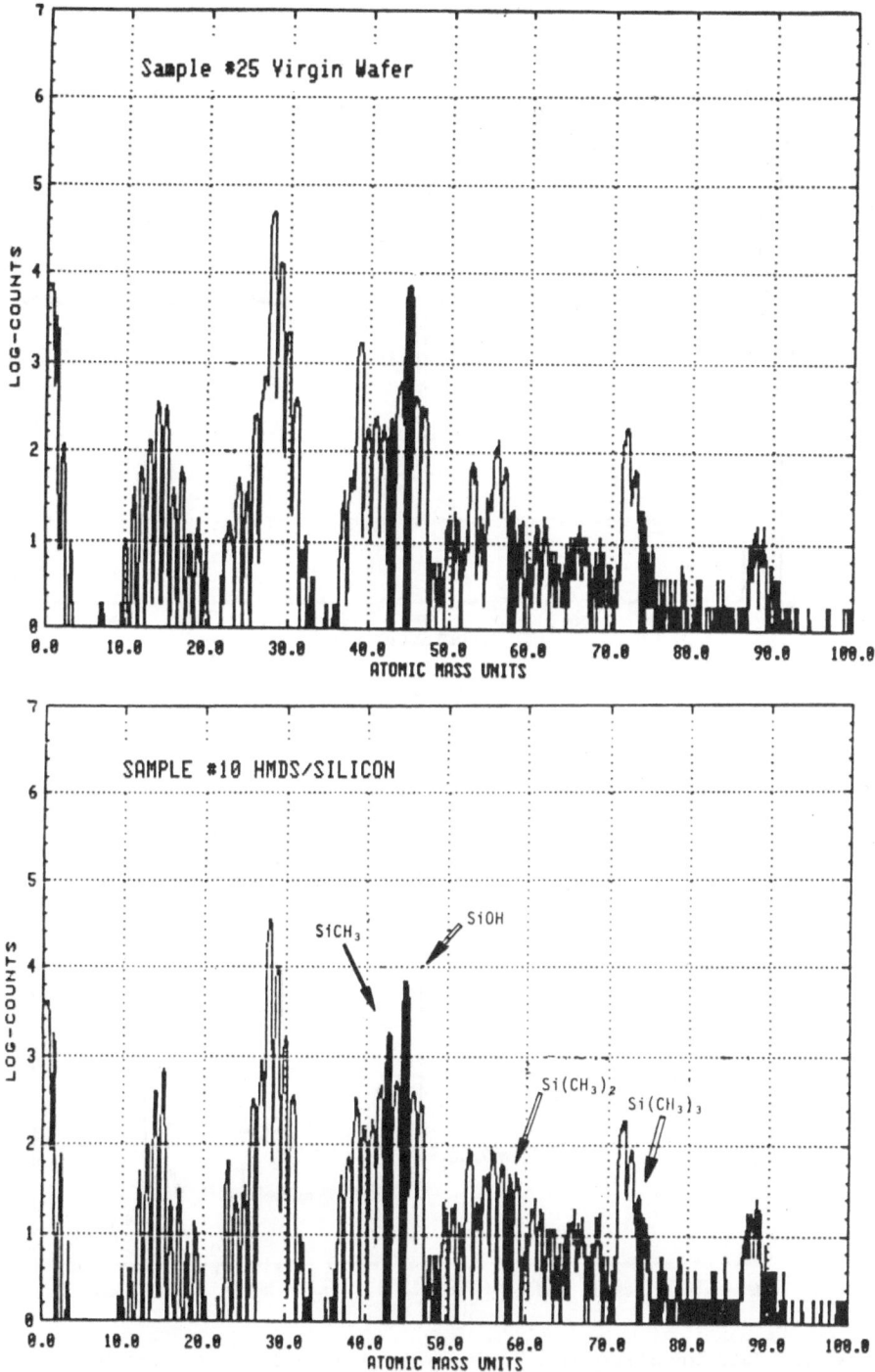

Figure 6. SIMS spectra of (a) an unprimed wafer and (b) a wafer primed with low-moisture HMDS.

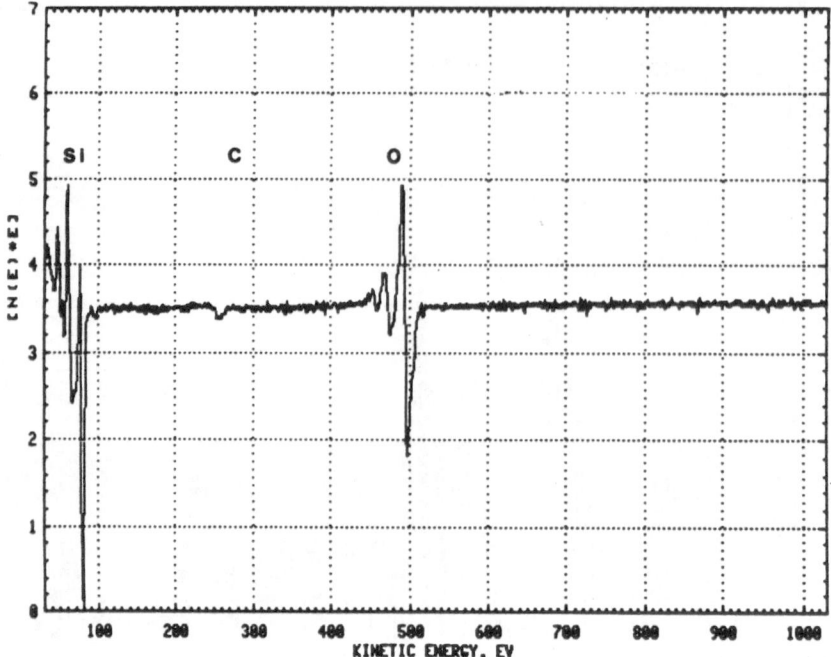

Figure 7. AES spectra of a wafer primed with HMDS containing 1000 ppm H_2O.

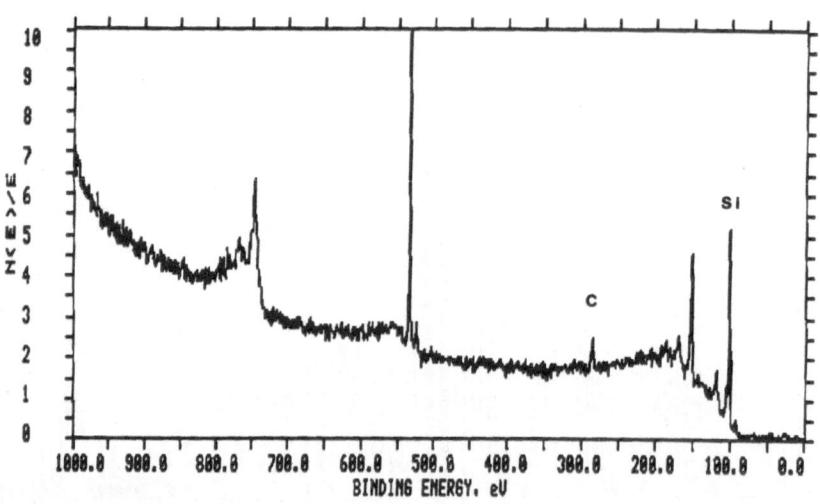

Figure 8. ESCA spectrum of a wafer primed with low-moisture HMDS. To determine carbon profiles, the carbon and silicon signals were compared as a function of the angle of incidence for various priming operations.

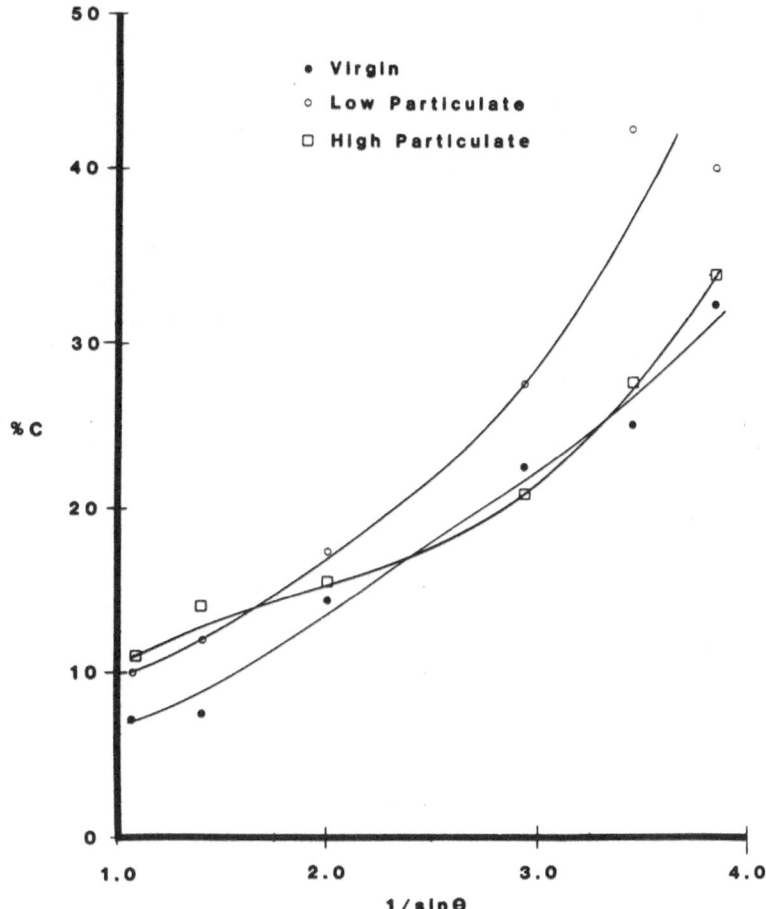

Figure 9. (a) Angle-resolved ESCA data showing the relative carbon signal strength as a function of reciprocal angle.

This reaction is consistent with the Si-Me groups seen in the SIMS spectra, as well as the carbon observed in the AES and ESCA spectra.

In the presence of water, which is known to degrade the adhesion properties of HMDS, the most likely reaction is the following:[2]

$$
\begin{array}{ccccccccc}
 & \text{Me} & & \text{H} & & \text{Me} & & & & & & \text{Me} \\
 & | & & | & & | & & & & & & | \\
\text{Me} - & \text{Si} & - & \text{N} & - & \text{Si} & - \text{Me} & + & 2\text{H}_2\text{O} & ---> & 2\,\text{H} - \text{O} - & \text{Si} & - \text{Me} & + \text{NH}_3 \\
 & | & & & & | & & & & & & | \\
 & \text{Me} & & & & \text{Me} & & & & & & \text{Me}
\end{array}
$$

In contrast to the relatively weak Si-N bond in HMDS, which readily allows reaction with the OH groups on the wafer surface, the strong Si-O bond in trimethylsilanol is much less likely to react with the hydrated wafer surfaces. As a result, once the HMDS reacts with H_2O to form trimethylsilanol, its photoresist adhesion properties are reduced.

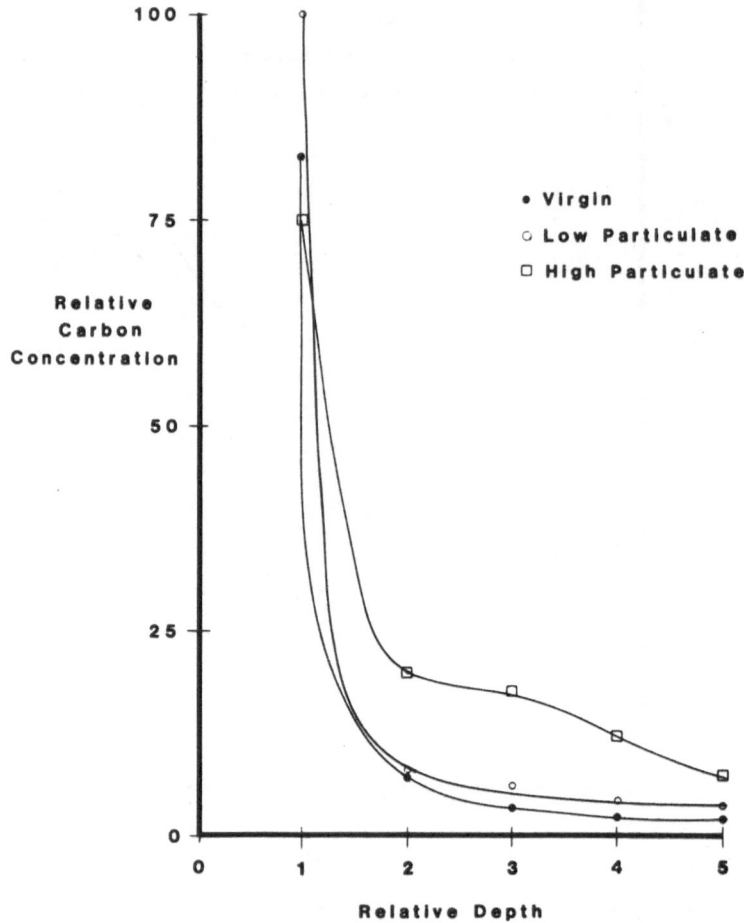

Figure 9. (b) Deconvolution of the ESCA data, showing relative carbon concentration profiles.

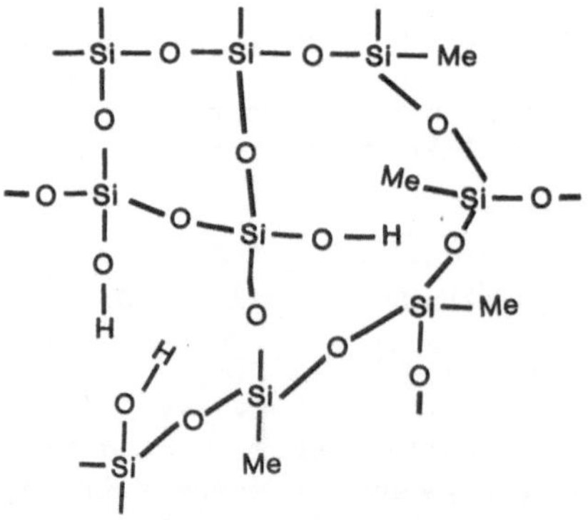

Figure 10. Proposed structure of particle contamination.

The same argument also holds if the trimethylsilanol is dehydrated, which it is reported[2] to do very readily:

```
          Me                        Me      Me
          |                         |       |
2 H - O - Si - Me    ----->   Me - Si - O - Si - Me    +  H₂O
          |                         |       |
          Me                        Me      Me
```

These reactions, however, are not sufficient to explain particle formation. Essentially, a polymerization reaction is necessary; the polymerization must be more extensive than the dimerization shown in the dehydration reaction.

For more extensive polymerization, methyl groups must be split from hexamethyldisiloxane, with substitution by hydroxyl groups:

```
     Me        Me                        Me        Me
     |         |                         |         |
Me - Si - O - Si - Me + H₂O -->   Me - Si - O - Si - O - H + CH₄
     |         |                         |         |
     Me        Me                        Me        Me
```

This reaction has been observed[3] to occur at 200°C in an alkaline solution. Conceivably, this reaction could also occur in the vapor phase at slightly lower temperatures, with NH_3 acting as a base-acceptor molecule. By continued reaction of this type with subsequent dehydration of the reaction products, an extensive silicate structure could be produced.

This silicate structure would be expected to contain a significant fraction of non-bridging methyl and hydroxyl groups. Fig. 10 shows the silicate structure we propose for the contaminating particles found on HMDS-primed wafers.

An alternative mechanism, yielding the same silicate end product, is cleavage of the methyl groups directly from the trimethylsilanol:

```
          Me                       Me
          |                        |
H - O - Si - Me + H₂O -->   H - O - Si - O - H   + CH₄
          |                        |
          Me                       Me
```

This reaction, followed by subsequent dehydration, yields the same final silicate structure. In fact, the addition of a second hydroxyl group may be more readily accomplished than the first,[3] making this a more probable route for silicate formation.

This hydrolytic cleavage of methyl groups could also occur on the primed silicon surface, yielding the following situation:

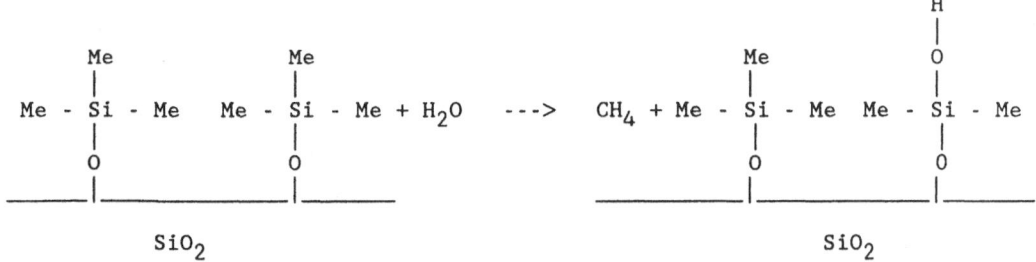

67

This reaction, which is consistent with the XPS data (Fig. 9a and 9b, showing a lower surface carbon concentration when the HMDS contains H_2O), could be instrumental in understanding the lift-off problem.

CONCLUSIONS

The proposed silicate structure, formed from the hydrolytic cleavage of the HMDS molecule with subsequent dehydration, is consistent with our findings. Since only trace amounts of H_2O are present, hydrolytic cleavage is not expected to occur for all of the methyl groups. Consequently, a substantial amount should remain in the silicate particle. Furthermore, a fraction of the hydroxyl groups would probably remain, forming additional non-bridging bonds in the final structure. The particulate contamination ultimately formed from these reactions would resemble a silica gel formed from organo-silicon precursors; the SEM micrographs (Fig. 4) clearly show this is true. The major difference is that with HMDS, nitrogen is initially present. The lability of the Si-N bond, however, would preclude the presence of nitrogen in the final structure, again consistent with the observed data.

A number of tentative reactions have been proposed in this paper. Clearly, such reactions would show sensitivity to various parameters such as temperature, H_2O concentration, and residence time at temperature. The dependence of particle formation on these parameters is the subject of another investigation;[4] however, from what is known, the deleterious nature of H_2O in HMDS wafer priming is obvious. This investigation has shown that under conditions of conventional HMDS wafer priming, there is ample opportunity for HMDS to react with H_2O, forming particulate contamination. By starting with low-moisture HMDS and not allowing contamination by atmospheric moisture, the particulate contamination level can be significantly reduced.

ACKNOWLEDGEMENTS

The authors wish to thank Yield Engineering Systems for use of their vapor prime system. They are also grateful to A. Lagendijk and S. Riahi of the J.C. Schumacher Company for stimulating discussions on the chemistry of the proposed reactions.

REFERENCES

1. K.L. Mittal, Solid State Technology, 22 (5), 89 (1979).
2. R.O. Sauer, J. Am. Chem. Soc., 66, 1707 (1944).
3. R.H. Krieble and J.R. Elliot, J. Am. Chem. Soc., 68, 229 (1946).
4. M.A. Logan, D.L. O'Meara, and J.R. Monkowski, Proceedings of the Microcontamination Conference and Exposition, pp. 111-122, Canon Communications, Santa Monica, CA (1986).

CONTAMINATION OF CHIP SURFACES BY PARTICLES DURING DESTRUCTIVE PHYSICAL ANALYSIS OF INTEGRATED CIRCUIT DEVICES

J. J. Weimer[*], J. Kokosinski, M. R. Cook and
M. Grunze[+]

Laboratory for Surface Science and Technology
9 Barrows Hall
University of Maine
Orono, ME 04469, USA

[*]Fritz Haber Institut der Max Planck Gesellschaft
 Faradayweg 4-6
 1000 Berlin 33/Dahlem
 West Germany

[+]Lehrstuhl fuer Angewandte Physikalische Chemie
 Institut fuer Physikalische Chemie
 Universitaet Heidelberg
 Im Neuenheimer Feld 253
 6900 Heidelberg, West Germany

The process of accessing a component or chip in a hermetically sealed, integrated circuit device by destructively cracking through the package is a major source of particulate contamination of the chip surface. If a polyimide layer exists on the chip surface and is subsequently removed by peeling, electrostatic charging during the peeling process results in an even greater particulate concentration both on the chip and on the side of the polyimide which had been adhering to the chip. The particles can be observed with a microscope and are characterized by X-ray Photoelectron Spectroscopy to be either lead oxide or aluminum oxide. They resist complete removal by such methods as blowing with an anti-static Freon duster or brushing away. The amount of contamination by the particles can be decreased but not completely eliminated by a modified cracking method.

1. INTRODUCTION

Integrated circuit (IC) devices are important and widely used components in electronic systems. The heart of an IC device is a chip containing a variety of fundamental elements integrated together to achieve the desired function of the device. The chip is sealed within the device package for protection and for ease of handling. In some cases a thin layer of polyimide (PI) material is coated onto the chip surface before sealing the device to act as an alpha particle barrier.

In the course of a study to characterize PI films on chip surfaces in IC devices[1], a problem of contamination by particles arose that is the subject of this report. The objective of the investigation described was to assess qualitatively the source and nature of this contamination. The findings are relevant to investigations such as testing and quality control programs which must access the chip in an IC device by cracking through the package in some manner. In particular, any work that probes the surface properties of a chip after opening an IC device in this way may need to consider the effects described here.

2. EXPERIMENTAL DETAILS

The major portion of the results deal with X-ray Photoelectron Spectroscopy (XPS) from the surfaces of chips in IC devices. The XPS analysis was performed in two different systems, both equipped with Al K_α x-ray sources. The analyzer in one system is a double pass cylindrical mirror and in the other a hemispherical sector. Although the transmission functions of the two analyzers differ, this is not significant for the results discussed in what follows. All of the spectra are aligned to position the major carbon peak at 284.4 eV in binding energy to correct for differences in charging from sample to sample.

Off-the-shelf IC devices were obtained from various industrial sources and a schematic of such a device is shown in Fig. 1. The industry standard encapsulation consists of an upper ceramic lid sealed with a low melting point lead glass to the lower body. Connecting pins run through the glass to the inner cavity of the device and fine wires join these pins to the chip at the center of the cavity. The package had to be opened to access the PI on the chip surface, and this was accomplished through the use of a knife-edge cracking tool. The tool had jaws that clamped firmly onto the device package on opposite sides at the points of the glass seal just above the connecting pins shown in Fig. 1a. A slight pressure on these jaws fractured the body of the device along the seal. The upper lid of the device package could then be lifted away to expose the chip, as shown in Fig. 1b. The PI film on the chip surface was removed by lifting a corner with a razor blade and pulling with tweezers, as discussed in a separate publication[1]. After the PI had been peeled, the devices were mounted on a stage with the chip surface facing out for insertion into the vacuum system through an introduction port. The peeled PI film was mounted on a separate stage with the surface that had been bonded to the chip facing the analyzer.

3. RESULTS AND DISCUSSION

The spectra in Fig. 2 are from an XPS analysis of a PI film peeled from the chip, the chip surface after peeling, and the same chip surface plus a mask described below. The carbon 1s peaks and silicon 2s and lead 4f peaks are on the left and right side of this figure, respectively. In Fig. 2a the carbon 1s region of the PI surface clearly shows two peaks. As discussed in the literature[1-3], the appearance of the smaller peak at higher binding energy is a characteristic of PI materials containing carbonyl functional groups. The silicon 2s peaks in Fig. 2a on the PI can arise from a number of sources such as silane adhesion promoters[4], silicon oxide carried with the film from the chip surface, or silicon in the lead glass, as discussed in ref. 1.

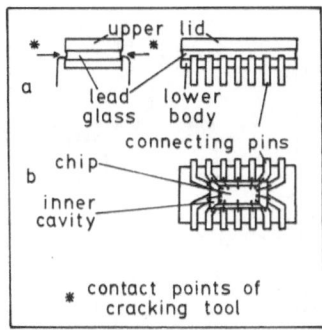

Fig. 1   Schematic representations of a 14 pin hermetically sealed,
         integrated circuit device. a) Side view of the device prior
         to cracking showing the upper lid, lower body, connecting
         pins, and lead glass sealing material. The points of con-
         tact to the knife-edge cracking tool are illustrated in one
         panel. b) Top view of the device after cracking the upper
         lid away to expose the inner cavity containing the chip.

     The important and initially unexpected finding is the presence of
a fairly strong lead signal on the peeled PI film surface, as seen in
Fig. 2a. The spectrum in Fig. 2b for the chip surface shows the same
lead signals. The lead was at first believed to arise from a contaminant in the PI in both cases, since, as explained in detail in a separate publication[1], a thin layer of PI or polymer-derived material
remains on the chip surface after the peeling. This is seen in Fig. 2b
which shows the same two peaks in the carbon 1s region that are characteristic of the PI film, even though the bulk of the PI film has been
peeled away.

     In order to exclude any possible contribution to the XPS spectrum
in Fig. 2b from the glass sealing material around the edges of the inner
cavity, the lower body of the device was masked with a sheet of tantalum
foil containing a window that exposed only the chip surface. The
resulting spectrum in Fig. 2c shows that this procedure eliminated a
major portion of the lead signal in this specific case. However, for
the same analysis of another sample, the intensity of the lead signal
was not affected by the mask, as seen in Fig. 3. For this sample, the
intensity of the lead signal from the chip plus mask (Fig. 3c) is essentially unaltered when compared to that from the chip alone (Fig. 3b).

     The results in Fig. 2c proved to be an exception in the XPS analy-
sis of over a dozen IC devices. In general, the spectra for the lead
region from XPS scans of chip plus mask samples always showed relatively
stronger lead 4f peaks than those in Fig. 2c. Lead and silicon were
also always observed on the PI film surface, as seen in Figs. 2a and 3a.
The amount of lead and silicon varied from sample to sample on the PI
films and on the chip surfaces.

     After a review of the preparation procedures for the analysis of
the devices and, in particular, the method of cracking the package to
open the device, the lead on the chip surfaces was determined to arise
from particles of the lead glass sealing material. Once fractured during the cracking, this glass crumbles easily into aggregates and small
particles which are scattered throughout the inner volume of the device
and across the chip. Although care was taken to avoid this and to clean

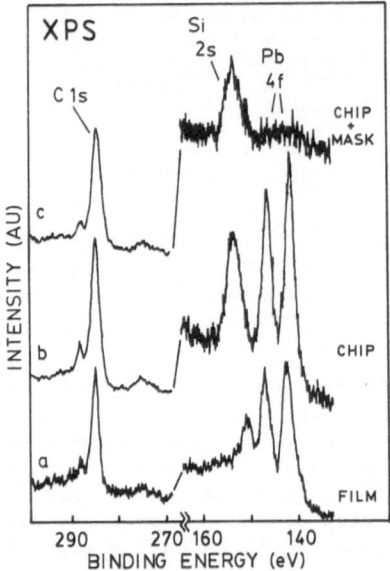

Fig. 2   Typical x-ray photoelectron spectra for the carbon 1s and
          silicon 2s regions after peeling the polyimide film from the
          chip in an integrated circuit device. a) The surface of the
          polyimide film previously bonded to the chip, b) the chip
          surface after peeling, and c) the chip surface when the rest
          of the device is masked by tantalum foil (see text).

the area around the chip before peeling the PI film, the particles were
not easy to remove.

     The lead (and possibly part of the silicon) signals on the PI film
surface after peeling are also from particles of the lead glass. The
process of peeling the PI leads to electrostatic charging of the film,
as evidenced by the tenacity with which it clings to metal tweezers
after being peeled from the chip. The electrostatic forces thus generated apparently draw particles to the peel interface. Since the lead
glass and PI film are insulators, no means of dissipation exists for the
charges once formed, and the particles adhere strongly.

     Supporting evidence for contamination by particles was subse-
quently found in examinations of the chips with an optical microscope.
Particles were seen in the range from about 10x magnification to above
30x scattered throughout the inner cavity and across the chip surface.
In addition, an analysis with a Scanning Electron Microscope (SEM)
equipped with Energy Dispersive X-ray Analysis (EDXA) identified lead-
containing particles to much smaller sizes on the chip surfaces. No
systematic determination of particle sizes was attempted in this work.

     An extensive amount of information exists concerning the strong
attraction of particles to surfaces and means for their removal[5-7].
Testing of different cleaning methods was, however, limited by the
necessity to maintain the peeled PI film and exposed chip surface undam-
aged and uncontaminated by foreign materials throughout the course of
the study[1]. Thus, opening the device in filtered, deionized water and
peeling under these circumstances could not be done due to the resulting
contamination of the PI film by water, for example.

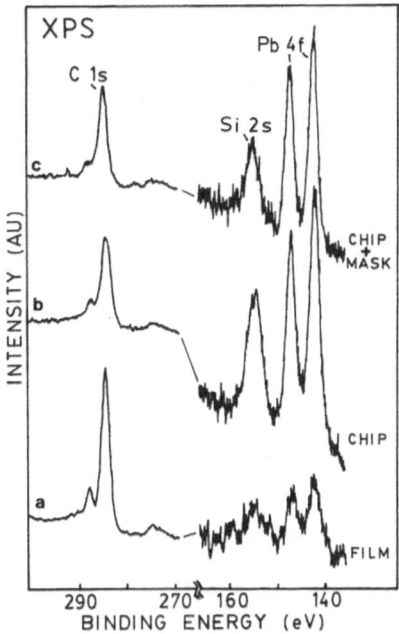

Fig. 3   The same results as in Fig. 2 for another device.

Two methods were used to try to remove the particles from the surfaces of the chips, although both were not without problems in respect
to damaging the PI film[1]. Since the particles are believed to adhere
mainly through electrostatic forces, an attempt was made to neutralize
the charges on a chip in the vacuum system. An electron beam of low
energy (about 50 eV) and beam current (1-2μA) was rastered across the
chip surface for about five minutes. If the residual surface charges
could be reversed or neutralized in this manner, the particles could
conceivably fall off of the chip, which was tilted at about 40° from the
horizontal in the vacuum system. An XPS scan of the silicon and lead
regions prior to such a raster is shown in Fig. 4a, and the spectrum
afterward in Fig. 4b. The ratio of peak intensities of lead to silicon
is 0.36 in Fig. 4a and is 0.33 in Fig. 4b, suggesting only partial if
any success with this method.

The second means tried for particle removal was to use an anti-
static spray duster. The device used for the XPS scans in Figs. 2a and
2b was removed from the chamber after the electron beam raster, and a
Freon spray duster equipped with an electrostatic neutralizer source was
used to try to blow the particles off of the chip. The scan in Fig. 4c
for the chip surface after the combined raster and anti-static clean
shows that the ratio of lead to silicon has decreased slightly to 0.30.
However, complete removal of the lead was still not effected.

One step performed during analysis of a number of the devices
required that the upper lid be punctured to sample the gases in the volume enclosing the chip prior to cracking. In subsequent analysis with
XPS of the chip and PI surfaces of these punctured devices less lead was
found compared with the amounts seen in unpunctured devices. This is
illustrated in Fig. 5, which shows an XPS scan of the lead regions for
an unpunctured and a punctured device. The ratio of lead to silicon
peak intensities is 2.4 for the data shown in Fig. 5a and is 0.36 in
Fig. 5b.

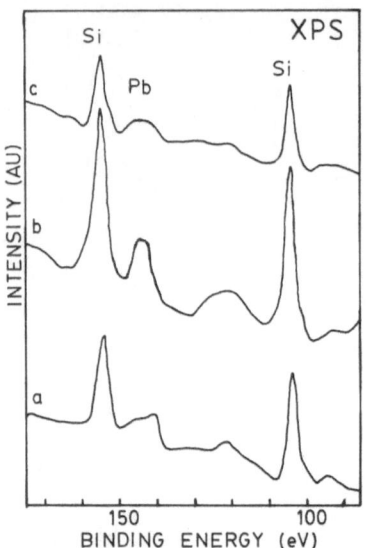

Fig. 4   The x-ray photoelectron spectra of the lead and silicon
regions for a chip subjected to cleaning treatments. a) The
chip as inserted in the system, b) after raster with an
electron beam, and c) after removal and spray with a Freon,
anti-static duster.

Two effects may be responsible for the lower lead signals in punctured devices. First, the volume of the inner cavity is generally at a
lower pressure relative to the outer room because the devices are heated
to 400°C to seal the package. After the device cools the pressure in
the sealed inner volume will be lower than it had been during sealing.
Cracking of the package will then cause an implosion that will carry
particles to the chip unless this pressure differential is first
removed. Thus, when the lid of the device is punctured prior to cracking, the lead glass particles will be carried only by the force of the
jaws on the knife edge, and this could result in a reduced amount of
particle contamination.

The second reason for lower lead signals in Fig. 5b relative to
those in Fig. 5a could be due to changes in the amount of moisture in
the device. The inner volume of the IC device is expected to be drier
than the room due to the heat treatment described above for sealing the
package or due to the use of an inert, dry gas during the sealing process. Puncturing the device prior to cracking may permit an equilibrium
to be established between the moisture content of the room and that of
the device cavity and PI material. An increase in the amount of moisture in the device and PI material can serve to reduce the generation of
electrostatic charges at the peel line or to decrease the extent to
which such charges interact with the lead glass particles. This would
act to lower the amount of lead pulled to the peel interface.

Since the particles from the lead glass material cannot be eliminated once the package is cracked open in the manner described above, an
effort was made to open a device by another technique that avoided
cracking the glass. The method involved going through the top of the
device package rather than cracking it off. The upper lid was ground
down so that a much thinner and more fragile ceramic wall existed above
the inner cavity of the device. Strongly adhering double-sided tape was
put across this lid and, by applying a gentle, even pressure to the

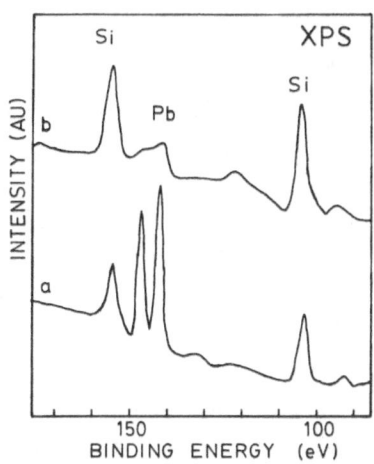

Fig. 5   A comparison of the silicon and lead regions for x-ray
photoelectron spectra of a chip in a device that was a)
unpunctured and b) punctured prior to cracking open.

upper lid, the thinner wall could be broken. The tape ensured that
major pieces of the ceramic did not fall onto the chip surface during
this process.

As seen in Fig. 6, the method of popping a device open as opposed
to cracking it open does not cause contamination by the lead glass. The
XPS spectrum in Fig. 6b shows no lead (or silicon) present on the chip
surface after peeling the PI film for a popped device. This spectrum
can be compared to Fig. 6a for a cracked device. The popping method is
to be recommended therefore to avoid such contamination.

Two problems are to be pointed out for the popping method, though.
One is that although contamination by lead glass particles from the
sealing material is avoided, contamination by ceramic particles from the
package lid is not. This is seen by the presence of an aluminum peak
from the ceramic in Fig. 6b. Therefore even the popping method is not a
"clean" way of opening a device package. The second problem is that, in
general, access to the chip surface is much more restricted for a popped
device as opposed to a cracked device. This should be considered if the
objective is to analyze some property of the chip surface, as was the
case in this work[1].

4. CONCLUSIONS

The method of opening an IC device by cracking through the glass
sealing material always causes contamination of the chip by particles of
lead glass. The particles can range in size from visible aggregates of
glass to particles invisible except to SEM. They adhere strongly to the
chip, probably through electrostatic interactions. If a PI film is on
the chip surface and is subsequently peeled, further contamination can
be expected because of the electrostatic charges generated during peeling.

As might be expected from the nature and strength of the forces
attracting the particles, they resist removal by attempts to neutralize
the electrostatic charges or by concurrent blowing or brushing techniques. A unique finding though is that the amount of contamination can

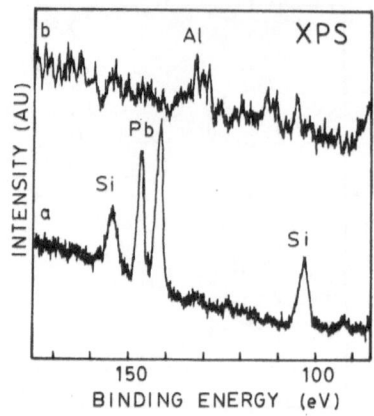

Fig. 6 A comparison of the silicon, lead and aluminum regions for x-ray photoelectron spectra of a chip in a) a cracked open device and b) a popped open device (see text for details).

be reduced by equilibrating the pressure and humidity in the inner cavity with the ambient some time prior to cracking away the upper lid. The technique of popping open the upper ceramic lid of the device does completely avoid contamination by lead glass particles but is also not a "clean" method because of ensuing contamination by ceramic from the device package.

5. ACKNOWLEDGEMENTS

This work was supported in part by IBM Federal Systems Division, Owego, New York.

REFERENCES

1. J. J. Weimer, M. R. Cook and M. Grunze, (to be published).
2. B. D. Silverman, P. N. Sanda and P. S. Ho, J. Polymer Sci. 23, 2857 (1985).
3. M. Grunze and R. N. Lamb, Chem. Phys. Letts. 133, 283 (1987).
4. K. L. Mittal, Ed. "Polyimides: Synthesis, Characterization and Applications", Vols. 1 and 2, Plenum Press, New York, 1984.
5. M. Coru, J. Air Pollution Control Assoc., 2, 567 (1961).
6. S. Bhattacharya and K. L. Mittal, Surface Technol., 7, 413 (1978).
7. J. M. Duffalo and J. R. Monkowski, Solid State Technol. 27, 3 (1984).

CALCULATION OF HAMAKER COEFFICIENTS FOR METALLIC AEROSOLS

FROM EXTENSIVE OPTICAL DATA

I. W. Osborne-Lee

Oak Ridge National Laboratory
Post Office Box X
Oak Ridge, Tennessee 37831

Studies of resuspension of aerosol particles from surfaces require consideration of van der Waals forces. Hamaker coefficients, which enable the calculation of van der Waals forces, for metals have been previously reported for only a few cases. The report presents application of the theory developed by Lifshitz to the calculation of Hamaker coefficients for metals. The theory requires data for the complex refractive index or dielectric function over a very wide range of frequencies. The data must undergo a Kramers-Kronig transformation, which involves integration of the data over the complete frequency range, prior to integration (again over the complete frequency range) for the Hamaker coefficient. An algorithm which performs numerical integration of unequally spaced data using Gaussian quadrature has been used, along with data available in the literature, to compute Hamaker coefficients for several metals of interest in core-melt aerosol studies including Ag, Al, Fe, Ni, and W. Hamaker coefficients fell between 10 and 50×10^{-13} ergs for all materials for which sufficiently extensive optical data were available. The available literature values show good agreement with Hamaker coefficients calculated by the method presented here. Temperature dependencies for the Hamaker coefficient are shown to be negligible for the materials studied.

INTRODUCTION

Aerosol behavior is a subject of importance in a great many fields of study. Of particular importance in some fields are those aspects of aerosol, or fine particle, behavior relating to deposition and resuspension. In the prediction of the radioactive compounds released from severe accidents at nuclear power plants, for example, aerosol deposition has long been accepted as a crucial factor impacting radionuclide release and thus harm to the environment. Recently, resuspension of aerosols has been recognized[1] as an important phenomenon affecting radionuclide release, and is now receiving increased attention.

This work derives from an interest in improvement of the radioactive compound release prediction capabilities of source-term prediction computer codes. TRAP-MELT[2], for example, is a computer code used to calculate the transport of aerosols in reactor coolant systems; it includes a variety of mechanisms for deposition of aerosols, but contains no mechanism for resuspension. The importance of resuspension in source-term prediction can be assessed through the incorporation of an appropriate mechanistic model for resuspension into TRAP-MELT for comparison of code predictions; such a model may be used to interpret test data.

It is expected that a model for resuspension should include the effects of adhesion. Neighboring objects experience a mutual force, normally effective over small separations, which can cause the bodies of matter to stick together. This force, termed the adhesion force, is actually a composite of several more basic forces. In general, the adhesion force includes such forces as van der Waals forces, electrostatic forces, magnetic forces, capillary forces, hydrogen bonding forces, and chemical bonding forces. Each force type can, under appropriate conditions, represent the dominant contribution to the adhesion force. For adhesion between uncharged, non-magnetic bodies in the absence of moisture, the van der Waals force usually predominates.

Of frequent concern are aerosols composed of metals or metallic oxides. In nuclear power plant accident sequences, for instance, reactor core and support materials can be vaporized and then condensed, or oxidized, to form aerosols. With such aerosols, in many cases, van der Waals forces may be expected to predominate in adhesion. Other effects, such as surface roughness, shape factors, and size distributions, are important but their consideration is beyond the scope of the present work. And, inevitably, quantification of the van der Waals force must be preceded by determination of some coefficient characterizing the adhering materials.

The van der Waals force, as will be shown later, depends on geometric factors, such as the size and separation of adhering objects, and a coefficient which depends on the material properties alone. The coefficient is frequently known as the Hamaker constant, or coefficient, but also as the Lifshitz-van der Waals coefficient. Historically, there have been two approaches to evaluation of the van der Waals force through the use of optical data, which differ in their determination of the Hamaker coefficient. Both approaches have been reviewed extensively before[3,4,5,6,7]. The first, simplest, and most approximate of the two approaches, is the microscopic approach of London (1930)[8] and, later, Hamaker (1937).[9] This approach is based on the assumption of additivity of intermolecular dispersion forces. This approach, however, even for crude estimates of the Hamaker coefficient, is limited to non-absorbing materials, which precludes its application to metals. The second approach, developed by Lifshitz[10] from considerations of only macroscopic properties of the media, has been shown to give more accurate determinations of the Hamaker coefficient and is applicable to all materials.

More recently, an alternative to the traditional approaches was developed by Johnson, Kendall and Roberts.[11] The method, which takes into account deformation of the adhering surfaces, involves the determination of the surface energy for the adhering surfaces through the measurement of the change of contact radius with pressing-on force for two surfaces in contact. The method has been used by Johnson et al.[11] for adhesion of gelatin and rubber spheres. However, while it is promising for such low modulus materials, the method may not be well suited to metal-metal adhesion where the changes in contact area with loading may be much smaller;

it is unclear whether such small changes are accurately measurable. Whether or not this new approach will be useful in modeling core-melt aerosol adhesion remains to be evaluated.

Hamaker coefficients have been reported for a variety of optical materials, some polymers, and for materials of interest to the semiconductor industries, including some metals[4,6,7,12]. Hamaker coefficients may be determined in a variety of ways, whether from direct measurement of adhesion force[6] or by derivation from colloid chemistry experiments,[4] surface and interfacial tension measurements,[13,14] optical data, or other means. However, there have been few reports of Hamaker coefficients for metals and metal oxides of interest in core-melt studies. This is so because it is difficult to know the surface energies of metals; because interfacial phenomena experiments are of limited use; and because, due to the complex nature of the interaction of metals with light, microscopic approaches fail seriously. There have, however, been some reports of calculations of Hamaker coefficients for metals from optical data by the macroscopic approach of Lifshitz[10,15] for the noble metals, in particular[4,12,16]. In recent years, methods using this approach to Hamaker coefficient determination have been discussed by several workers.[16,17,18,19,20]

The object of this paper is to present the method and results of calculations by a simple and expedient method using the approach of Lifshitz, for previously unreported values for the Hamaker (Lifshitz-van der Waals) coefficient for some metals of interest in core-melt aerosol studies. The method used herein has provided for quick, inexpensive, and easy determination of Hamaker coefficients from finite optical data existing in the literature by use of a numerical integration algorithm by Gill and Miller.[21] Also, Hamaker coefficients are reported for some previously studied materials in order to allow comparison with literature data.

THEORY

The Macroscopic Theory of Lifshitz

The van der Waals force is a quantum mechanical phenomenon. In neutral atoms, it arises due to instantaneous dipole moments. The electric field about an atom fluctuates from one instant to the next as the electron positions change. The atoms of two bodies in close proximity can interact through their electric fields. Phase correlation of the fluctuating electic fields reduces the energy of the system and is the origin of the dispersion force between the two bodies.

Lifshitz[10] used a perturbational approach to the calculation of the fluctuating electromagnetic fields in two planar masses, separated by a vacuum, for which the separation is small compared with the dimensions of the masses. Formulation of the interaction between the fluctuating fields showed that the van der Waals force between two macroscopic bodies can be computed from their dielectric spectral properties. Experimental validation of the Lifshitz-theory has been made by various workers, the work of Shih et al.[22] being of particular notice. Dzyaloshinskii, Lifshitz, and Pitaevskii[15] extended the work of Lifshitz to two half spaces separated by a third medium using quantum field theory methods. The expression for the van der Waals force, F_{VDW}, between the two parallel semi-infinite masses may be written[3] as shown in Eqn 1.

$$F_{VDW} = \frac{\hbar\bar{\omega}}{8\pi^2 z_o{}^3} \tag{1}$$

$$\hbar\bar{\omega} = h \int_0^\infty \sum_{n=1}^\infty \frac{1}{n^3} \left(\frac{\varepsilon_1(i\xi) - \varepsilon_3(i\xi)}{\varepsilon_1(i\xi) + \varepsilon_3(i\xi)} \right)^n \left(\frac{\varepsilon_2(i\xi) - \varepsilon_3(i\xi)}{\varepsilon_2(i\xi) + \varepsilon_3(i\xi)} \right)^n d\xi \tag{2}$$

Other workers have developed analogous expressions for contact involving different geometries and multilayer or composite bodies.[40-42]

Equations 1 and 2 represent the nonretarded case, where the two masses are separated by small distances. In these equations, z_o is the separation distance, $h\omega$ is the Lifshitz–van der Waals coefficient, h is Planck's constant h divided by 2π, and n is a summation index. ε is the dielectric function, or permittivity, and represents the ratio of the electric field density flux of the material to that of a vacuum; it is a real function of the complex frequency ($\omega = \omega' + i\,\xi$). ξ, the variable of integration, represents the imaginary portion of the complex frequency.

The subscripts 1, 2, and 3 denote the two masses and the intervening medium, respectively. According to Eqn. 1, the van der Waals force may be calculated from knowledge of the dielectric function (or, as will be seen later, refractive index) over the complete electromagnetic spectrum range. The Hamaker coefficient is more frequently reported than the Lifshitz–van der Waals coefficient. The two coefficients are related by Eqn. 3, where A_{132} is the Hamaker coefficient.

$$A_{132} = \frac{3\hbar\bar{\omega}}{4\pi} \tag{3}$$

Equations 1–3 are valid for small separations, that is, separations much smaller than the characteristic absorption wavelengths of the media. At larger separations, where z_o is of the order of magnitude of the wavelength at which the imaginary portion of the dielectric function $[\varepsilon''(\omega)]$ is particularly large, a relativistic effect is observed. The phase correlation between the fluctuating electric fields of atoms separated by the gap is diminished due to the finite time required by an electromagnetic wave to cross the gap. Hence, the attractive force between the adhering bodies is diminished. The van der Waals force in this case is said to be retarded, and is given by

$$F_R = B_{132}/z_o{}^4 \tag{4}$$

where B_{132} is the retarded Hamaker coefficient, given by

$$B_{132} = \frac{\pi^2 \hbar c}{240} \frac{1}{\sqrt{\varepsilon_3(0)}} \left(\frac{\varepsilon_1(0) - \varepsilon_3(0)}{\varepsilon_1(0) + \varepsilon_3(0)} \right) \left(\frac{\varepsilon_2(0) - \varepsilon_3(0)}{\varepsilon_2(0) + \varepsilon_3(0)} \right) \phi\left(\varepsilon_1(0), \varepsilon_2(0), \varepsilon_3(0) \right) \tag{5}$$

The constant c is the speed of light and ϕ is a function depending on the dielectric functions evaluated at zero wavelength, $\varepsilon_1(0)$. For very large $\varepsilon_1(0)$, as for metals, ϕ tends to unity. Hence, for interaction between metallic media

$$B_{132} = \frac{\pi^2 \hbar c}{240} \frac{1}{\sqrt{\varepsilon_3(0)}} \left(\frac{\varepsilon_1(0) - \varepsilon_3(0)}{\varepsilon_1(0) + \varepsilon_3(0)} \right) \left(\frac{\varepsilon_2(0) - \varepsilon_3(0)}{\varepsilon_2(0) + \varepsilon_3(0)} \right) \tag{6}$$

Expressions for both the retarded and non-retarded van der Waals force have been reported for several different geometries.[3,7,12,40] In each case, the force can be written as proportional to two terms, one depending only on the geometry of the media, and a term depending only on the optical properties of the media. The latter is embodied in the Hamaker coefficient.

The Kramers-Kronig Relation

Normally, it is the complex dielectric function ($\varepsilon(\omega) = \varepsilon'(\omega) + i\,\varepsilon''(\omega)$), as a function of the real frequency that is known, rather than the real dielectric function $\varepsilon(i\xi)$ as a function of the imaginary frequency. An important quantity here is the imaginary portion of the dielectric function, $\varepsilon''(\omega)$, which represents that portion of the energy of an electromagnetic wave of frequency ω which is dissipated by the medium. The desired quantity $\varepsilon(i\xi)$ can be obtained through the use of the Kramers-Kronig relation, given in Eqn. 7.

$$\varepsilon(i\xi)-1 = \int_0^\infty \frac{\omega\varepsilon''(\omega)d\xi}{\omega^2+\xi^2} \tag{7}$$

The expression in Eqn. 7 is derived using methods from the theory of functions of a complex variable[23] and is true for functions $\varepsilon(\omega)$ which are single-valued and finite for all $\omega > 0$. The quantity $\varepsilon(i\xi)$ contains all of the information required for calculation of the Hamaker coefficient for use in predicting the van der Waals force; it depends only on the imaginary portion of the dielectric function--that part of incident radiation which is absorbed.

Alternatively, it may be the complex refractive index $n(\omega) = n'(\omega) + i\,n''(\omega)$ that is known. In that case, the imaginary portion of the dielectric function can be obtained from the complex refractive index by the following relation.

$$\varepsilon''(\omega) = 2n'(\omega)n''(\omega) \tag{8}$$

The imaginary portion of the refractive index is known as the absorptivity.

Equations for Adhesion

Equations 1 and 2 give the attractive force for the case of two plane surfaces composed of different materials, separated by a third material. The first term in Equation 2 is the most important; the second term is rarely more than a few percent.[3] If the intervening material is a low density gas, such as air at normal conditions, the absorptivity is small and the function ε_3 may be set to 1.

Taking the first term in the series, for this case, the expression for the Hamaker coefficient reduces to

$$A_{12} = \frac{3\hbar}{4\pi} \int_0^\infty \left(\frac{\varepsilon_1(i\xi)-1}{\varepsilon_1(i\xi)+1}\right)\left(\frac{\varepsilon_2(i\xi)-1}{\varepsilon_2(i\xi)+1}\right)\ d\xi \tag{9}$$

where the subscript 12 denotes the fact that the intervening material is considered to be transparent, as for a vacuum.

Equations for Cohesion

If the two bodies of material are similar, as in the case of cohesion, Eqn. 9 can be further reduced. Eqn. 10 gives the relation for the Hamaker coefficient in the case where the materials are considered to be identical.

$$A_{11} = \frac{3\hbar}{4\pi} \int_0^\infty \left(\frac{\varepsilon_1(i\xi)-1}{\varepsilon_1(i\xi)+1}\right)^2 d\xi \tag{10}$$

It is useful to note the behavior of the integrand in equation 10. For free-electron metals, the imaginary portion of the dielectric function has an infinite asymptotic value at zero frequency and in the limit of unbounded frequency approaches the value of zero. Likewise, according to Eqn. 7, $\varepsilon(i\xi)$ ranges from positive infinity at zero frequency to unity at infinite frequency. Hence, the integrand, which will be denoted Y, may be expected to range from unity, for $\varepsilon(i\xi) \gg 1$, to zero where $\varepsilon(i\xi) = 1$. A complete set of optical data should produce a set of Y values which range, asymptotically, to 1 at the lower frequencies reported, and to 0 at the highest frequencies reported, while completely defining the shape of the function Y between these limits.

Effect of Temperature

At finite temperatures, the above expressions for the van der Waals force must be corrected by the addition of another term F_T which depends on the temperature. According to Dzyaloshinskii, Lifshitz and Pitaevsky[15], this term is given by

$$F_T = \frac{kT}{8\pi z_0^3} \sum_{n=1}^\infty \frac{1}{n^3}\left(\frac{\varepsilon_1(0) - \varepsilon_3(0)}{\varepsilon_1(0) + \varepsilon_3(0)}\right)\left(\frac{\varepsilon_2(0) - \varepsilon_3(0)}{\varepsilon_2(0) + \varepsilon_3(0)}\right) \tag{11}$$

where k is the Boltzmann constant and T is the absolute temperature. It should be noted that the above are corrections to the method used in calculation of the dispersion force. The need for correction arises out of the assumption made in the derivation of equation 2 that $z_0 kT/hc \ll 1$, which for temperatures near absolute zero is certainly satisfied, but for finite temperatures may not be. The correction does not account for any temperature dependence which may be inherent in the dielectric functions of the adherents. Equation 11 is valid at all separations.[3] Taking only the first term in the series, for an intervening vacuum or dilute gas, the equations for the cases of adhesion and cohesion are, respectively,

$$F_T = \frac{kT}{8\pi z_0^3} \left(\frac{\varepsilon_1(0)-1}{\varepsilon_1(0)+1}\right)\left(\frac{\varepsilon_2(0)-1}{\varepsilon_2(0)+1}\right) \tag{12}$$

$$F_T = \frac{kT}{8\pi z_0^3} \left(\frac{\varepsilon_1(0)-1}{\varepsilon_1(0)+1}\right)^2 \tag{13}$$

The correction to the Hamaker coefficient, for the case of cohesion, due to temperature dependence is given by equation 14.

$$A_T = 6\pi z_0^3 F_T = \frac{3kT}{4}\left(\frac{\varepsilon_1(0)-1}{\varepsilon_1(0)+1}\right)^2 \tag{14}$$

Combining Rules

Equation 9 for the Hamaker coefficient may be used to compute the van der Waals force between two different materials separated by an optically transparent medium. Its use requires the execution of a double integral of optical parameters for the two materials over a very large range of frequencies. Coordination of optical data for numerical integration can present some difficulty, since the data may occur at different intervals and over different ranges of frequency. Alternatively, the Hamaker coefficient can be computed for the simpler case of cohesion, by use of Eqn. 10, for each material. The resulting values of A_{11} and A_{22} can be combined to yield a value for A_{12} by use of eqn. 15.

$$A_{12} = (A_{11}A_{22})^{1/2} \tag{15}$$

This combining law is an approximate expression. It has been shown to overpredict the Hamaker coefficient for adhesion, but the error associated with its use is generally not more than a few percent.[3] For convenience, Eqn. 15 along with Eqn. 10 will be used to determine A_{12}, rather than Eqn. 9.

HAMAKER COEFFICIENT CALCULATIONS

It was established in the preceding section that Hamaker coefficients may be calculated by use of equation 10 for cohesion between like materials, and, in the case of unlike materials, by use of equation 15 following application of equation 10 for both materials. This section includes a discussion of the sources of optical data, required in equation 10, which are available in the literature. Also, the computational details are discussed prior to presentation of the results.

Source of Data

A large amount of optical data has been reported, but the most extensive data have been reported mainly for optical materials, such as glass, and for elemental metals--especially those of importance to the semiconductor industries, such as silicon and germanium. The data may take the form of refractive index, dielectric function, reflectance or transmittance, and absorbance.

The Handbook of Chemistry and Physics[24] has tabulations of the complex refractive index for 25 metallic elements. Weaver et al.[25] have reported an even more comprehensive compilation, including optical data for more than forty metallic elements. Hageman et al.[26] have reported optical data for Mg, Al, Cu, Ag, Au, Bi, C, and Al_2O_3. In addition, there have been many less-comprehensive reports of optical properties for various metals, including the measurements of Arakawa et al.[27] and Petrakian et al.[28] on Sn and other metals, Jellison and Modine[29] on Si, Smith et al.[30] on Al, and others.[31-37]

The frequency range over which the optical data are reported varies for different reports and also for different materials within the same report. In accordance with the findings of Parsegian and Weiss,[16] it is anticipated that the more extensive the data range used in computations of the Hamaker coefficient, the larger will be the result. The important optical quantity $\varepsilon(i\xi)$, however, vanishes in the limit of unbounded frequency, and so data which range to large enough energies should yield accurate values for the Hamaker coefficient. The work of Hageman et al.[26]

represents the most extensive data sets for the materials reported. This source of data was used whenever possible. The data sets reported by Weaver et al.[25] are less extensive in terms of the frequency ranges used, but were found to be adequately extensive for some of the materials used here, and was used for the materials not reported on in the work of Hageman et al.

Computational Method

The data used consisted of up to 292 (ω, $\varepsilon''(\omega)$) data pairs, for each material. The integral in equation 2, whose integrand depends on the computed $\varepsilon(i\xi)$ values, is over the dummy variable ξ. In theory, this integration should be from zero to positive infinity. In practice, the range of ξ values chosen must be finite, since the functional form for $\varepsilon''(\omega)$ and thus $\varepsilon(i\xi)$ is unknown. The ξ values, which are arbitrary, were chosen to be identical to the ω values in the data set, since it is already assumed that the data set adequately approximates the semi-infinite range required. The transformation was carried out using equation 7, which also depends on ξ, by numerical integration of the complete data set for each value of the dummy variable ξ. Integration of the resulting set of (ξ, $\varepsilon(i\xi)$) data pairs over the range of ξ, yielded a value for the Lifshitz-Van der Waals coefficient, $\hbar\bar{\omega}$, which is proportional to the Hamaker coefficient, as shown in equation 3.

The calculations of A_{11} values were performed with the use of a numerical integration algorithm due to Gill and Miller.[21] This quadrature method is suited to integration over a finite range of a function whose analytical form may be unknown and for which data exist at intervals which may be unequally spaced. The code exists as part of the NAG FORTRAN Library as the subroutine D01GAF. The calculations were programmed in FORTRAN and carried out at Oak Ridge National Laboratory on a DEC Systems PDP-10 mainframe computer.

The value of the dielectric function at zero frequency is necessary in a number of calculations of interest. Aside from extending the range of the numerical integrations, which lead to values for the unretarded Hamaker coefficient, and thus increasing the accuracy of those calculations, extrapolation for $\varepsilon(\omega=0)$ allows calculation of retarded Hamaker coefficients and also the contribution to the Hamaker coefficient due to temperature dependency. Linear extrapolation of $\varepsilon''(\omega)$ to zero frequency was done in the calculations for each material reported. Equation 13 was then used to evaluate the importance of the temperature dependent contribution to the Hamaker coefficient.

It is feasible to further extend the finite data sets by extrapolation to infinite frequency using a model of electronic behavior, such as the Drude oscillator model.[38]

$$\varepsilon' = 1 - \frac{\omega^2 \zeta^2}{(1+\omega^2\zeta^2)} \tag{16}$$

$$\varepsilon'' = \frac{\omega_p^2 \zeta}{\omega(1+\omega^2\zeta^2)} \tag{17}$$

Equations 16-17 apply to free-electron metals. The plasma frequency, ω_p, and the mean time between electron collisions, ζ, can be determined by fitting the optical data with Eqns. 16-17 over an applicable frequency range. However, this was not done in this report.

Results

Calculations of the Hamaker coefficient were performed for several metals of interest in aerosol studies. With respect to radioactive compound releases resulting from light water reactor accidents, several metals and oxides have been identified previously[39] as being of particular importance, including those shown in Table I. Figure 1 shows the plots of Y, the integrand in Eqn. 10, versus the natural logarithm of the frequency (Hz). Hamaker coefficient calculations have been made for those metals for which available optical data were deemed to be sufficient including, iron, manganese, nickel, and tungsten. The results are shown in Table II. Also shown in the table are calculations which were made, for aluminum and the noble metals, silver, copper, and gold, in order to enable comparison with recently reported values for Hamaker coefficients. Table III shows Hamaker coefficients previously reported by other workers, for the materials in Table II, where available.

Table I. Chemical form of aerosols in pressure vessel prior to failure[a]

Clearly oxides	Clearly metals	Could be either oxide or metal
ZrO_2	Ag	$Fe/FeO/Fe_3O_4$
Al_2O_3	Te	Sn/SnO_2
RE_2O_3	Ni	In/In_2O_3
UO_2	Cd	Mo/MoO_2
BaO	Rh	
SrO	Pd	
	Ru	

[a]Taken from Wichner and Spence.[39]

Table II. Hamaker coefficients computed from optical data (this work).

Material	A^{a} $(J \times 10^{20})$	Source of data	#Data points
Silver	44	Hageman et al.[26]	148
Aluminum	33	Hageman et al.[26]	148
Gold	48	Hageman et al.[26]	150
Copper	46	Hageman et al.[26]	150
Iron	26	Weaver et al.[25]	181
Manganese	15	Weaver et al.[25]	49
Nickel	32	Weaver et al.[25]	181
Tungsten	40	Weaver et al.[25]	292

[a]Hamaker coefficient calculated using equations 10 and 15.

Table III. Hamaker coefficients from various sources.

Material	A^a ($J^{12} \times 10^{20}$)	Reference
Silver	50.0	Parsegian[16]
Silver	40.0	Visser, 1972
Gold	40.0	Parsegian, 1981
Gold	45.5	Visser[4], Van den Tempel[12]
Copper	40.0	Parsegian 1981
Copper	28.4	Visser[4]
Iron	21.2	Visser[4]
Germanium	30.0	Visser[4]
Mercury	43.4	Visser[4]
Lead	21.4	Visser[4]
Selenium	16.2	Visser[4]
Silicon	25.6	Visser[4]
Tin	21.8	Visser[4]
Tellurium	14.0	Visser[4]

[a]Hamaker coefficient.

The correction to the Hamaker coefficient due to its dependency on temperature is shown in Table IV. These values have been estimated using Eqn. 14 along with values of $\varepsilon(i\xi)$ extrapolated from the available data to zero frequency. Since the function $\varepsilon(i\xi)$ is nonlinear, there is some error inherent in the extrapolation; however, the effect of the inaccuracy is to cause an overprediction of the correction to the Hamaker coefficient. Hence, the values in Table IV may be considered to represent an upper limit. The calculations were made with an arbitrarily assumed temperature of 1000K.

Table IV. Contribution to Hamaker coefficient of temperature dependence at 1000K.

Material	$\varepsilon''(0)^a$	A_T^b ($J \times 10^{20}$)
Silver	1150000	1.05
Aluminum	3120000	1.05
Gold	1390000	1.05
Iron	720	1.04
Manganese	73	.99
Nickel	2830	1.05
Tungsten	16500	1.05

[a]Imaginary portion of the dielectric function evaluated at zero frequency.
[b]Temperature dependent portion of the Hamaker coefficient, arbitrarily calculated at 1000K.

DISCUSSION OF RESULTS

Early calculations, in which attempts were made to use a microscopic approach in determination of Hamaker coefficients for the metals, iron, tin and tungsten, resulted in Hamaker coefficient values which were several orders of magnitude lower than previously reported values. These results are not reported here, but serve to illustrate the inadequacy of the microscopic approach when applied to metals, which show strong absorption bands, and the need for a macroscopic approach utilizing full spectral data.

Hamaker coefficients have been reported previously for some metallic materials. The most comprehensive report was made by Visser[4] and included Hamaker coefficients for ionic crystals, water, hydrocarbons, silicon dioxides, carbons, and some polymers, as well as some metals and metal oxides. The Hamaker coefficients reported by Visser were obtained by a variety of methods, including those derived from colloid chemistry experiments, surface tension measurements, rheological data, and optical data. It is important to note that, in Visser's assessment of Hamaker coefficients derived from the different methods, the most reliable results were found to be those derived from optical data using the method of Lifshitz. More recently, Parsegian and Weiss[16] reported Hamaker coefficients for the noble metals, silver, gold and copper, derived from optical data from various sources, with the finding that the data of Hageman et al.[26] are the more reliable by virtue of the more extensive range which is covered. The data in Table III from these sources represent the more reliable of the values reported.

Of the calculations made for this report, only Hamaker coefficient values for which the available data have been determined to be sufficiently extensive are reported. Sufficiency is determined with reference to Fig. 1; it is apparent that the data for manganese and tin do not range to high enough frequencies, while the data for aluminum oxide is inadequate at the lower frequencies. From the curves in Fig. 1, it is clear that the data used for all others but iron and tungsten, are adequate. For iron and tungsten, however, the high-frequency asymptote is not quite reached, but the error due to this is not significant. This can be shown if Y is plotted versus frequency (Hz) in a linear plot for iron and tungsten, respectively. It can be shown that the increase in area under such a curve as would be obtained by extending the curve to higher frequencies (assuming its continued decay) would be a small fraction of the total area, which is directly proportional to the Hamaker coefficient. A least squares polynomial fit of degree 3 was performed on the last 12 data points, for iron and tungsten. Extrapolation of the polynomial to frequencies just beyond the range of the data set showed the integrand of equation 10 reaches zero at approximately $6x10^{16}$ Hz and $8.5x10^{16}$ Hz for Fe and W, respectively. The corresponding area increases were estimated at less than 3%.

The findings of this report, as shown in Tables II and III, agree quite well with those of previous workers. No Hamaker coefficient calculations are reported for oxides, due to lack of sufficiently extensive optical data. The Hamaker coefficients for copper, silver, gold, and iron agree to within 20% with values reported by Visser[4] and by Parsegian and Weiss.[6] Although the comparison indicates significant uncertainty in the Hamaker coefficient, it also shows considerable improvement in the agreement between Hamaker constants reported by different workers, which have historically differed by as much as an order of magnitude.[4] In fact, the magnitude of the uncertainty is not unexpected in light of the fact that a portion of the uncertainty, perhaps 10%,[24] is inherent in the optical data. Also, Parsegian and Weiss[16] have found that the method of treatment of the data affects the result by some amount less than 10%.

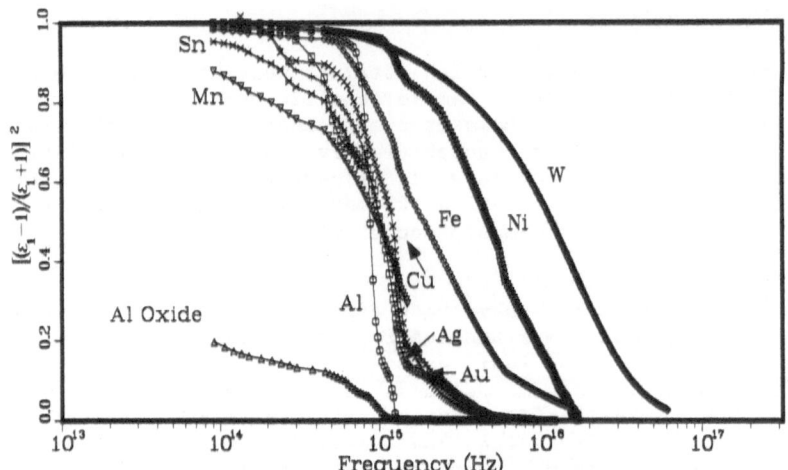

Figure 1. Transformed spectral data for ten metallic compounds. Semi-log
plot of the integrand of Equation (10) as a function of fre-
quency in Hertz using optical data from Hageman, Gudat and Kunz,
Deutsches Elektronen-Synchrotron, DESY SR-74/7, 1974(Ag, Al,
Al_2O_3, Au, Cu and Fe) and also data from Weaver, Kraftka and
Lynch, Physik Daten, 18-1 and 18-2, 1981 (Ni, Mn, Sn and W).

     The temperature dependency of the Hamaker coefficient is seen to be
small for all materials reported, for two bodies interacting across an
optically transparent medium. As seen in Table IV, the correction to the
Hamaker coefficient is much less than the uncertainty in the Hamaker coefficient itself. This is in agreement with previous estimations.[3,4]

     It is interesting to note that the Hamaker coefficients all fall
within a relatively narrow range of values, between $10-50 \times 10^{-13}$ ergs,
for the materials considered. This is surprising, since the materials all
exhibit somewhat different absorption spectra. However, such finding
could be useful for rough predictions of van der Waals forces for metals
for which adequate optical data do not yet exist. Also, the similarity in
Hamaker coefficient values for different materials lends additional validity to combining relations, such as Equation 15.

SUMMARY

     Hamaker coefficients have been reported for several metals of
interest in core-melt aerosol studies, including, aluminum, iron, manganese, nickel and tungsten. These included previously unreported values
for which were obtained by use of the macroscopic theory of Lifshitz and
extensive optical data already available in the literature. The calculations included a Kramers-Kronig transformation of the imaginary portion of
the dielectric function and utilized a numerical analysis routine based on
a Gaussian quadrature algorithm by Gill and Miller. Hamaker coefficients
were similarly obtained for materials studied previously, including the
noble metals, silver, copper, and gold, the values being slightly larger
than the previously reported values, due to the more accurate calculations. All Hamaker coefficents obtained were in the range of $10-50 \times 10^{-13}$
ergs. It was found that temperature effects could be neglected in Hamaker
coefficient calculations for the materials studied.

The approach used here to determine the Hamaker coefficients for metals provides a quick, relatively simple and reliable means for obtaining Hamaker coefficient estimates.

ACKNOWLEDGEMENTS

This report describes work performed as part of the TRAP-MELT Validation Project, which is sponsored by the Fuel Systems Behavior Research Branch of the U.S. Nuclear Regulatory Commission Division of Accident Evaluation.

REFERENCES

1. R. Wilson, "Report to the Americal Physical Society of the Study Group on Radionuclide Release from Severe Accidents at Nuclear Power Plants," Rev. Mod. Phys. 57, (3) (1985).
2. H. Jordan, J. A. Gieseke, and P. Baybutt, "TRAP-MELT User's Manual," Nuclear Regulatory Commission, NUREG/CR-0632, BMI-2017, R3, 4, 1979.
3. J. N. Israelachvili and D. Tabor,"Van der Waals forces: theory and experiment," Prog. Surf. Membr. Sci. 7, 1 (1973).
4. J. Visser, "On Hamaker coefficients: A comparison between Hamaker and Lifshitz-Van der Waals," Adv. Colloid Interface Sci. 33, 311 (1972).
5. J. N. Israelachvili, "Calculation of Van der Waals dispersion forces.. macroscopic bodies," Proc. Royal Soc. A 331, 39 (1972).
6. J. Gregory, "Calculation of Hamaker coefficients," Adv. Colloid Interface Sci. 2, 396 (1970).
7. H. Krupp, "Particle adhesion, theory and experiment," Adv. Colloid Interface Sci. 1, 111 (1967).
8. Von F. London, "Zur theorie and systematik der molekularkrafte," Z. Physik 63, 245 (1930).
9. H. C. Hamaker, "The London-van der Waals attraction between spherical particles," Physics 4 No. 10 (1973).
10. E. M. Lifshitz, "Theory of molecular attractive forces between solids," Sov. Phys. 2, 73 (1956).
11. K. L. Johnson, K. Kendall, and A. D. Roberts, "Surface energy and the contact of elastic solids," Proc. Royal Soc. A 324, 301 (1971).
12. M. Vanden Tempel, "Interaction forces between condensed bodies in contact," Adv. Colloid Interface Sci. 3, 137-159 (1972).
13. F. M. Fowkes, "8. Attractive forces at interfaces," Ind. Eng. Chem. 56, 40-52 (1964).
14. C. J. van Oss, R. J. Good, and M. K. Chaudhury, J. Colloid Interface Sci. 111(2), 378(1986).
15. I. E. Dzyaloshinskii, E. M. Lifshitz, and L. P. Pitaevskii, "General theory of van der Waals forces," Adv. Phys. 10 165 (1961).
16. V. A. Parsegian and G. H. Weiss, "Spectroscopic parameters for computation of van der Waals forces," J. Colloid Interface Sci. 81, 285 (1981).
17. M. K. Chaudhury and R. J. Good, Langmuir 1(6), 673(1985).
18. Ya. I. Rabinovich, "Calculation of Hamaker coefficient from depth of secondary potential well," Kolloidnyi Zhurnal (see Colloid J. of USSR for translation) 42(1), 75 (1980).
19. Ya. I. Rabinovich, "Approximate method for calculation of Hamaker coefficient..effect of lag.," Kolloidnyi Zhurnal (see Colloid J. of USSR for translation) 41(6), 1146 (1980).
20. R. Pashley, "Van der Waals interaction for liquid water: comparison of..(methods)," J. Colloid Interface Sci. 62, 344 (1977).

21. P. E. Gill and G. F. Miller, "An algorithm for the integration of unequally spaced data," Computer J. <u>15</u>, 80 (1972).
22. A. Shih and V. A. Parsegian, Physical Review A <u>12(3)</u>, 835(1975).
23. L. D. Landau and E. M. Lifshitz, "Electrodynamics of Continuous Media," Sect. 122, Pergamon Press, New York, 1960.
24. R. C. Weast, editor, "CRC Handbook of Chemistry and Physics," 64th edition, CRC Press, Boca Raton, FL (1983).
25. J. H. Weaver, C. Krafka, and D. W. Lynch, "Optical properties of metals," Physik Daten, <u>18-1</u> and <u>18-2</u> (1981).
26. H. J. Hageman, W. Gudat, and C. Kunz, Deutsches Elektronen-Synchrotron, DESY SR-74/7 (1974).
27. E. T. Arakawa, T. Inagaki, and M. W. Williams, "Optical properties of metals by spectroscopic ellipsometry," Surface Sci. <u>96</u>, 248 (1980).
28. J. P. Petrakian, A. R. Cathers, and E. T. Arakawa, "Optical properties of liquid tin between 0.62 and 3.7 eV.," Phys. Rev. <u>B 21</u>, 3043 (1980).
29. G. E. Jellison and F. A. Modine, "Optical Constants for Silicon at 300 K and 10 K from 1.64-4.73eV by Ellipsometry," ORNL/TM-8002, Oak Ridge National Laboratory, Oak Ridge, Tennessee, 1982.
30. D. Y. Smith, E. Shiles, and M. Inokuti, "Optical Properties and dielectric function of aluminum from 0.04-10,000 eV," ANL-83-24, Argonne National Laboratory, Argonne, Illinois (1983).
31. H. Ehrenreich, H. R. Philipp, and B. Segall, Phys. Rev. <u>132</u>, 1918 (1963).
32. H. Ehrenreich and H. R. Philipp, Phys. Rev. <u>128</u>, 1622 (1962).
33. R. A. McRae, E. T. Arakawa, and M. W. Williams, "Optical properties of vacuum-evaporated white tin," Phys. Rev. <u>162</u>, 615 (1967).
34. A. I. Golovashkin and G. P. Motalevich, "Optical properties of tin at helium temperatures," Sov. Phys.-JETP <u>20</u>, 44 (1965).
35. H. R. Philipp and H. Ehrenreich, Phys. Rev. <u>129</u>, 1550 (1963).
36. H. R. Philipp and E. A. Taft, Phys. Rev. <u>136A</u>, 1445 (1964).
37. H. R. Philipp and H. Ehrenreich, Phys. Rev. <u>131</u>, 2016 (1963).
38. F. Wooten, "Optical Properties of Solids," Sect. 3.2., Academic Press, New York, 1972.
39. R. P. Wichner and R. D. Spence, "1983 Quantity and Nature of LWR Aerosols Produced in the Pressure Vessel during Core Heatup Accidents - A Chemical Equilibrium Estimate," NUREG/CR-3181, ORNL/TM-8683, Oak Ridge National Laboratory, Oak Ridge, Tenn.
40. D. Langbein, J. Adhesion <u>1</u>, 237 (1969).
41. J. E. Keifer, V. A. Parsegian, and G. H. Weiss, J. Colloid Interface Sci. 57, 580(1976).
42. W. H. Marlow, "Aerosol Microphysics I: Particle Interaction," pp. 133-154 Springer-Verlag, New York, 1980.

SOILING MECHANISMS AND PERFORMANCE OF ANTI-SOILING SURFACE COATINGS

Edward F. Cuddihy

Jet Propulsion Laboratory
4800 Oak Grove Drive
Pasadena, California 91109

Physical examination of surfaces undergoing natural outdoor soiling suggests that soil matter accumulates in up to three distinct layers. The first layer involves strong chemical attachment or strong chemisorption of soil matter on the primary surface. The second layer is physical, consisting of a highly organized arrangement of soil creating a gradation in surface energy from a high associated with the energetic first layer to the lowest possible state on the outer surface of the second layer. The lowest possible surface energy state is dictated by the chemical and physical nature of the regional atmospheric soiling materials. These first two layers are resistant to removal by rain. The third layer constitutes a settling of loose soil matter, accumulating in dry periods and being removed during rainy periods.

Theories and evidence suggest that surfaces that would be naturally resistant to the formation of the first two rain-resistant layers should be hard, smooth, hydrophobic, free of first-period elements, and have the lowest possible surface energy. These characteristics, evolving as requirements for low-soiling surfaces, suggest that surfaces or surface coatings should be of fluorocarbon chemistry.

This article describes the three soil layer concept, soiling adhesion mechanisms, and the positive performance of candidate fluorocarbon coatings on glass and transparent plastic films after 56 months of outdoor exposure.

INTRODUCTION

The accumulation of dust, dirt, pollen, and other atmospheric contaminants and particles on the surfaces of solar energy devices such as solar-thermal collectors and photovoltaic (PV) modules results in a loss of performance due to a decrease in transmitted sunlight. This accumulation of a diversity of deposited atmospheric materials, hereafter referred to as soil reduces light transmission by a combined action of absorption and scattering. To minimize performance losses caused by soiling, solar devices should have surfaces or surface coatings that have low soil retention, maximum susceptibility to natural cleaning by wind, rain and snow, and which are readily cleanable by simple and inexpensive maintenance techniques.

From considerations of known and postulated mechanisms of soil retention on surfaces, it became possible to develop theoretical definitions[1,2] of the requirements of low-soiling surfaces or surface coatings, and of the requirements for low-soiling environments. The requirements for low-soiling surfaces appear to be:

1) hardness
2) smoothness
3) hydrophobicity
4) low surface energy
5) nonstickiness (chemically clean of sticky materials, both surface and bulk)
6) cleanliness (chemically clean of water-soluble salts and first-period elements, both surface and bulk)

and the requirements for low-soiling environments appear to be:

1) low to zero airborne organic vapors
2) frequent rains, or generally dry (low dew, low RH)
3) few dew cycles or occurrences of high RH between heavy rain periods.

Further, of the expected natural cleaning agents--wind, rain, and snow --only rain is really effective. With respect to snow, observations have been made[3] that the surfaces of photovoltaic modules and mirrors are noticeably quite clean after a heavy snow pack has slid off the tilted modules. The presumption is that cleaning is accomplished by a combination of abrasive action and the presence of liquid water at the module surface/snow pack interface. However, accumulated snow that is removed by melting and not sliding is not effective[4].

Wind is also not an effective cleaning agent. The aerodynamic lifting action of wind can remove particles greater than about 50 μm from surfaces[5], but is ineffective for smaller particles. Thus, the particle size of soil matter is generally found to be less than 50 μm, and predominantly to be less than 5 μm[6,7].

In general, rain appears to be the primary natural outdoor cleaning agent, but rain is not necessarily efficient at all times in removing all accumulated soil on a surface. This article will describe known and speculated mechanisms of soil adhesion to surfaces which can result in resistance to soil removal by rain. Therefore, the above mentioned six requirements of low-soiling surfaces are theoretically considered as required to minimize rain-resistant adherence of soil matter on surfaces. Based on these theoretical considerations, candidate low-soiling surface coatings based on fluorocarbon chemistry could be identified. These coating materials and their positive performance after 56 months of outdoor exposure are herein reported.

In addition, activities related to the removal of soil specimens from the surfaces of soiled photovoltaic modules, using transparent adhesive tape, for later chemical and physical analysis, in combination with detailed observations of the maintenance washing and cleaning of these soiled surfaces, resulted in a theoretical speculation that soil accumulates in tiers of up to three distinct layers. These layers are designated, outward from the surface, as A, B, and C. Layer C always forms during dry periods, and is removed during rainy periods. Layers A and B, which are resistant to removal by rain, may or may not form, but if they do, will be in the sequence A followed by B (followed by C), or B only (followed by C). Therefore, refinement of the soiling theory suggests that the six requirements for low soiling surfaces are those for preventing the formation of layers A and B, or B alone, but will have no influence on layer C. The observational evidence for the natural formation of up to three soil layers are also described.

SOILING MECHANISMS

Cementation

Adherent particles clinging to the surfaces of video disc interfere with both record and playback signal quality. This problem was encountered in the development of flat surface video discs. RCA[8] identified a specific mechanism causing the problem, which involved soluble salts and relative humidity.

Atmospheric dust contains a distribution of inorganic and organic particulates. The inorganic particulates in turn contain a distribution of water soluble and insoluble salts. At high humidities, water-soluble dust particles on the surface formed microscopic droplets of salt solutions which also retained any insoluble particles. As the relative humidity decreased, the drops of salt solution dried out leaving the precipitated salt to function as a cement to anchor insoluble particles to the surface (Figure 1). RCA reported that a cross-linked surface layer of a General Electric silicone product designated SF-1147 aided in reducing the affinity of the surface to retain particles by the cementation mechanism.

McDonnell-Douglas published a report[9] on the soiling of mirror surfaces, which also identified cementation as a retention mechanism.

The weathering of soda-lime glass is schematically illustrated in Figure 2, which is reproduced from Reference 10. Observe that one result of glass weathering is the formation and deposition of water soluble sodium salts on the surface. The source of water soluble salts for cementation can originate not only as a component of atmospheric particles, but in the surface material itself, either inherently or by weathering.

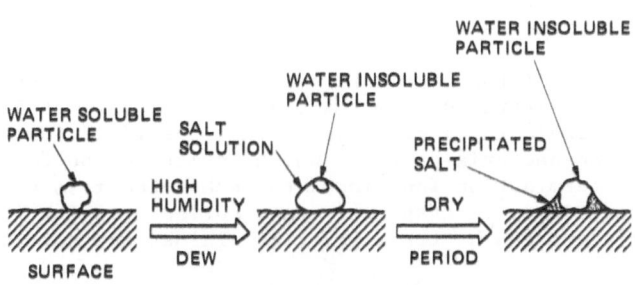

BUILDUP OF LAYERS OF CEMENTED PARTICLES

Figure 1. Cementation process.

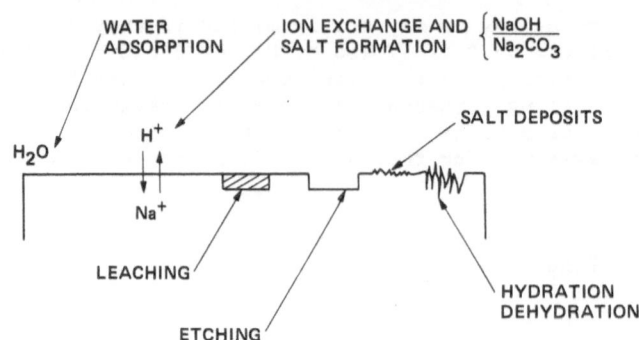

Figure 2. Weathering of soda-lime glass (taken from Reference 10).

Organic Deposition

An investigation into the mechanisms by which salt deposits built up on desalinization membranes revealed[11] the deposition sites were first coated with an ultra-thin layer of organics, presumably organic colloids present in the water. Cleaning of the salt deposits was facilitated by the use of surfactant and/or detergent solutions to remove the organic layer, but the efficiency of cleaning was time-dependent. This was because the salt buildup became so thick that the cleaning solutions could not effectively reach the organic layer.

With respect to cementation, this same characteristic of time-dependency on the efficiency of cleaning was mentioned in the McDonnell-Douglas report[9]. Repetitive cycles of dew formation and evaporation and high relative humidity (RH) presumably resulted in a gradual buildup of cemented layers, as illustrated in Figure 1, and the eventual inability of cleaning agents and rain to penetrate and dissolve the cement.

Surface Tension

Early microscopic examination of bacteria revealed them to be flat and saucer shaped, contrary to expectation. In preparation for microscopic studies, fluid dispersions of bacteria were deposited onto a slide and the fluid allowed to evaporate away. Eventually an alternative preparative technique called "freeze drying" was employed and bacteria were then observed in their expected shape.

An investigation of the evaporative technique revealed that the bacteria were being flattened by mechanical forces associated with surface tension. In the final stages of fluid drying, the fluid formed an enveloping film around the bacteria particle which terminated on the surface of the slide. This termination functioned to anchor the film around the particle. As fluid continued to evaporate away, surface tension forces were sufficient to flatten the bacteria particle against the surface of the slide.

If instead of bacteria, the particle was soil with irregularly shaped sharp points, then surface tension forces acting on these sharp points could become enormous and drive the particle into a soft surface. Visual examination of the soiled surfaces of soft silicone elastomers revealed that the surfaces acquired "craters," having the appearance typical of an impact crater. As a speculation, the action of surface tension forces pushing down on particles could conceptually squeeze up the compliant elastomer around the particle perimeter and generate the observed "crater" appearance.

Particle Energetics

The energetics of particles, and therefore "particle-particle" attraction is a function of particle size[5,12]. The attractive forces are considered to be van der Waals forces. One experimental method which has been successfully used to measure the relative magnitude of "particle-particle" attraction forces as a function of particle size is sedimentation. For this experimental method, particles of a known and narrow size distribution are dispersed in an inert liquid and the sedimentation volume ϕ_m measured after centrifuging. The sedimentation volume ϕ_m decreases as particle-particle attraction[12] becomes more dominant than separation forces in the fluid medium.

Figure 3 is reproduced from Reference 12, and is a plot of experimentally measured values of ϕ_m as a function of particle size, for a variety of chemically different particles. The results show that particle-particle attractions which limit the sedimentation volume begins at a nominal particle size of 10μm.

It can be suggested that the natural attractive forces which develop between small size particles might also develop between a small particle and a surface. During the development of the Mars landing probe for the Viking mission, there was a requirement that the probe be decontaminated so as not to carry earth bacteria into the Mars environment. One carrier of bacteria is atmospheric dust, which would settle freely on the surfaces of the landing probe. Efforts to remove the settled dust with a revolving brush vacuum cleaner were fruitless. Dust particles of a size greater than 50 μm were readily removed, whereas particles of a size 10 μm and less could not be removed at all. There was some removal of particles whose size was intermediate between 10 and 50 μm.

The vacuum cleaner generates a wind velocity across the surface which produces a force on the dust particles. It was decided to experimentally

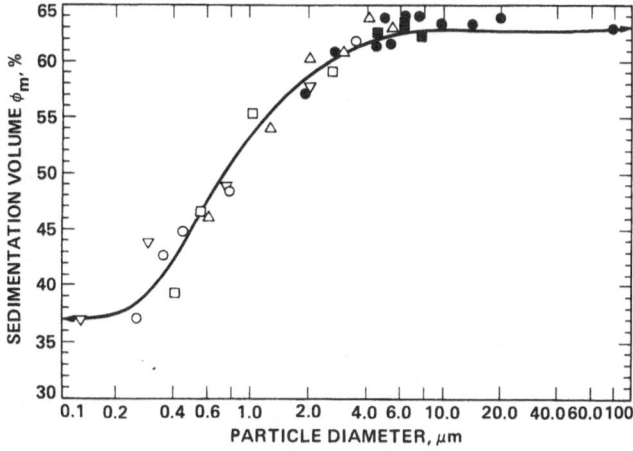

Figure 3. Effect of particle-particle interactions on sedimentation volume.

measure the efficiency of particle removal as a function of wind velocity
and particle size. The experimental surface was selected to be a glass
slide. Details of the experimental technique are described in Reference 5.
The dust, referred to as "facility dust" consisted of 90% silicon based
materials, with the remaining 10% being essentially equal parts of non-
magnetic metals, cellulose, Teflon and some unidentified colored materials.

The efficiency of removal of the facility dust from dry, oil-free
glass slides as a function of particle size and air velocity is shown in
Figure 4, which is reproduced from Reference 5. The measurements were
made in a room having an average relative humidity of 40%. Note the
dramatic resistance of the facility dust against removal by wind forces,
with the resistence significantly increasing as the particle sizes decrease
below 50 μm. It is pointed out in Reference 5 that no particles of size
10 μm and smaller were removed at air velocities below 25 m/s (≈55 miles
per hour).

Figure 5, also reproduced from Reference 5, is a plot of the wind
forces acting on particles, as a function of particle size, and also the
attractive forces between particles and a surface as a function of particle
size at 3 levels of relative humidity. Note that both the attractive forces
and wind forces decrease with decreasing particle size, but that wind forces
decrease faster. Thus for particle sizes greater than about 50 μm, wind
forces are greater than the attractive forces, but as particle sizes become
smaller, the attractive forces become greater than wind forces. Therefore
for particle sizes less than 10 to 50 μm, wind forces become ineffective for
removing these small particles from surfaces.

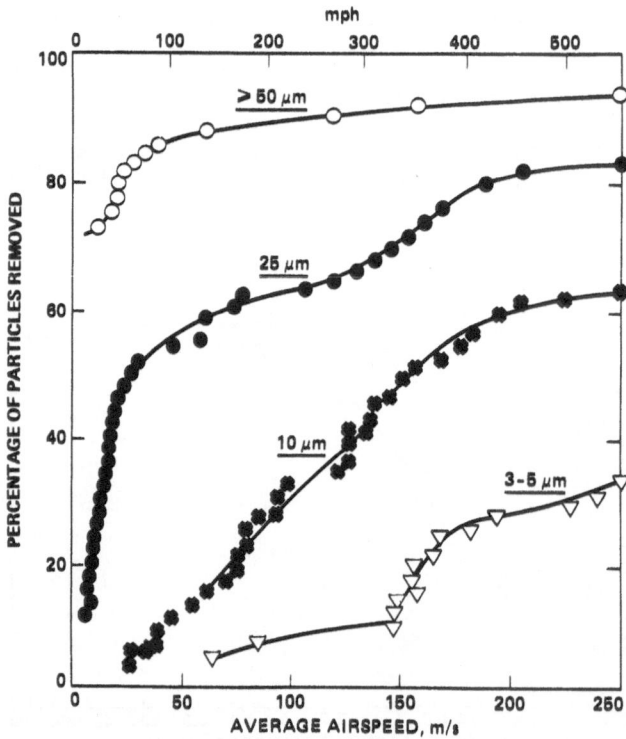

Figure 4. Removal by wind of dust particles from the surface of a clean,
oil-free glass slide (40%RH) (taken from Reference 5).

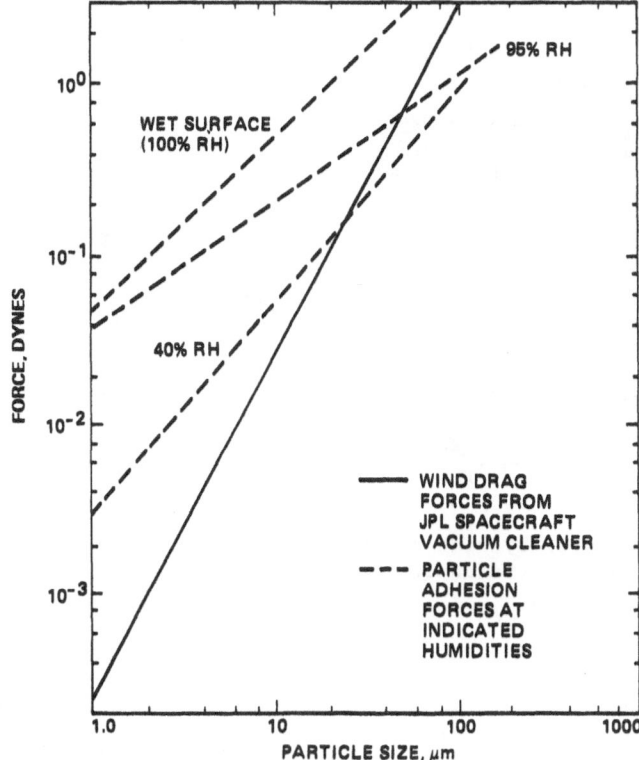

Figure 5. Comparison of wind forces on particles, and the attractive
forces between particles and a surface, as a function of
particle size (taken from Reference 5).

Two additional experiments with dust removal by wind forces were carried
out. The first was to allow facility dust to settle on a glass slide whose
surface had been touched by human fingers in order to generate a thin oily
layer. The second consisted of placing a pre-cleaned glass slide with
settled facility dust into a refrigerator and then back out into the ambient
air to allow the cold surface to fog with condensed moisture. The test
results for both slides are shown in Figures 6 and 7, which are reproduced
from Reference 5. The effect of an oily layer and one dew cycle are indeed
spectacular.

ANTI-SOILING REQUIREMENTS

The considerations relative to particle energetics and the experimental investigation of particle removal by wind forces strongly suggest that
for small particles (<50 μm), wind is nonfunctional as a natural cleaning
agent.

The cementation mechanisms identified by RCA and McDonnell-Douglas and
the significant increase in resistance of particles to be removed by wind
after just one dew cycle strongly suggest for outdoor soiling that relative
humidity and dew are involved in soil retention. For the cementation
mechanism, frequent rains should dissolve the salts and remove the trapped
particulates. However, for low rain locations, the buildup of cemented

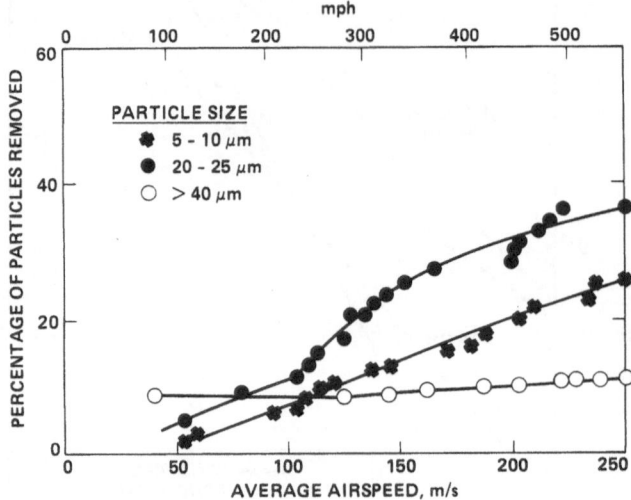

Figure 6. Removal by wind of dust particles from an oily glass slide surface (taken from Reference 5).

particles should begin to resist water penetration to lower layers. The RCA article suggests that surfaces such as generated with General Electric's SF-1147 silicone product may have a reduced affinity for soiling caused by cementation. The surface can be characterized as hard and hydrophobic.

In the case of the surface tension mechanism of drying water, this could cause particulates to be pressed into surfaces for permanence. Hard surfaces are suggested in order to counter this postulated mechanism of soil retention. Additionally, it would seem that both particles and the

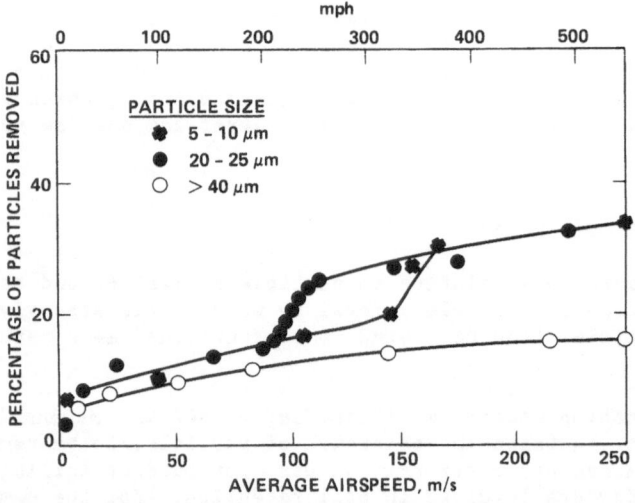

Figure 7. Removal by wind of dust particles from a glass slide surface after one dew cycle (taken from Reference 5).

surface should be wetted by water in order to set up the surface tension mechanism. This would not be expected with hydrophobic surfaces.

Sub-surface deposition of particulates may be caused by permeation and/or solubility of salt solutions. During a dry cycle, the water dries out of the material and leaves behind sub-surface precipitated salts, i.e., water spots. An opposing consideration is that sub-surface water dissolves impurities and "processing additives" present in the bulk of the material and draws these agents to the surface (i.e., by elution). Some of the agents may be sticky and tacky.

An additional argument for low surface energy involves the attractive forces between surfaces and particles which cause clinging. It can be speculated that these forces, basically van der Waals in origin, arise from both the level of surface energy and the energetics of particles. Employing surfaces having lower surface energy may shift downward the sizes of particles which tend to naturally cling to surfaces.

Defining soil retention mechanisms as those which result in resistance of soil particles to be removed by rain, particle-surface attraction would not appear to be a mechanism. Rather, particle-surface attraction keeps the small particles (<<50 μm) "clinging" to the surface until rain-resistant mechanisms occur.

In summary, the evolving requirements for low-soiling surfaces appear to be:

1) hard
2) smooth
3) hydrophobic
4) low-surface energy
5) chemically clean of sticky materials
6) chemically clean of water soluble salts

and the evolving requirements for low-soiling environments appear to be those having:

1) low to zero airborne organic vapors
2) frequent rains, or generally dry (low dew, RH)
3) few dew cycles or occurrences of high RH between heavy rain periods.

NATURE OF THE SOIL LAYERS

The concept that soiling behaves as though it were accumulating in tiers of up to three distinct layers resulted serendipitously from an effort to lift surface soil from test surfaces with a transparent adhesive tape for analysis. The soil samples were to be taken from the surfaces of photovoltaic modules exposed outdoors at test sites in Panama and Alaska. At each site three modules of different designs, with three different surfaces, were accumulating soil. The three surfaces were:

1) A soft silicone elastomer (GE RTV 615)
2) A semihard silicone coating (Dow Corning Q1-2577)
3) Hard soda-lime window glass.

Test Sites

At the Panama site each of the three surface types was soiled. Hand washing with a solution detergent in water cleaned the glass surfaces thoroughly, but cleaned the soft and semihard silicone surfaces only partially. The residual soil on these silicone surfaces, which could not be removed by hand washing, appeared to constitute a thin, uniform, tan-gray coating over the whole of the module surfaces. This coating could only be removed by vigorous and abrasive scrubbing which was conveniently performed with a slightly moistened thumb. The scrubbed area, cleaned of the tenacious soil coating, exhibited the crystal-clear, water-white appearance associated with fresh, unsoiled silicone.

The Panama observation suggested the existence of two presumably distinct layers of soil on the silicon surfaces: a primary layer directly on the silicon surface, very tenacious, which could not be removed by washing and on top of this primary layer, a secondary layer that could be readily removed by washing. The glass surfaces, on the other hand, could be washed clean of all soiling matter. In the interest of establishing definitions, the soil layer on glass was comparable to the secondary soil layer on the silicones, in the operational sense that they both could be removed by hand washing. Thus, when observed at Panama, glass was covered with a secondary layer of soil, while the silicones were covered with both a primary and a secondary layer of soil.

Conversations with persons who operate the Panama test site revealed that rain occurs there almost daily over an eight-month period from April to November and almost none at all, or very little, occurs during a dry period from December to March. They noted that more soil can be observed on the modules during the dry period than during the rainy period. When the rains begin in April, some soil is removed. The inspection of the Panama modules was done in mid-May, when presumably only rain-resistant soil remained on the module surfaces. However, the observation of additional soil during the dry months suggested the existence of yet a third chracteristic surface layer of soil, probably the outermost layer, which is removable by rain. Since this third type of soil layer builds up during dry periods and is depleted during rainy periods, it probably is a fluctuating surface layer whose quantity of soiling matter fluctuates in some sequence with rain patterns.

At the Alaska site in mid-August, it was observed that the glass surfaces were extremely clean with no visual evidence of any soil on their surfaces. The soft and semihard silicones, however, were soiled. Frequent rains in the local area preceded the inspections.

Hand washing did nothing for the glass modules, removed essentially all soil from the semihard silicone and cleaned the soft silicone surface only partially. The tenaciously adhesive soil layer remaining on the soft silicone surfaces had the same appearance as that observed in Panama and could be removed by abrasive scrubbing. The scrubbed area became crystal clear.

For the Alaska site, it is observationally inferred that a rain-removable layer settles on all surfaces during dry periods, that a rain-resistant secondary soil layer forms on the semihard silicone, and that both primary and secondary rain-resistant soil layers form on the soft silicone.

Transparent Tape Sampling

At Panama, two modules with soft silicone surfaces (GE RTV 615) were visibly quite dirty and had never been washed. They had been outdoors for about three years. When transparent adhesive tape was placed on the dirty surface of one of the modules for soiling sampling, two observations were made: the tape piece readily and immediately adhered to the soiled surface, and visibility through the tape into the silicone interior was strikingly improved over visibility of the silicone interior when viewed through the untaped soiled surface. The enhanced clarity of the taped area afforded a dramatically clearer and sharper view of the underlying solar cell and substrate panel, as compared with the obscurity of the untaped surface.

However, the taped area did not acquire the crystal-clear appearance associated with a clean, brand-new soft silicone surface; there was a tan-grey color under the tape, which appeared to be uniform over the entire taped area.

When the tape was peeled off the surface, some soil adhered to the tape, but in the tape-sampled area, the thin, uniform, tan-grey soil coating remained. This coating was virtually identical in appearance with the tenanciously adhering coating that could not be removed by hand washing. The residual coating in the tape-sampled area could be removed by abrasive scrubbing, exposing the crystal-clear appearance associated with a clean silicone surface.

At Panama, the tape overlay acted optically to eliminate or reduce the light obscuration associated with the secondary soiling layer. The tape also caused optical disappearance of the secondary soiling layer; thus, the next lower layer was revealed through the tape, which for each module was the tenaciously adhering, tan-grey primary soil layer.

The "Hide-a-Layer, Reveal-the-Next" effect of the tape overlay was also observed at the Alaska site. A tape overlay on the unwashed soft silicone modules revealed the tan-grey primary soil layer, and after the modules were washed, a tape overlay on the primary soil layer offered a crystal-clear view into the interior of the module. A tape overlay on the unwashed semihard silicon modules resulted in a crystal clear effect and these modules, when washed, were observed to be free of a primary soil layer.

Three Soil Layers

The three soil layers whose existence was indicated from field observations are illustrated in Figure 8. They can be designated and defined for descriptive convenience as follows:

1) Layer A, a primary surface layer of soil that is resistant to removal by rain, washing, and adhesive tape. This layer can only be removed by abrasive scrubbing. This layer is speculated to be attached by strong chemical bonds or strong chemisorption.

2) Layer B, a secondary surface layer of soil that is resistant to removal by rain, but can be readily removed by washing or adhesive tape. This layer is speculated to be of a cemented construction.

3) Layer C, a top surface layer of dirt that can be readily removed by rain. The depth of layer C fluctuates with rain patterns.

The field observations suggest that if layer A forms, it will do so directly on the material surface, and then layer A will be over-coated with layer B, which in turn will be overcoated with layer C.

If layer A does not form, then layer B will form directly on the material surface and then will be overcoated with layer C.

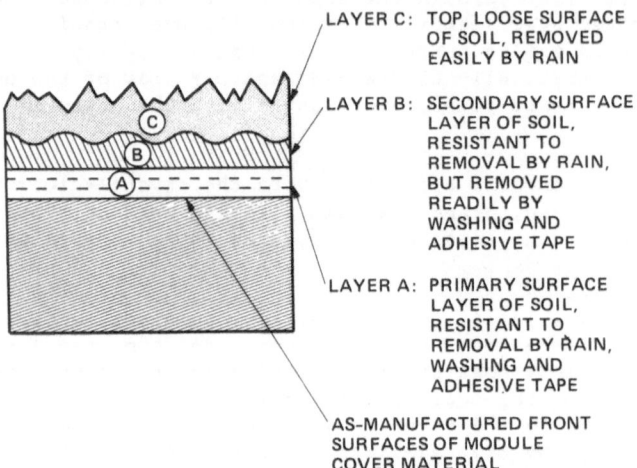

LAYER C: TOP, LOOSE SURFACE OF SOIL, REMOVED EASILY BY RAIN

LAYER B: SECONDARY SURFACE LAYER OF SOIL, RESISTANT TO REMOVAL BY RAIN, BUT REMOVED READILY BY WASHING AND ADHESIVE TAPE

LAYER A: PRIMARY SURFACE LAYER OF SOIL, RESISTANT TO REMOVAL BY RAIN, WASHING AND ADHESIVE TAPE

AS-MANUFACTURED FRONT SURFACES OF MODULE COVER MATERIAL

Figure 8.   The three soiling layers.

If layer B does not form, then only layer C will reside on the material surface. The field observations did not indicate in any way that layer C will reside directly on layer A without the intermediary layer B.

In other words, the soiling possibilites seem to be:

    1)   A + B + C
    2)   B + C
    3)   C only

Jet Propulsion Laboratory Soiling Studies

JPL carried out measurement of the decrease in light transmittance of seven different transparent materials being soiled naturally at 7 climatically different locations[13,14]. The seven materials included three different glasses: an aluminosilicate glass, a borosilicate glass, a soda lime glass; and four polymers: Korad acrylic film, Tedlar fluorocarbon film, the semihard silicone surfacing material, and the soft silicone elastomer (RTV-615). The method of measuring soiling accumulation involved the measurement of the short-circuit current from a standard solar cell positioned behind the transparent materials. The short-circuit current of a solar cell is directly proportional to light transmittance, and decreases with increasing quantities of soil on the surfaces of the transparent material. Test results are reported as a percentage using the equation

$$\text{Loss from soiling, \%} = \frac{I_c - I_s}{I_c}$$

where I_c is the short-circuit current measured with the clean transparent material over the cell and I_s is the short-circuit current measured with the soiled transparent material over the cell.

The test materials had been outdoors for more then two years, unwashed, and with soiling measurements made on these materials at intervals of two

102

to three months. The time-dependence of the outdoor soiling behavior of
the materials generally follows the pattern schematically illustrated in
Figure 9. The oscillating solid line traces the time-dependent magnitude
and behavior of the surface soiling, which increases during dry periods and
decreases during rainy periods.

Accepting the soil-layering concept, the curve in Figure 9 should
reflect the existence of rain-resistant and rain-removable soil layers.
The dotted line connecting the minimums, therefore, is associated with
the light obscuration caused by the development of the rain-resistant
layers, either layers A and B, or layer B alone; and the solid, oscillating line riding on the dotted line, is therefore associated with the
light obscuration caused by the rain-controlled layer C. A general characteristic illustrated in Figure 9 is that the dotted line associated with
A and/or B formation rises rapidly for the first 30 to 60 days and there-
after slows dramatically, ranging for various combinations of materials
and sites from a virtual asymptote to a perceptibly detectable slope.

The published JPL soiling data[14] can be de-coupled into light obscuration
caused by the rain-resistant layers (A and/or B), and layer C. To do
this, the minima of the experimental data curves, as schematically illustrated in Figure 9, are associated with the light obscuration caused by
rain-resistant layers (A + B, or B only), and the difference then between
this rain-resistant minimum baseline and the maximum peak to have been
observed is associated with the maximum light obscuration caused by layer C.
The latter calculation is arbitrary as there were other intermediate highs
in the soiling data. Thus, the calculated valued to be allocated to
layer C represents the maximum quantity of layer C soil to have been
present on the surface during the outdoor exposure period.

The de-coupled light obscuration values are described in Table I using
the JPL soiling data[14] from the seven different sites. The available data
do not permit further de-coupling of the minimum into separate values for
A and B: therefore the minimum is considered the sum of A and B as
indicated in the column heading of Table I.

As expected, the data indicate that the largest quantity of rain-
resistant soil (Column A + B) is found on the soft silicone followed by
the simihard silicone, and last, by the remaining five harder materials.
Although the numbers for these five materials are small, there is an
indicated ranking. Comparing the plastic films, the fluorocarbon (Tedlar)
is slightly better than the acrylic (Korad). Comparing the glasses, the
ranking (in improving order) is soda-lime, aluminosilicate and borosilicate.

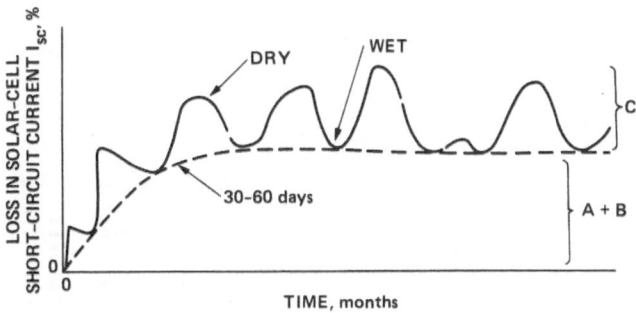

Figure 9. Behavior of natural outdoor soiling.

Table I. JPL Soiling Data: Percent Reductions in Short-Circuit Current From Soiling Layers A, B, and C(a).

Site Materials	Torrance (A+B)	C	Point Vicente (A+B)	C	Goldstone (A+B)	C	Table Mountain (A+B)	C	Pasadena[b] (A+B)	C	JPL 34-deg Site[c] (A+B)	C	JPL 45-deg Site[c] (A+B)	C
						Reduction, %								
Soft Silicone RTV 615	20	10	?	?	?	?	?	?	25	8	24	6	24	7
Semihard Silicone, Q1-2577	14	8	5	2	6	2	1	3	17	15	16	12	15	8
Acrylic Film, Korad 212	3	8	0	8	1	2	2	1	5	14	3	13	3	11
Fluorocarbon Film, Tedlar	1	8	0	5	0	2	0	2	3	13	1	16	2	12
Soda-Lime Glass	2	6	1	4	2	2	0	2	3	9	4	12	3	9
Aluminosilicate Glass	1	12	1	5	0	2	0	2	2	12	2	13	2	11
Borosilicate Glass	0	7	0	5	0	2	0	2	1	11	1	15	1	13
Average for layer C		8.4		4.8		2		2		12.3		13.5		10.6

a Data from Reference 14
b Pasadena station of South Coast Air Quality Management district.
c 34 deg and 45 deg tilt angles from ground.

As was observed for glass in Alaska, the JPL soiling data indicate for some combinations of sites and materials that neither layer A nor layer B has formed (the minima of the soiling curves are zero). The data suggest that the formation of the rain-resistant soil layers are both material- and site-dependent, but that material dependency dominates. The trend of surface properties for minimizing the formation of A and B layers, as revealed in Table I, supports the concept of the theoretically derived six requirements of low-soiling surfaces as described above in the section on soil theory.

There is a strong indication in the data that the magnitude of layer C soiling is site-dependent and material-independent. This is understandable, given the development of layer B; it is on this surface rather than the natural material surface that layer C resides. Thus, the development of layer B leads to material independence. For those materials that do not form a layer B, their natural surface must have properties similar to those of layer B.

The site dependency of layer C relates to the atmospheric concentrations of soiling materials, their types and rain cycles. The average of the six or seven values of light obscuration by layer C is also included in Table I. If the average value for layer C is treated as a measure of the soiling characteristics of an environment, then (of the sites listed in Table I) JPL and Pasadena are the dirtiest and Goldstone and Table Mountain are the cleanest.

The two JPL sites designated as 34 deg. and 45 deg. are at the same location, differing only in the angles at which test modules are tilted from the ground. The tilt-angle dependence implied for layer C is a reduction in layer C accumulation with increasing tilt toward the vertical. Figure 10 is a linear plot of tilt-angle data for layer C. Although there are only two data points, it is interesting to observe that a linear extrapolation would suggest no layer C deposition at a near vertical alignment. There appears to be no tilt-angle effect on the formation of rain-resistant soil layers A and B.

From the perspective of developing maintenance-cleaning strategies and techniques, the soiling studies suggest that for hard surfaces, light obscuration by rain-resistant layers A and B is low, typically much less than 4%. The real problem is the three layers that develop on soft and semihard surfaces and layer C on hard surfaces. Because soft and semihard surfaces are not favored for solar devices, requirements for establishing maintenance-cleaning methods should probably be related to layer C behavior on hard surfaces.

It is being suggested that maintenance cleaning techniques for hard surfaces should not be designed for layers A and B, which generate the least light obscuration and which would require the most demanding cleaning approaches, such as extremely high-pressure water. Rather, cleaning strategies should be developed for layer C, perhaps a low-pressure water spray (rain simulation) during dry cycles.

LOW-SOILING COATINGS

Field soiling observations and JPL soiling data indicate that there are two distinct soiling problems to be dealt with to achieve low soiling. The first is to have top surfaces that resist the formation of the rain-resistant soil layers; the second is related to the rain-removable layer.

As previously mentioned, there are six characteristics of low-soiling

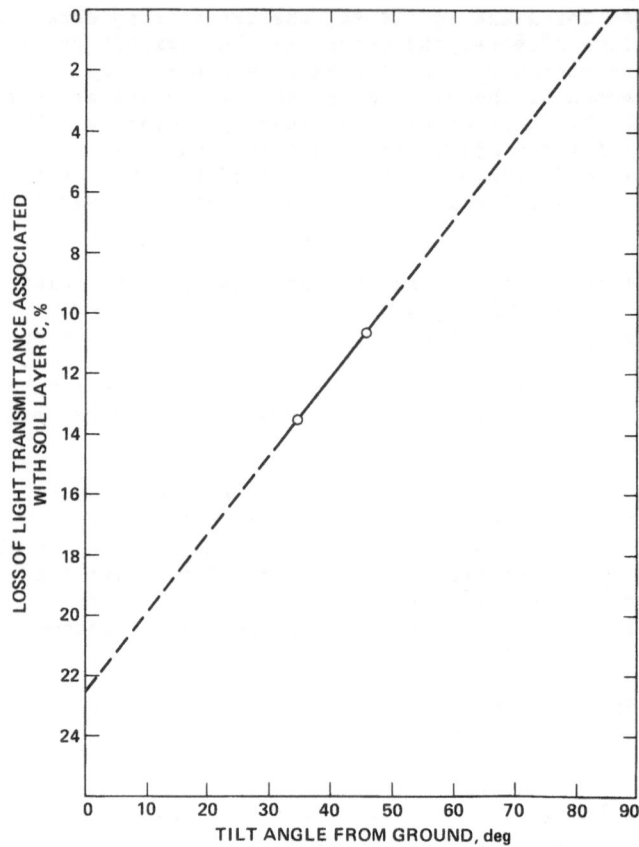

Figure 10. Effect of tilt angle on accumulation of soil layer C (Data from Table I, JPL sites).

surfaces,

1) hardness
2) smoothness
3) hydrophobicity
4) low surface energy
5) nonstickiness (chemically clean of sticky materials, both surface and bulk)
6) cleanliness (chemically clean of water-soluble salts and first-period elements, both surface and bulk)

These six postulates relate to surface requirements for resisting the formation of rain-resistant soil layers and, taken in total, lead to the conclusion that the top surface must be a hard, smooth, fluorocarbon material, or a very thin (micrometers) coating of a fluorocarbon on a hard, smooth backing (i.e., glass).

Candidate materials for the outer surfaces of photovoltaic modules currently consist of glass, Tedlar fluorocarbon film (DuPont Co. 100BG3OUT) and a biaxially oriented acrylic film, Acrylar (3M Co. X-22417)[15]. These materials are all relatively hard, smooth, and free of water-soluble residues. Experiments were conducted to determine if an improvement in soiling resistance could be obtained by the application of low surface-energy treatments. A survey[4] of coating materials showed that very few commercial materials exist that could be useful for this purpose. Nevertheless, two candidate fluorocarbon coating materials were identified:

1) L-1668, an experimental fluorochemical silane produced by 3M Corp. that is used to impart water and oil repellency to glass surfaces. This material will bond chemically to glass surfaces.

2) Dow Corning Corp. E-3820-103B, an experimental treatment consisting of perfluorodecanoic acid chemically reacted with a commercial Dow Corning silane, Z-6020. This experimental compound will bond chemically to glass surfaces.

In a trial test, each of these two fluorocarbon coatings, which are supplied in solvent solutions, was brushed onto the surfaces of the three outer surface candidate materials and allowed to dry in air and react chemically for 24 hours. The treated materials were then soaked in water, simulating rain, to determine if they were adequately attached chemically. The criterion for judging this attachement was whether water would bead up or wet and spread on the surfaces when the materials were periodically removed from the water bath. By this criterion, both coatings were judged to have become permanently attached to glass; the E-3820 to have become attached to the Tedlar, and the L-1668 to have become attached to the Acrylar. Droplets of liquid water on the Tedlar treated with L-1668 and on the Acrylar treated with E-3820, tended to wet and spread.

To promote chemical attachment of the L-1668 on Tedlar and the E-3820 on Acrylar, the surfaces of both of these films were first activated by exposure to ozone, to generate surface polar groups that would react chemically with the silanes followed by brushing on the fluorocarbon coating solutions. This technique worked excellently. Therefore, as an additional experiment effort, E-3820 was also applied to an ozone-treated Tedlar surface and L-1668 was also applied to an ozone-treated Acrylar surface, even though the earlier trial testing did not indicate such a need. The concept was that the ozone treatment may also enhance the adhesion of these fluorocarbon coatings on the plastic films. Glass and the plastic films coated with the fluorocarbon coatings were then mounted in outdoor racks on the roof of Springborn Laboratories' facilities in Enfield, Connecticut. Evaluation was performed monthly and the surfaces of these test specimens were not washed or touched with the hands.

The degree of soiling on the test specimens was monitored by measurement of the percentage of decrease in the short-circuit current (I_{sc}) output of a standard silicon solar cell positioned behind the soiled test specimens as described above for the JPL soiling studies.

For Acrylar, the better coating was found to be E-3820 in combination with ozone and Figure 11 compares the soiling behavior of uncoated Acrylar (control) and the E-3820-ozone-coated Acrylar specimen. For Tedlar, the better coating was found to be E-3820 and Figure 12 compares the soiling behavior of uncoated Tedlar (control) and E-3820-coated Tedlar. For glass, little difference was found between E-3820 and L-1668, but E-3820 might be slightly better. Figure 13 compares uncoated glass with the E-3820 coated glass.

Comparing the uncoated controls, glass has the least tendency to retain natural soil followed by Tedlar and then Acrylar, both having greater tendencies to retain natural soil. This difference in soiling behavior between glass and plastic films had been observed earlier[13,14]. However, with the fluorocarbon antisoiling coatings, the soiling behavior of all three materials becomes essentially the same.

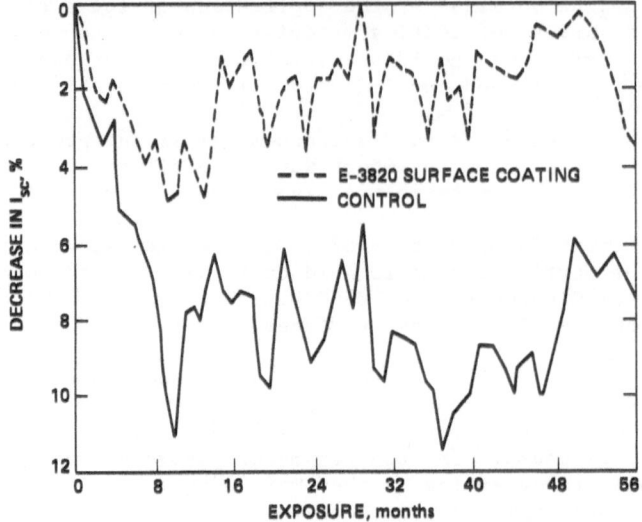

Figure 11. Outdoor soiling behavior of Acrylar X-22417 plastic film, with
and without a fluorocarbon antisoiling coating.

The soiling data were averaged over the 56 month period and the time-
averaged values are given in Table II. The glass control sample realized
an average optical loss of about 2.7% over the 56-month period, whereas the
Tedlar control realized an average loss of about 5.4% and the Acrylar
control specimen realized an average loss of about 7.2%. Soiling data
measured on similar materials in Pasadena (Table I), showed about 3% loss
for glass, 3% loss for Tedlar, and 5% loss for Korad acrylic film, which
is similar in chemistry to Acrylar.

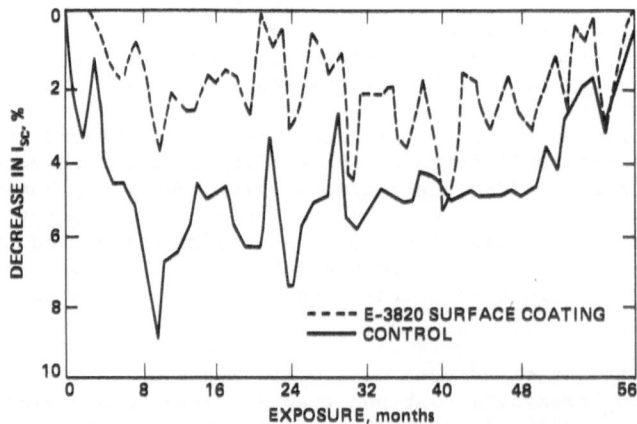

Figure 12. Outdoor soiling behavior of Tedlar 100BG30UT plastic film,
with and without a fluorocarbon antisoiling coating.

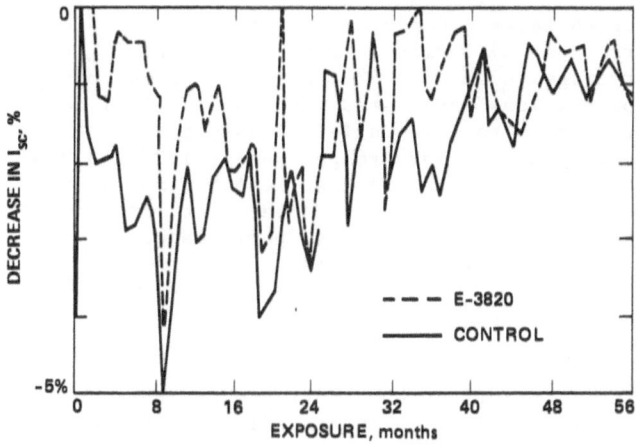

Figure 13. Outdoor soiling behavior of glass, with and without a fluorocarbon antisoiling coating.

The data in Table II also indicate that the better-performing fluorocarbon coating for all three materials is E-3820. On glass, the E-3820
coating resulted in a reduction of soiling-related optical losses from 2.7%
to 1.5%, for an average performance gain of nearly 1.2%. Similarly, on
Tedlar, the E-3820 results in an improvement from 5.4% to 1.7% for an
average performance gain of nearly 3.7%, and Acrylar realized an average
performance gain of nearly 4.6%. These performance gains can be econom-
ically important to the electrical power output of a photovoltaic module.

Table II.   Time-Averaged Values of the 56-Month Soiling Data.

Materials	Optical Losses, %
Glass	
Control	2.7
with E-3820	1.5
with L-1668	1.6
Tedlar	
Control	5.4
with E-3820	1.7
with L-1668/ozone	4.3
with L-1668	4.4
with E-3820/ozone	4.7
Acrylar	
Control	7.2
with E-3820/ozone	2.6
with L-1668	4.2
with E-3820	4.4
with L-1668/ozone	5.2

With respect to layer C behavior, if it can be assumed that no rain-resistant layers (A or B) formed on the E-3820 coated glass specimen, then the time-average value of 1.5% (Table II) can be considered as the time-average optical loss associated with the cyclical deposition and rain removal of layer C soil in Enfield, Connecticut. The higher minimums observed for Tedlar and Acrylar, respectively, 1.7% and 2.6%, may reflect the formation of some lower layers that resist removal by gentle rainfall but not intense rainfall as occurred in the 21st month of outdoor exposure. This suggests possibilities for further performance gains from the use of improved fluorocarbon coating materials.

CONCLUSIONS

In conclusion, low-surface-energy treatments based on fluorosilane chemistry appear to be effective in retarding the accumulation of dirt on the candidate outer surfaces of interest. The most effective soil retardant treatments identified to date are: for glass, E-3820; for Acrylar, ozone activiation followed by E-3820; and for Tedlar, treatment with E-3820.

ACKNOWLEDGMENTS

The research described in this paper was carried out by the Jet Propulsion Laboratory, California Institute of Technology, and was sponsored by the U.S. Department of Energy through an agreement with the National Aeronautics and Space Administration.

REFERENCES

1. E. F. Cuddihy, "Theoretical Considerations of Soil Retention," Solar Energy Materials, 3, 21-22, (1980).
2. E. F. Cuddihy, and P. B. Willis, "Antisoiling Technology: Theories of Surface Soiling and Performance of Antisoiling Surface Coatings," DOE/JPL Publication 1012-102, JPL Publication 84-72, FSA Project Report No. 5101-251, Jet Propulsion Laboratory, Pasadena, California, November 15, 1984.
3. J. M. Freese, "Effects of Outdoor Exposure on the Solar Reflectance Properties of Silvered Glass Mirrors," Sandia National Laboratories Report 78-1649, September 20, 1978.
4. P. B. Willis and B. Baum, "Investigation of Test Methods, Material Properties and Processes for Solar Cell Encapsulants, "Annual Report, Springborn Laboratories, Inc., Enfield, Conn., DOE/JPL Document No. 954527-82/23, Jet Propulsion Laboratory, Pasadena, California, July, 1982.
5. H. Schneider, "Mechanical Removal of Spacecraft Microbial Burden," Subtask I of Spacecraft Cleaning and Decontamination Techniques, Chapter 6 of Planetary Quarantine, Annual Review, Space Technology and Research, JPL TR-900-597, Jet Propulsion Laboratory, Pasadena, California, February, 1973.
6. E. P. Roth and A. J. Anaya, "The Effect of Natural Cleaning on the Size Distribution of Particles Deposited on Silvered Glass Mirrors," paper presented at the Second Solar Reflective Materials Workshop, sponsored by the Department of Energy; held in San Francisco, California, February 12-14, 1980.
7. R. G. Draft, R. J. Mell, and E. C. Segers, "Investigation of Soiling Characteristics of Polymeric Film," IIT Research Institute, Chicago, Illinois, Final Report No. LS-72-7451 for the Jet Propulsion Laboratory, February, 1982.

8. D. L. Ross, "Coatings for Video Discs," RCA Review, 39, 136 (1978).

9. M. B. Sheratte, "Cleaning Agents and Techniques for Concentrating Solar Collectors," Final Report, No. MDCG 8131, McDonnell-Douglas Astronautics Co. West, September 1979.

10. P. B. Adams, "Glass Containers for Ultrapure Solutions," Chapter 14 in Ultrapurity, M. Zeif and R. Speights, Editors, Marcel Dekker, New York 1972.

11. M. C. Porter, "Membrane Filtration," Section 2.1 in Handbook of Separation, pp. 2-3, McGraw-Hill, New York (1979).

12. B. G. Moser and R. F. Landel, "A Theory of "Particle-Particle Interaction Describing the Mechanical Properties of Dental Amalgam," Jet Propulsion Laboratory SPS No. 37-40, Vol. IV, p. 84 (1966).

13. A. R. Hoffman and C. R. Maag, "Airborne Particulate Soiling of Terrestrial Photovoltaic Modules and Cover Materials," in Proceedings of the Institute of Environmental Sciences, Philadelphia, Pennsylvania, May 11-14, 1980, Institute of Environmental Sciences, Mount Prospect, Illinois, 1980.

14. A. R. Hoffman and C. R. Maag Photovoltaic Module Soiling Studies, May 1978 to October 1980, JPL Document No. 5101-131, JPL Publication 80-87 DOE/JPL-1012-49, Jet Propulsion Laboratory, Pasadena, California, November 1, 1980.

15. E. F. Cuddihy, W. Carroll, C. Coulbert, A. Gupta, and R. Liang, Photovoltaic Module Encapsulation Design and Materials Selection: Volume I, JPL Document No. 5101-177, JPL Publication 81-102, DOE/JPL-1012-60, Jet Propulsion Laboratory, Pasadena, California, June 1, 1982.

IMPLICATIONS OF PARTICULATE CONTAMINATION IN THE
PERFORMANCE OF FLOPPY DISKS

A. F. Lewis and R. J. Rogers

Data Resource Group
The Kendall Company
Research Department
Walpole, MA 02081

Flexible diskettes have been deliberately contaminated with various particulate materials in an attempt to determine which may be responsible for the generation of transient missing pulse errors in flexible disk magnetic information transfer. It was learned that deliberate contamination of diskettes with debris vacuumed from nonwoven fabric jacket liner materials does not have a high propensity to induce transient errors. On the other hand, magnetic media burnishing dust was found to be a most detrimental contaminant for diskettes; large numbers of transient errors were observed in diskettes contaminated with this material. It is hypothesized that the most likely cause of transient missing pulse errors in flexible disk computing is magnetic media burnishing dust generated on the diskette surface during manufacture and operation of the disk.

INTRODUCTION

During the use of flexible magnetic disks, from time to time errors are encountered in the transfer of information to or from the recording media. These errors are commonly the result of loss of signal (i.e., a missing bit or dropout) and may be divided into two classes according to their characteristics. Those errors which are permanently located in one physical location on the diskette surface are referred to as "hard" or permanent errors. These are understood to be caused by defects in the magnetic coating such as coating flaws, scratches, foreign materials permanently imbedded in the diskette surface and others. A hard error in the transfer of digital information generally renders a diskette useless for information transfer purposes. Disks observed to have hard errors are therefore labeled as manufactured rejects during their certification process.

Relative to the overall problem of computer software process errors, few events are more frustrating to the users and manufacturers of flexible magnetic disk media than the occurrence of transient errors during the testing and/or functional use of flexible disks. Transient or "soft" errors are errors that result in failure to transfer information during a read or write operation which may not be present during subsequent attempts

113

to repeat the operation. These errors may appear, then disappear if the disk is reinserted into the drive or if the system is re-booted. It is surely a problem that has been encountered by all users of flexible disks. In some cases it may not be recognized as a magnetic media error problem; it is a non-repeating, fleeting condition. For a manufacturer of floppy disks, if these soft or transient errors occur during the certification operation, diskette certification yields are reduced. If these transient errors occur during user copying operations or general certification by the purchaser of the diskette, the product is labeled unqualified for use, time and money are lost; the reputation of the particular diskette manufacturer is damaged as a result. The manufacturer is faced with the unpleasant task of explaining why his product was 100% certified when it left his plant yet the customer found it necessary to reject it.

MANIFESTATION OF TRANSIENT ERRORS

The reality and capriciousness of the "soft" dropout errors (transient missing bit) are easily shown. An illustration experiment was conducted whereby ten 5.25" commercially available diskettes of identically specified quality (single side, double density) were purchased from five different manufacturers. Each diskette of this test sample grouping was inserted in the Shugart SA-460L disk drive interfaced to a diskette analysis system (Cloutier Design Services Flexible Disk Analysis System, Model 1200, Eden Prairie, MN 55344). The missing pulse test[1] was performed ten times (ten separate insertions) for each of the fifty diskettes in the test grouping. On their initial insertion, all fifty diskettes were free of missing pulses and were error-free to a clipping level of 75% for all tracks. Nevertheless, during the 500 insertions of the test, five transient missing pulses were observed. In no case did these errors appear twice on the same disk and there was no apparent correlation of missing pulses to track number. In this test, transient errors seem to have appeared randomly in about 1% of all diskette insertions. It should be noted that these transient errors were not characterized by marginal signal level reductions. Instead, each of the transient missing pulses observed showed complete loss of signal in the region of the error. In each case, the error appeared only once; on subsequent insertions, the diskette was fully functional.

Transient errors on a disk can also be observed during missing bit testing without reinserting the disk. Table I presents Cloutier error analysis results on a diskette that was continuously rotated in a Shugart Type SA-455 disk drive as part of a real time wear test experiment. The head was loaded on the outer tracks of the disk throughout the duration of the test. After 6 weeks of this wear cycling (representing over 18 million revolutions of the disk) the diskette was tested for missing bits using the Cloutier FDAS (interfaced with a Shugart SA-460 disk drive). Referring to Table I, test pass Number 1 shows that the diskette failed to certify; two missing pulses occurred at the 50% threshold level. Without reinserting the disk, a second test pass was conducted. As shown, the two error pulses at the 50% threshold level disappeared. The diskette also continued to "clean-up" error-wise in the third pass. At about the fourth pass, the diskette showed a profile of error count versus threshold level similar to the non-wear cycled disk (zero wear time). Table I clearly demonstrates the essentials of transient error behavior and how they can be observed and followed using the Cloutier FDAS.

Obviously there is much concern throughout the flexible disk industry (manufacturers and users) over transient errors. To this day, the exact cause of soft or transient errors in floppy disk magnetic media is not known. However, some speculation has been made on this subject. Most of this speculation centers on separation of the media from the head by the

114

interposition of loose particles, but other causes are possible. Table II lists some of the reported possible causes of these soft errors.

Table I. Example Of Cloutier Error Analysis Data
Relative To The Demonstration Of Missing
Pulse Transient Error Occurrence As A
Function Of Successive Test Passes (a)(b)(d)

PERCENT OF TRACK AVERAGE AMPLITUDE

(c) is the Certification threshold level marker above the 50 column.

	90		80		70		60		50 (c)		40		30		20		10	
ZERO WEAR TIME →	28	1																
TEST PASS NUMBER (d) 1	31	5	4	4	4	3	2	2	2	2	2	1	1	1	1			
2	26	3	1	1	1	1												
3	35	4	1	1	1													
4	30	3	1															
5	34	5	1															
6																		

(a) Test results are of a diskette with Rayon/Polypropylene blend liner, wear tested by continuously running it in a Shugart Drive Type SA-455 (head loaded on outer tracks) for over 18 million revolutions.

(b) Data show two transient error pulses (50% threshold level) on first test pass. Second test pass shows the disk has no errors at 50% threshold. After the fourth or fifth pass, disk error pulse patterns were similar to the unworn disk.

(c) Certification threshold level is 50%.

(d) Values refer to number of signal pulses failing to exceed the indicated percentage of the track average amplitude.

Table II. Some Possible Causes
Of Flexible Disk Magnetic Media
Transient Errors

1. Speck of dust or liner debris momentarily between the R/W head and the medium.

2. Magnetic Coating or R/W head wear debris particles in transient motion on medium.

3. Electrostatic discharge effects due to the dynamics of the media / R/W head / jacket liner nonwoven fabric system.

4. Micro-scratches on the medium which are polished away by the action of the head.

5. Extraneous/random vibration of medium due to mechanical defects at head/media interface.

Much of the present speculation as to the cause of transient errors has been focused on the R/W head - media separation due to loose particles generated within the diskette or deposited from the ambient environment being interposed between the R/W head and the media causing a separation loss. Such separation losses were first treated experimentally and theoretically by Wallace[2]. The subject is still of interest and has been discussed and studied more recently by Jorgensen[3], Jeffers[4], and Bertram[5].

From the Wallace equation, RSL = 55 d/λ dB, where RSL is the reproduce separation loss, d, the R/W head separation distance and λ , the fundamental wavelength or twice the bit length, B , one can get an estimate of the loss of signal caused by a momentary separation or loss of signal between the head and the medium such as what could occur when the R/W head rides over a speck of dirt or debris on the media. According to the Wallace equation, one or more magnetic signals formatted onto the medium could be blanked out resulting in the occurrence of missing bit errors. Taking into consideration a 5.25" floppy at a track density of 96 tpi, imposing a 250K FTPS (2F) signal (imposing 125K bits per second at 300 rpm) on both the outer and inner tracks of the disk, from the system's geometry, bit lengths, B , of approximately 14 and 8 microns result for the outer and inner tracks respectively. From RSL = 55 d/λ, the smallest R/W head-media separation distance, d, necessary to cause sufficient reduction in signal for a missing bit error to appear should be about 3 microns for the outer tracks and about 2 microns for the inner tracks. Thus, it is shown theoretically that transient errors can possibly be caused by mobile particles, of a sufficient size, momentarily interacting with R/W head-media contact in such a way that information transfer is impaired.

EFFECT OF NONWOVEN FABRIC LINER DEBRIS ON TRANSIENT ERROR OCCURRENCE

In the construction of all floppy disk systems, the magnetic media disk (magnetic coating onto polyester plastic film) is enclosed in an envelope composed of a polyvinyl chloride sheet stock outer structural shell lined with a nonwoven fabric. This nonwoven fabric liner is a critical part of the diskette system. In addition to its functioning as a mechanical protective cushion for the magnetic media disk element in the jacket, it also serves the vital function of wiping the disk surface while it is being rotated in the disk drive. Thus the liner works to keep the R/W head-media contact zone clean, free of dirt and other foreign particles which, if allowed to migrate, could possibly cause head-disk separation errors.

The choice of a nonwoven fabric to perform this wipe function is itself a paradox since all nonwoven fabrics currently used in floppy disk system manufacture contain, by their very nature, some level of lint, fiber fragments or particulate polymeric matter. Nonwoven fabrics are essentially produced by consolidating, in a controlled manner, a random web of staple (1-2 inch length) textile fibers. A mat-like assembly of fibers is formed. Consolidation takes place by thermal, chemical and/or mechanical means. Since staple (non-continuous) fibers are involved in their manufacture, lint and fiber fragments or particles are an inherent part of the final material because, among other reasons, such materials must be mechanically slit to a proper width before use. While a good portion of this liner debris is considered to be firmly entangled and embedded in the depth of the nonwoven fabric, some proportion of liner surface debris must always be considered to be present in such nonwovens.

In addition to the lint and fiber debris inherent with the liner material itself, debris can also be produced as the result of the edge slitting and hole cutting operations that are needed in the manufacture of the diskette jackets. It is believed that liner debris from the jacket manufacture cutting operation can be a major source of liner surface debris and it is recommended that diskette jacket manufacturers should vacuum all cutting edges of the jacket liner materials before the jacket folding operation.

Experiments have recently been conducted in an effort to determine what effect, if any, does liner fabric debris have on the certification and function of diskette magnetic media. In this work, commercially available diskettes (5.25" diameter, DS, DD, 96tpi, Reinforced Hub) were deliberately contaminated with various types of liner debris and tested for missing pulse errors by the Cloutier FDAS Model 1200 interfaced to a Shugart SA460 disk drive. The particular jacket used in these studies was one lined with a nonwoven fabric composed of a blend of rayon and polypropylene fibers. Debris was collected by vacuuming the surface of various nonwoven fabrics. For example, Figure 1 is a SEM (Scanning Electron Microscope) photomicrograph of liner debris collected from the surface of a rayon/polypropylene nonwoven fabric. Note that it is characterized by lint

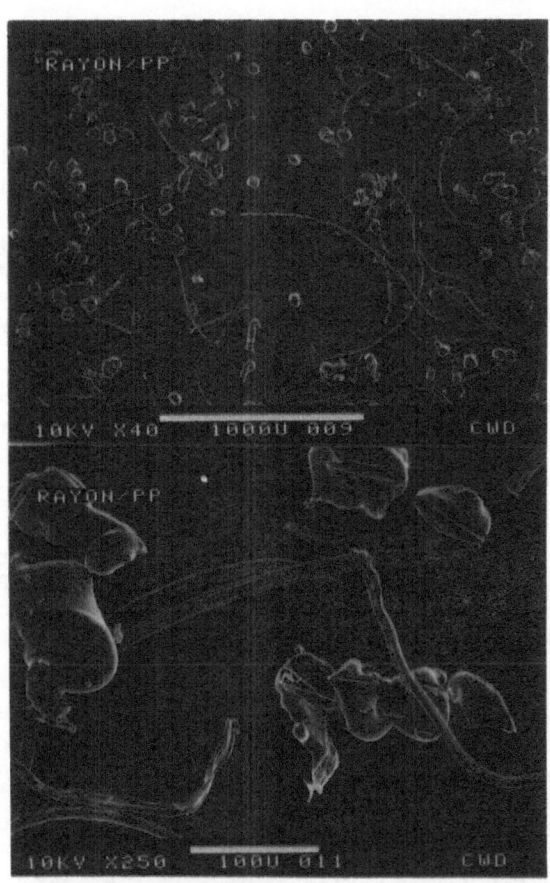

Figure 1.   SEM photomicrographs of debris vacuumed from Rayon/Polypropylene blend nonwoven fabric diskette liner material.

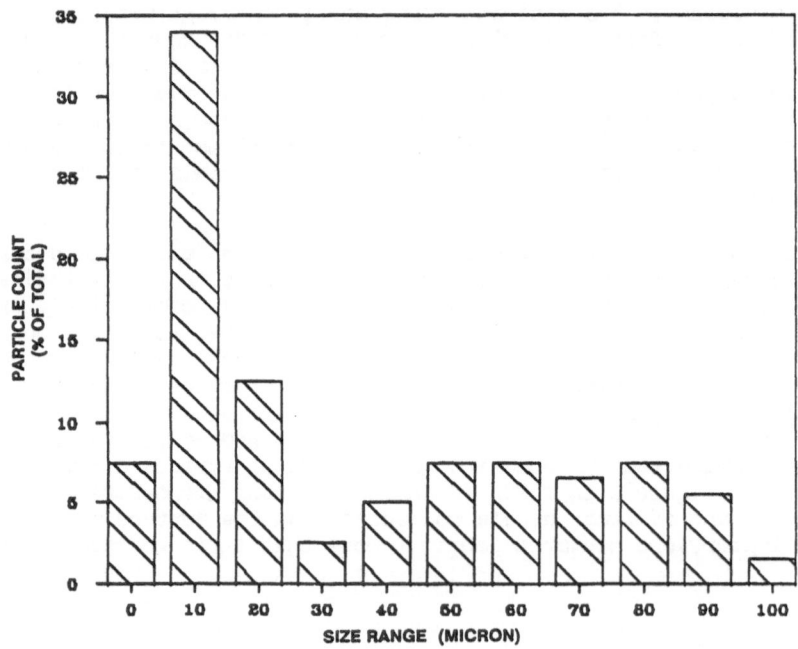

Figure 2.   Particle size distribution of Rayon/PP liner
debris.

fibers, fiber fragments and polymer globules. The lint fibers are 1 to 4
millimeters in length. To further characterize this debris, a size
analysis was performed on the particulate material in this debris using an
OPTOMAX IV Image Analysis system (Optomax Company, Hollis, NH 03049). A
histogram describing the particle size distribution of the particulate
matter in the debris is shown in Figure 2. As shown, more than 30% of the
particulate matter in the liner debris from the rayon/polypropylene blend
liner fabric is less than 20 microns in size.

Having characterized the particle size distribution of the liner debris, a
procedure for deliberately contaminating the diskettes with this debris was
devised. Here, the diskette was first cut open at the label edge and the
disk was carefully removed from its jacket. The disk was weighed on an
analytical balance and then a quantity of collected debris, similar to what
is shown in Figure 1, was randomly sprinkled onto the disk. The disk was
then reweighed and carefully inserted back into the jacket. Using this
procedure, several diskettes contaminated with from 3 to 8 milligrams of
debris were prepared. This series of debris contaminated diskette samples
was then error count tested using the Cloutier FDAS 1200 system. The
results of these experiments are presented in Table III. As shown, very
few transient R/W errors were observed, even though these disks were
drastically contaminated with as much as two hundred times the amount of
liner debris inherent in an uncontaminated disk. In the one case where a
transient error was observed, the error disappeared on the second test pass
(same behavior as illustrated in Table I). It can, therefore, be generally
concluded that liner debris has no drastic immediate effect on the
existence of transient or permanent errors in diskette magnetic media R/W
systems. The suspicion that mobile liner debris particles being interposed
between the R/W head and the media causing momentary transient, soft or
casual errors in magnetic information transfer has not been substantiated
by these experiments.

Table III. Transient Errors Encountered In 5.25" Diskettes
Deliberately Contaminated With Various Types
of Liner Debris

LINER DEBRIS CONTAMINANT(a)(e)	WEIGHT OF DEBRIS ADDED TO DISKETTE(b) (MILLIGRAMS)	TRANSIENT ERRORS ENCOUNTERED(c)	TEST PASS WHEN ANSI CERTIFIED DISK BEHAVIOR WAS ACHIEVED
None (d)	(0.04(d)	0	First
Rayon/PP	3.7	0	First
Rayon/PP	4.5	0	First
Rayon/PP	4.8	2	Second
Rayon/PP	8.5	0	First
PP	4.4	0	First
PP	8.4	0	First
PET	3.8	0	First
Rayon/PET	3.6	0	First

(a) All diskettes tested had jackets lined with Rayon/Polypropylene blend
liner material; this material served as the "carrier" for the liner
debris particles introduced into the system.

(b) Only Side 1 of the disk was contaminated with the debris and tested.

(c) Number of missing bits (50% clipping level) occurring on first
insertion of the diskette in the Cloutier FDAS Model 1200; Shugart
SA-460 (96tpi) disk drive.

(d) From surface vacuuming studies on unsupported liner material, the
inherent debris content of the particular Rayon/Polypropylene liner
material used in the jackets of the diskettes employed in this study
measured less than 0.04 milligrams of debris per disk side.

(e) Abbreviations: PP - Polypropylene fiber
PET- Polyester fiber

DELIBERATE CONTAMINATION OF FLEXIBLE DISKS WITH VARIOUS POWDER MATERIALS

To further investigate the effect of particulate matter on the
Cloutier error analysis of floppy disks, an additional series of diskettes
was deliberately contaminated with powders of different material properties
including dust collected from the disk burnishing operation. Table IV
presents a list of the materials studied. In each case, the diskette was
first weighed and then a quantity of powder contaminant was deposited in
the head access slot (Side 1) of the diskette. The diskette envelope was
then carefully lifted up at the head access slot section and the disk was
manually rotated in its jacket to capture the powder contaminate inside the
jacket on the "down-stream" side of the disk/jacket liner contact area.
The powder, therefore, was inserted in the diskette in a manner which
enabled it to completely contaminate the media/liner interface before it
again reached the head access slot. The contaminated diskette was then
spun for five minutes (at 300rpm) on a turntable-like wear test stand. A
pad loading of 2.4 kPa (0.35psi) was set in position near the "upstream"
side of the head access slot imposed on the diskette during this
preparation period. This step was included in order to assure that the

powder contaminate was somewhat distributed in the diskette system. The diskette was then re-weighed in order to determine the amount of powder contaminant in the system. The deliberately contaminated diskettes were then subjected to Cloutier Error Analysis; the results of these experiments are presented in Table IV. As shown, the electrically conductive powders, copper and aluminum metal as well as the inorganic (dielectric, non-conducting) powder, aluminum oxide, are found not to produce any significant errors. However, the magnetically active powders, iron oxide (Fe_2O_3) and the media burnishing dust, showed a high incidence of missing pulse error occurrence. The burnishing dust showed a high number of read/write errors even at low diskette contamination levels, as low as two milligrams.

In the interest of characterizing this burnishing dust, SEM photomicrographs were obtained on this material as well as an Optomax IV particle size analysis. The results of this work are presented in Figures 3 and 4. Note from Figure 3 that the burnishing dust particles are of two types (a) "curled" slivers (which resemble debris from a plowing or cutting tool operation) and (b) fine particles or fragments. Supplementing Figure 3, the histogram in Figure 4 shows that the burnishing dust particles are very small; over 50% of them are 5 microns or less.

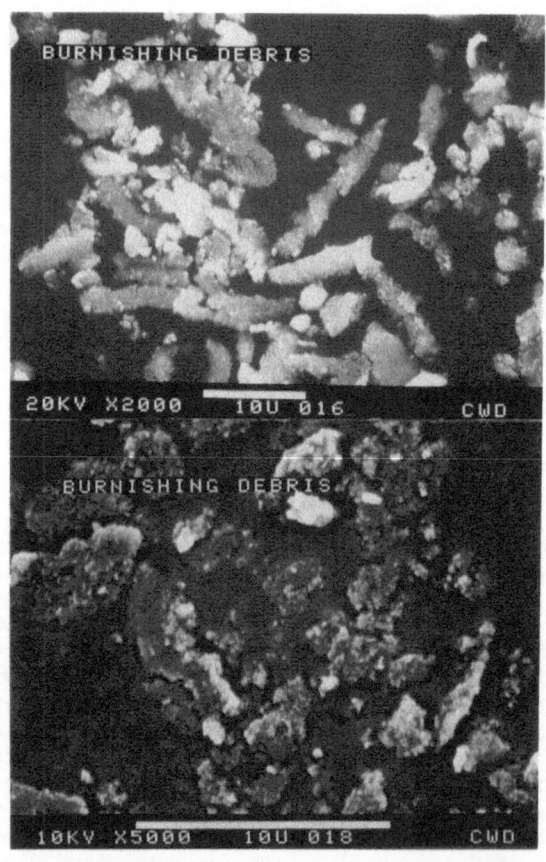

Figure 3. SEM photomicrographs of magnetic media
           burnishing dust.

120

The observations that magnetic media burnishing dust is a most virulent agent in causing errors to appear in the operation of a floppy disk is an important one. This is because, among all the powders listed in Table IV, the most likely one to be encountered in the operation of floppy disks is media dust. Two potential sources of media dust exist. First, during manufacture, all floppy disks are polished (burnished) with an abrasive tape before certification. Here the surface asperities of the magnetic coating are smoothed down so that intimate contact between the disk drive read/write head is assured and no surface "chatter" due to surface roughness occurs. If the residues from this polishing process are not

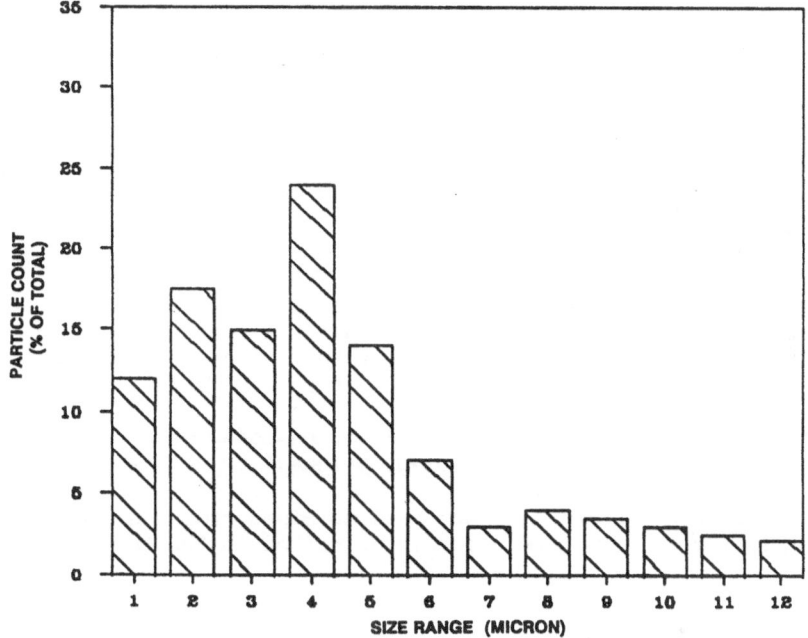

Figure 4. Particle size distribution of magnetic media burnishing dust.

completely removed from the media surface, i.e., remain occluded at the surface, their presence could eventually cause transient errors to appear sooner or later in the operation of the floppy disk. The second source of magnetic media dust could result from the wear process between the ceramic read/write head and the magnetic media surface. Media wear particles can, therefore, be generated during the functional operation of a floppy disk. This is especially true if for some reason the internal lubrication mechanism of the floppy disk material system has failed. Regardless of the source of media dust, as shown in Table IV, its existence in floppy disk systems is most detrimental.

121

Table IV. Effect Of Various Powder Materials On The
Occurrence Of Missing Pulse Errors In
Deliberately Contaminated Diskettes (a)

POWDER CONTAMINANT	AVERAGE PARTICLE SIZE microns	POLYESTER LINER FABRIC		RAYON/PP (b) LINER FABRIC	
		mg. contam.	M.P. errors	mg. contam.	M.P. errors
COPPER METAL	11 +/- 7	25	0	26	0
ALUMINUM METAL	10 +/- 6	111	0	91	0
ALUMINUM OXIDE	6 +/- 3	15	0	18	0
FERRIC OXIDE	5 +/- 4	17	16	16	27
MAGNETIC (c) MEDIA BURNISHING DUST	5 +/- 3	4	4	2	11

(a) In all cases, report of missing pulse errors refers to number of
errors appearing at 50% threshold level at first test pass; on second
test pass, the error pulses disappeared indicating the transient
nature of such an error occurrence (Refer to Table I)

(b) The diskettes studied contain a liner composed of a nonwoven fabric
blend of 75% Rayon and 25% Polypropylene.

(c) Magnetic media burnishing dust was obtained from a diskette
manufacturer. Its generation is described in the text.

MAGNETIC MEDIA LINER FABRIC WIPE EFFECTIVENESS TEST

The finding that the presence of magnetic media burnishing dust is the
primary cause of transient error occurrence in floppy disk operation allows
one to postulate that the general procedure used to generate the data in
Table IV might well serve as the foundation for a test to evaluate the
effectiveness of magnetic media liner wipe fabrics. To this end, a
procedure has been developed involving the deliberate contamination of
diskettes with media burnishing dust and subsequent error analysis testing
of these contaminated diskettes. It is presumed that the number of errors
encountered will reflect the wiping effectiveness of the particular liner
fabric. The fewer the errors observed, the more effective a liner fabric
would be in wiping up media burnishing dust debris.

As the result of numerous experimental trials, a procedure was
established relative to this proposed test. Several steps are involved.

1. Test disks are checked for certification/error analysis using the
Cloutier FDAS instrument to make sure these test disks are performing up to
specification.

2. A series of diskette jackets having fabric liner of the type under study is prepared; one (flap) end is left open for subsequent insertion of the media disk.

3. Media dust is generated by abrading an unburnished media disk with super fine, 400 grit carborundum paper to generate some dust. Here a 35mm (1 3/8") wide strip of the carborundum paper is pressed against the media disk at 4.8 kPa (0.7 psi) for 3 minutes at 300 rpm. The disk is rotated on a turntable like assembly and the carborundum paper is pressed against the rotating disk that is fabric backed to provide a cushion; the pressure pad was 25 X 13 mm (1" X 1/2").

4. The media dust generated in Step 3, is found to collect on the surface of the carborundum paper. It can easily be brushed off onto the surface of a certified disk (Step 1). In practice, the media dust is weighed directly onto the certified disk. In most instances, the use of multiple "injections" of media dust is required to obtain the proper amount of media dust contaminant on the disk to be studied. Four "injections" results in a contamination level of approximately 1 mg of burnishing dust.

5. The disk prepared in Step 4 is then carefully inserted into the fabric lined test jackets. Studies were carried out and it was found that the general position of the media dust contaminant deposit was not critical in the generation of results. However, to assist reproducibility, care was taken to place the deposit of media dust contaminant "up-stream" of the read/write head hole of the diskette.

6. The diskette, as prepared in Step 5, is then tested for missing pulse errors. Data are recorded in Table V on the occurrence and number of missing pulse errors occuring at the 50% threshold level.

The liner wipe evaluation test procedure was applied to evaluate the relative effectiveness of polyester and rayon/polypropylene fabrics in preventing transient errors in diskettes. Data for these two systems are presented in Table 5; results of tests conducted on diskette systems prepared with jackets with no liner are included for comparison.

Table V. Summary Of Missing Pulse Error Analysis Data For Various Diskettes Deliberately Contaminated With Media Burnishing Dust. (a)

LINER FABRIC	PASS RATIO (b)	PERCENT PASS	AVERAGE (c) NO. OF ERRORS AT 50% THRESHOLD ON 1ST PASS
RAYON/PP	16/22	73%	1
POLYESTER	4/22	18%	6
NO LINER	0/22	0%	34

(a) Contamination level approximately 1 mg per diskette

(b) Ratio refers to the number of deliberately contaminated diskettes that showed no errors at the 50% threshold level divided by the total number of diskettes tested.

(c) This value refers to the total number of missing bit signals recorded at the 50% threshold level during the first test pass for all the diskettes tested divided by the number of diskettes tested that had errors at the 50% threshold level.

The results indicate that the rayon based liner is more effective in preventing errors caused by media dust than the polyester based liner jacket system. In addition, fabric lined jacket diskettes are much better than unlined diskettes. The data in Table V are definitive in that they show that liner fabric effectiveness with respect to wiping burnishing dust from media follows: Rayon/PP > Polyester>> no liner. This order is determined by the percentage of diskettes that pass the "zero errors at the 50% threshold level" test and column three of Table V which presents an index of the intensity of the error condition at the 50% threshold level when errors do occur. As shown, with unlined diskette jackets, many missing pulse errors were observed giving support to the need (and function) for liner fabrics in diskette assemblies.

From the limited results presented, the described liner wipe effectiveness test appears to have promise and should be worthy of further development. Some questions remain, however. Referring to Table IV, it is observed that rayon/PP lined diskettes show a higher number of transient errors than the polyester fabric lined system. This trend is opposite to what is presented in Table V. The exact reason for this is not known. Consider that two different diskette contamination procedures were used to generate the data in Tables IV and V. The procedure used to obtain the Table V data was much more deliberate and controlled. Table IV data involved higher contamination levels; furthermore, the media burnishing dust used in the Table V experiments was freshly abraded. More research work remains on proving the efficacy of the proposed liner fabric wipe effectiveness test. The results to date indeed show promise.

CONCLUDING REMARKS

In some experiments where diskettes were deliberately contaminated with excessive amounts of nonwoven fabric liner debris, certification tests did not exhibit any occurrence of missing pulse errors. The proposal that liner debris particles getting between the R/W head and the media causing separation loss (missing pulse or dropout errors) in magnetic media recording has not been substantiated in these studies. From experiments where various powders and dusts were used to deliberately contaminate diskette systems, it was found that magnetic media coating burnishing dust is a detrimental contaminant for flexible disks. These present experiments point to the strong indication that burnishing dust particles are a primary culprit in generating transient errors in flexible disk magnetic media operations. Based on this hypothesis, it appears that it is possible to develop a magnetic media liner fabric wipe effectiveness test. The scope and reliability of this test remain the subject of future studies, but initial results show that rayon based liner materials are superior to PET liners with respect to their ability to remove burnishing dust.

The overall implication of this present study is that magnetic media burnishing dust is a contaminant particle of primary consideration in causing transient errors during the operation of floppy disks. The precise mechanism of transient error occurrence is not known. However, from this work, it is believed that burnishing debris that remains occluded at the media surface during manufacture and becomes dislodged during the functioning of the disk initiates transient error phenomenon. Once a speck of burnishing dust (could also be considered in the realm of a media coating wear particle) becomes mobilized, the possibility exists that the particle can interfere with signal transfer causing an error to appear. In subsequent operation, if the particle then proceeds to get wiped from the media surface by the liner material and becomes immobile, the error condition disappears. A transient error sequence has thus been generated.

ACKNOWLEDGEMENTS

The authors would like to thank Carol DuPuis of Kendall's Lexington Laboratories for the SEM photomicrographs presented in this paper. Also, the assistance of Chris Cavanaugh who performed the particle size analysis work is appreciated.

REFERENCES

(1) American National Standards Institute - Proposed Standard X3B8/82-51 - for two sided double-density unformatted 5.25 inch 96 tpi, flexible disk cartridge - general, physical and magnetic requirements (for 7958 BPR use) - Signal Quality Section 7.2.1 - missing bit Standards Date August 1983.

(2) R. Wallace, Bell Tech. J. 1145 (October 1951)

(3) F. Jorgensen "Complete Handbook of Magnetic Recording", P. 218 TAB Books, Inc. Blue Ridge Summit, PA., 1980.

(4) F. Jeffers. IEEE Trans. on Magnetics, Mag. 18, #6, 1146 (1982)

(5) H. Bertram IEEE Trans. on Magnetics, Mag. 18, #6, 1206 (1982)

PART II. PARTICLE-SUBSTRATE INTERACTION AND
PARTICLE ADHESION

A THEORETICAL REVIEW OF PARTICLE ADHESION

R. Allen Bowling

Texas Instruments Incorporated
Materials Science Laboratory
P.O. Box 655936, MS 147
Dallas, Texas 75265

This paper constitutes a theoretical review of the forces of adhesion of small particles to surfaces. The primary force of adhesion of small, less than 50 micron diameter, particles on a dry surface are van der Waals forces. These van der Waals forces can be increased due to particle and/or surface deformations which increase the particle contact area. Particles less than 1 micron diameter can be held to surfaces by forces exceeding 100 dynes, which corresponds to forces per unit area of 10^9 dyn/cm^2 or more. Total forces of adhesion for 1 micron diameter particles can exceed the gravitational force acting on that particle by factors greater than 10^6. Electrostatic forces, although they only become important and predominate for particles larger than 50 microns diameter, may play a significant role in bringing particles to surfaces for adhesion. The presence of a liquid between the particle and surface, due to either immersion and subsequent removal from a liquid or due to high humidity conditions, can add a very large capillary force to the total force of adhesion. This capillary force is known to remain, in some cases, even after baking at above the liquid boiling point for many hours. Although some possibilities exist for particle removal, removal of small particles has been in practice extremely difficult due to these large forces. Each of these forces will be discussed theoretically with some attempts at quantification versus particle diameter. Some data will also be presented on attempted particle removal as a function of removal method. Clearly, emphasis should be placed on prevention of particle deposition on surfaces rather than relying on achieving subsequent removal.

INTRODUCTION

Particles are present all around us in the atmosphere in great abundance. They include, for example, pollens, dusts, fibers, metals, metal oxides, etc. They are generated by several means including mechanical abrasion, chemical reactions, and combustion processes. Particles are known to have a very dramatic effect on the performance of integrated circuits and are a major yield loss factor. Particles will only become more important as the size of individual semiconductor operational units decreases.[1] People are also large sources of particles, from their clothes, from their skin and hair, and from their breath, particularly if they smoke. Particles adhere to surfaces with great tenacity. One needs only to invert a surface with adhered particles to realize that, even for particles that are quite large, the adhesion force is greater than the gravitational force on those particles. Similarly, vigorous blowing of the surface only manages to dislodge relatively few of the particles.[2,3] The total adhesion force on a particle, as will be shown later in this paper, decreases approximately linearly as a function of the diameter of the particle. On the other hand, the weight of a particle decreases as a function of the diameter cubed, i.e., volume reduction. For a 1 micron diameter particle, the force of adhesion easily exceeds the force due to gravity by a factor greater than 10^6.[4] A basic understanding of particle adhesion is vital to the search for ways to insure the particle free semiconductor processing needs of the future. Information about particle adhesion mechanisms can aid in more effective particle prevention.

There are, to date, practically no physical models of adhesion which relate adhesion strength to established fields of physics. The difficulties arise from the fact that adhesion strength is a combination of physical and chemical forces, and mechanical strains and stresses, all at and around that adhesion interface. Also, adhesion forces can only be measured destructively. An adhesion measurement, therefore, does not represent an equilibrium situation; thus a kinetic model of adhesion is required. Adhesion must also normally be approximated geometrically by a sphere on a flat surface.[5] The discussions below on adhesion mechanisms all assume a spherical particle on a flat surface. The discussions are also primarily theoretical because of the problems in making measurements of adhesion force for very small particles. A typical measurement method is to apply a force for removal by spinning the surface with the attached particles and monitoring the rotation rate at which particles are removed. Removal of 1 micron diameter particles, however, is impossible by this method; it would require forces greater than 10^7 g, which corresponds to rotation rates greater than 10^6 rotations per minute, a rate not obtainable for known materials. Total forces of adhesion of small particles to surfaces can range from a total of 10^{-5} to 10^2 dynes. This total force can correspond to a tremendous force per unit area of up to $>10^9$ dyn/cm^2 for micron size particles.

Interactions between solids which bring about adhesion can be classed into several groups. The first group includes long-range attractive interactions which act to bring the particle to the surface and establish the adhesion contact area. These include van der Waals forces, electrostatic forces, and magnetic attractions. Electrostatic attractions include both bulk excess charge image forces and electrostatic contact potentials, also known as electrical double layer forces. A second group includes other forces which, along with the first group of forces, establish the adhesion area. This group of interfacial reactions includes sintering effects such as diffusion and condensation, diffusive mixing, and mutual dissolution and alloying at the interface. This category also includes the establishment of liquid and solid bridges between particle and surface, and the consequential capillary forces associated with these phenomena. The third group includes very short range interactions which can add to adhesion only after the establishment of an adhesion contact area. These include chemical bonds of all types and intermediate bonds such as hydrogen bonds.[5,6]

In general, the quantitative treatment of the second and third groups of adhesion forces are very difficult because they are primarily specific to each case, being dependent upon the particle and surface materials. The sections below will thus treat primarily the first group only. The effects of capillary forces will also be discussed and quantitatively treated. This paper essentially ignores a general treatment of complex chemical bonding to surfaces. We do know, however, that chemical bonds play an important role in particle adhesion on silicon surfaces.[7] For example, it has been shown that silicon particles on a silicon surface can oxidize along with the surface and become entrapped in the oxide; HF cleaning can be important to the removal of the oxide, thus allowing the particle to be removed by other means. The outer surface of silicon is also known to have a large number of "dangling" hydroxide groups, and is referred to as a silanol surface.[8] Molecules which themselves have hydroxide groups can form strong estersil bonds with this silanol surface. The remaining Si-O-Si surface of silicon can also hold many species by hydrogen bonding at the electronegative oxygen atoms. All these effects must be considered when trying to remove particles, but none of these bonding factors are general enough to be treated in a quantitative manner as a function of particle size. One important factor in ignoring these in a general treatment is that bonding and interfacial reactions are not generally very active at room temperature.

The primary forces which act to bring particles to a surface and then hold them there are van der Waals forces and electrostatic forces. Electrostatic forces predominate for large particles, i.e., greater than about 50 microns diameter. Van der Waals forces, however, predominate for smaller particles. Electrostatic forces are comprised of two types of forces, excess charge image forces and electrostatic contact potentials, also known as electrical double layer forces. For dry uncharged particles on a dry uncharged surface, only van der Waals and electrical double layer forces will act to hold the particles on the surface, i.e., specifically if no bonding or other interfacial reactions can occur for the materials. Charged particles and/or charged surfaces then add an additional electrostatic image force. Wet systems can then have an additional capillary force acting to hold the particles, and liquid immersed systems may experience a shielding of each of these forces so that the total force holding the particles is reduced. As follows, the forces for dry particle systems will be described followed by a treatment of wet and immersed systems. This paper is an update of a previous paper on the same subject with some additional experimental evidence which has been obtained.[9]

VAN DER WAALS FORCES

Van der Waals forces can be understood as follows. Even at absolute zero temperature, solids can contain local electric fields which originate from polarizations of the constituent atoms and molecules. Above zero degrees, additional contributions come from thermal excitations of the atoms and molecules. As explained by quantum theory, the electrons of an electrically neutral solid do not occupy fixed states of a sharply defined minimum energy which results in spontaneous electric and magnetic polarizations varying quickly with time.[5] Van der Waals forces include forces between molecules possessing dipoles and quadrapoles caused by the polarizations of the atoms and molecules in the material. This can include both natural as well as induced instantaneous dipoles and quadrapoles. It, more importantly, also includes nonpolar attractive forces.[10] The nonpolar van der Waals forces are also referred to as London-van der Waals dispersion forces because London associated these forces with the cause of optical dispersion, i.e., spontaneous polarizations.[10-12] This dispersion force will make the major contribution to the intermolecular force except in the case where the polarizability is small and the dipole moment is large. These forces are postulated by some to be additive for assemblies of atoms and molecules. This assumption has been used to calculate the van der Waals force using a microscopic approach which starts from interactions between individual atoms or molecules and calculates the attraction between larger bodies as an integration over all pairs of atoms and molecules.[2,3,10] This method has been used by Hamaker.[13] His method uses the so-called Hamaker constant, A. This approach has severe shortcoming because linear additivity is not correct, as it does not consider cross-correlation of charge.[5,14] One author has stated that the Hamaker approach is in certain cases "not only inaccurate but also downright misleading as to the laws of force".[15] A more satisfactory macroscopic approach was developed by Lifshitz who started directly from the bulk optical properties of the interacting bodies.[16] In this approach, the decisive material value is the Lifshitz-van der Waals constant, h, which is defined as an integral function of the imaginary parts of the dielectric constants of the adhering materials.[5,14,15,17]

This constant, h, depends only on the materials involved, provided the separation distance is very small. Under some conditions the Lifshitz-van der Waals constant may be related to the Hamaker constant[4] by the equation, $h = 4\pi A/3$. This Lifshitz-van der Waals constant generally ranges from about 0.6 to 9.0 eV, depending on the materials combination.[14] Qualitatively, since this constant is related to the optical absorptivity of a material, materials with strong optical absorption have strong spontaneous fields and thus should be bound by greater forces of adhesion due to van der Waals attractions. The van der Waals force, F(vdW), per unit area between two parallel flat surfaces in contact can be exactly calculated to be

$$\frac{F(vdW)}{cm^2} = \frac{h}{8\pi^2 z^3} \; ,$$

where z is the atomic separation between the surfaces.[5,6,14] The van der Waals adhesion force between a spherical particle and a flat surface reduces in approximation to

$$F(vdW) = \frac{h\,r}{8\pi z^2} \; ,$$

where r is the particle radius. This formula treats the force of adhesion as the force necessary to remove the particle, and as such is a maximum force under ideal conditions for a perfectly spherical particle on a flat surface. This formula can be further reduced by substituting d/2 for r, and assuming an adhesion distance, z, of about 4 angstroms. In approximation, the formula becomes

$$F(vdW) = 2\,h\,d \;\; mdyn \; ,$$

where d is the particle diameter in microns. The constant, h, ranges from about 0.6 eV for polymers to about 9.0 eV for metals such as silver and gold. In Table I are shown some typical particle/surface van der Waals constants. From a 1 micron diameter particle with a constant of 0.6eV to a 100 micron diameter particle with a constant of 9.0 eV, the van der Waals force can range from about 1.2 mdyn to 1800 mdyn. This corresponds to forces per unit area of from about 2×10^4 to 1.5×10^5 dyn/cm^2. These are tremendous pressures, and most particles and/or surface can be deformed by such forces. The amount of deformation depends on the hardness of the particle and surface. The additional van der Waals force due to deformation, F(vdW deform.), is a function of the increased contact area caused by this deformation and is given by

$$F(vdW\,deform.) = \frac{h\rho^2}{8\pi z^3} \; ,$$

where ρ is the radius of the adhesion surface area. This reduces, for z of 4 angstroms, to approximately

$$F(vdW\,deform.) = 9.96 \times 10^3 h\rho^2 \;\; mdyn \; ,$$

for ρ in microns. The total van der Waals force, F(vdW total), is then

$$F(vdW\ total) \ = \ F(vdW) + F(vdW\ deform.)$$

If the contact of a 5 micron diameter particle is increased from point contact to a 0.05 micron radius area by deformation, and assuming h is 7 eV, the F(vdW deform.) can be 174 mdyn, which is much greater than the 70 mdyn F(vdW) of the system. This is only a 2% deformation of the particle, i.e. the 5 micron diameter particle is flattened only to the point of having a 0.1 micron diameter contact diameter. The F(vdW deform.) would increase by a factor of 5 if the deformation were increased to a 0.5 micron diameter area, or 10%. This new contact area, and thus the F(vdW deform.), depends on the hardness of the particle and surface materials. Figure 1 is a graph of F(vdW) and F(vdW deform.) versus particle diameter for materials combinations of 4 and 8 eV and 1 and 5% deformations. It shows the F(vdW) without deformation for h equal to 4 eV and 8 eV, plus the additional force due to deformation of 1% and 5% for h equal to 8 eV. It clearly shows that deformation can add tremendously to the total force of adhesion. A first look at these plots should not be incorrectly interpreted that the forces of adhesion simply decrease with decreasing particle diameter. Although the total force does decrease, the force per unit area increases with decreasing particle size, and the force of adhesion also increases relative to the gravitational force acting on the particle. These relationships will be further discussed later.

TABLE I. Empirically Determined
Lifshitz-van der Waals Constants
(Data from references 4 and 5)

PARTICLE	SURFACE	h (eV)
polymer	polymer	0.6-0.9
KBr	KBr	2.0
alumina	alumina	4.0
Ge	Ge	6.6-7.6
Si	Si	6.8-7.2
Ge	Si	7.5
graphite	graphite	7.2
graphite	Si	6.8
Cu	Cu	8.5
Ag	Ag	9.0

ELECTROSTATIC FORCES

Two types of electrostatic forces may act to hold particles to surfaces. The first is due to bulk excess charges present on the surface and/or particle which produce a classical coulombic attraction known as an electrostatic image force. This electrostatic image force, F(i), is given by the equation:

$$F(i) = \frac{q^2}{4\pi\,E\,e\,l^2}$$

133

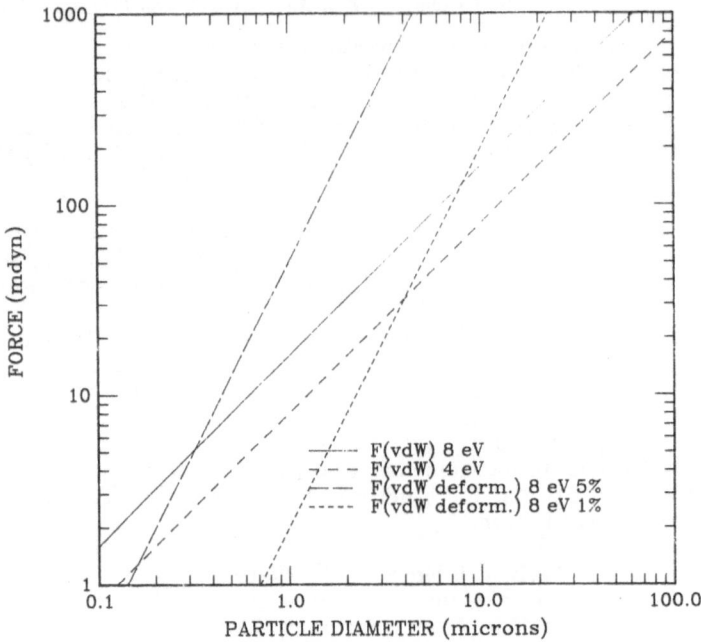

Figure 1. F(vdW) and F(vdW deform.) as a function of particle diameter.

where e is the dielectric constant of the medium between the particle and surface, E is the permittivity of free space, q is the charge, and l is the distance between charge centers. The distance between charge centers is in this case approximately equal to 2r, where r is the particle radius, and q can be expressed as a function of the particle radius by the expression:

$$q = C U = 4 \pi E r U ,$$

where C is capacitance, and U is the potential in volts. The capacitance has been expressed as a function of r via an approximation of the Euler equation. The expression for total image force then reduces as a function of particle diameter, d, in approximation to:

$$F(i) = 3 \times 10^{-2} d^2 \; mdyn ,$$

where d is the particle diameter in microns. This equation assumes a charge density of 10 electronic charges per square micron, what might be considered a typical large charge. The maximum charge density normally possible is 100 electronic charges per square micron, which corresponds to a flashover potential of about 20,000 volts/cm. It is important to note that for conductors, these excess charges are balanced by contact charge flow so that adhesion force by electrostatics is small. On the other hand, for nonconductors electrostatic attraction is significant.

The more important electrostatic force for very small particles is electrostatic contact potential induced electrical double layer forces. Two different materials in contact develop a contact potential caused by differences in the local energy states and work functions. Electrons are transferred from one solid to another until an equilibrium is reached where the current flow in both directions is equal. The resulting potential difference is called a contact potential difference, U, which generally ranges from zero to about 0.5 volts. It sets up a so-called double layer charge region. In the case of two metals in contact, only the surface layer carries contact charges. For semiconductors and insulators, these regions may extend into the bulk up to 1 micron or deeper. For a particle on a surface, this electrostatic double layer force, F(el), can be calculated as

$$F(el) \; = \; \frac{\mathrm{n}\,E\,r\,U^2}{z} \; dynes,$$

which reduces in approximation to

$$F(el) = 4\,d\,U^2\;mdyn,$$

where d is in microns and U is in volts. For a maximum potential difference of 0.5 volts, F(el) is approximately equal to d mdyn.[4,5,18] Figure 2 is a comparison of the van der Waals forces and electrostatic forces of particle adhesion versus particle diameter. It is clear, at least for these ideal calculations, that the van der Waals forces predominate over electrostatic forces for very small particles. Double layer electrostatic forces also generally predominate over electrostatic image forces for small particles. Materials that can carry charges high enough to allow electrostatic forces to contribute significantly to the total force of adhesion are generally polymers of poor conductivity or other extreme insulators.

It is very important in this case to point out again that these first considerations have been for spheres on flat surfaces. Electrostatic forces can in fact predominate over van der Waals forces in the case where surface asperities of the particles are significant enough to remove the bulk of the particle form the contact point. This is due to the rapid decrease in F(vdW) with increasing distance between the adherents. In the presence of surface asperities, calculation of the van der Waals force should not be based on the full radius of the particle, but rather on the radius of curvature of the surface elevation at

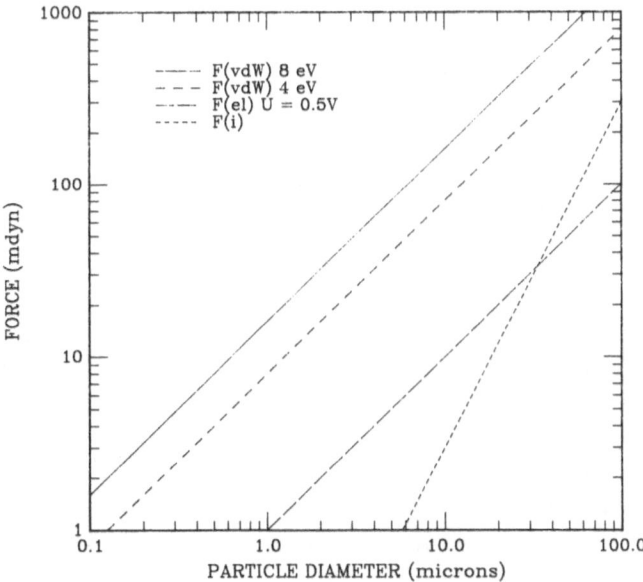

Figure 2. F(vdW) in comparison to electrostatic forces of adhesion as a function of particle diameter.

which the adherents are in contact. Conversely, electrostatic forces do no change much with asperities; at not too large deviations from spherical, the effective radius for electrostatic forces is approximately equal to the true radius. This comparison can be quite complex, especially when considering that multiple asperities may be present and thus multiple contact point may be formed so that F(vdW) once again predominates. Also, deformation of particles and/or surfaces may change these asperities at the contact point so that F(vdW) again also predominates.[4,5,18]

For a dry system, the total forces of adhesion is thus a combination of van der Waals forces and electrostatic forces. As stated before, for particles less than 50 microns diameter, this total adhesion force greatly exceeds the force due to gravity for a given particle size. For a one micron diameter silicon particle on a dry silicon surface, the total force of adhesion exceeds the gravitational force by a factor of greater than 10^7. A comparison of the total force of adhesion for a dry system versus the ratio of total force of adhesion to gravitational force is made in Figure 3. The total force of adhesion line is calculated for a van der Waals constant of 8 eV and 1% particle deformation, assuming a contact potential of 0.5 volts and an image force from Figure 2. The gravitational attraction, F(grav.) is calculated for a particle density of silicon (2.33 g/cc) based on the equation

$$F(grav.) = \frac{4}{3}\pi r^3 \rho g$$

where ρ is the density, g is the gravitational acceleration, and r is the particle radius. For particles less than about 20 micron diameter it can be seen that the total force of adhesion exceeds the gravitational force by a factor of at least 10^5.

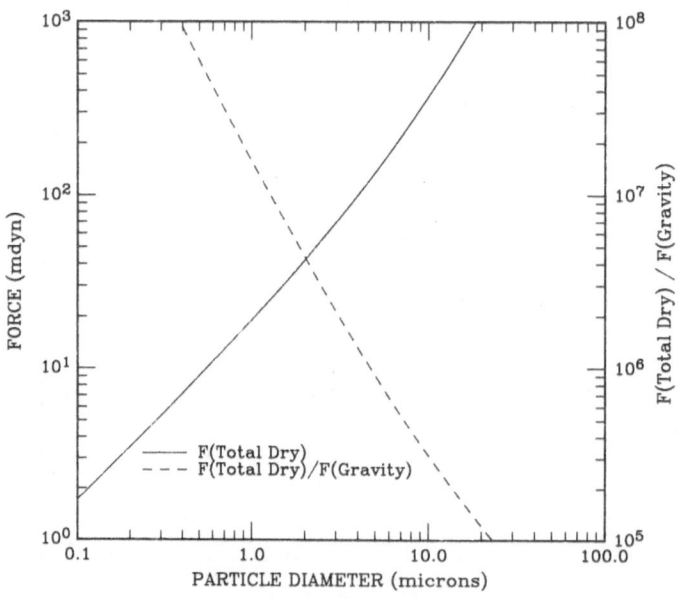

Figure 3. Total force of adhesion of particles on a dry surface as compared to the ratio of total force to gravitational force as a function of particle diameter.

EFFECTS OF HUMIDITY AND LIQUID IMMERSION ON ADHESION

Due to high humidity or to an adhered particle/surface system having been immersed and then withdrawn from a liquid, a liquid film can be formed by capillary condensation or capillary action between the particle and surface. The resulting capillary force can make a large contribution to the total force of adhesion. The capillary force is a function of the particle radius and liquid surface tension as shown by the following formula:

$$F(cap.) = 4\pi r\gamma \quad dynes,$$

where r is the particle radius, and γ is the liquid surface tension. The surface tension of water at 18°C is 73 dyn/cm. This reduces to

$$F(cap.) = 0.63\,d\,\gamma \quad mdyn,$$

where d is the particle diameter in microns. There is evidence that a capillary force may remain even after baking at above the boiling point of the liquid even as long as for 24 hours.[6] If the liquid contains substances that can crystallize upon evaporation a solid crystalline bridge might also form during drying. It should be pointed out that hydrophobic surfaces are not generally affected by capillary forces. Figure 4 shows the magnitude of a capillary force, for water as the liquid, in comparison to the van der Waals forces for h equal to 4 eV and for 1% deformation of the particle. It can be seen that the capillary force can be clearly predominant over other forces for small particles. A comparison of the total wet force of adhesion, as a combination of reduced van der Waals force and

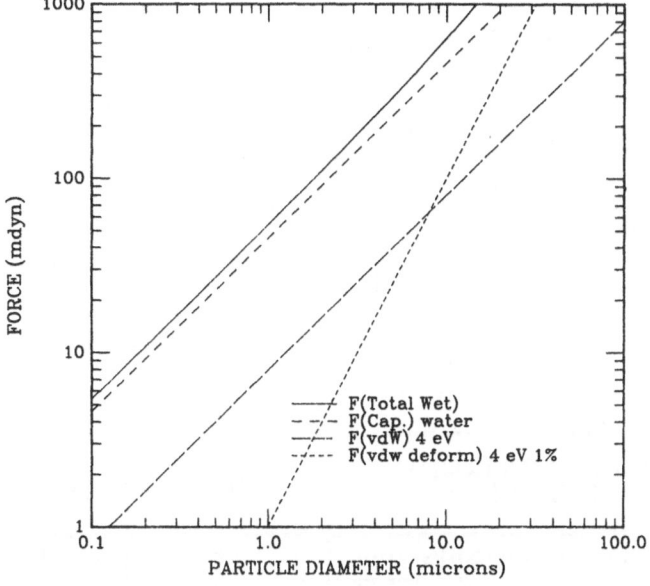

Figure 4. Capillary Force and total wet force of adhesion as a function of particle diameter.

capillary force, to the gravitational force is given in Figure 5. As will be discussed later, particle forces may be reduced by immersion in liquids, particularly for large particles, but if the particles are not removed in the liquid then the added capillary forces which remain upon removal from the liquid can increase the total adhesion force by an order of magnitude or more.

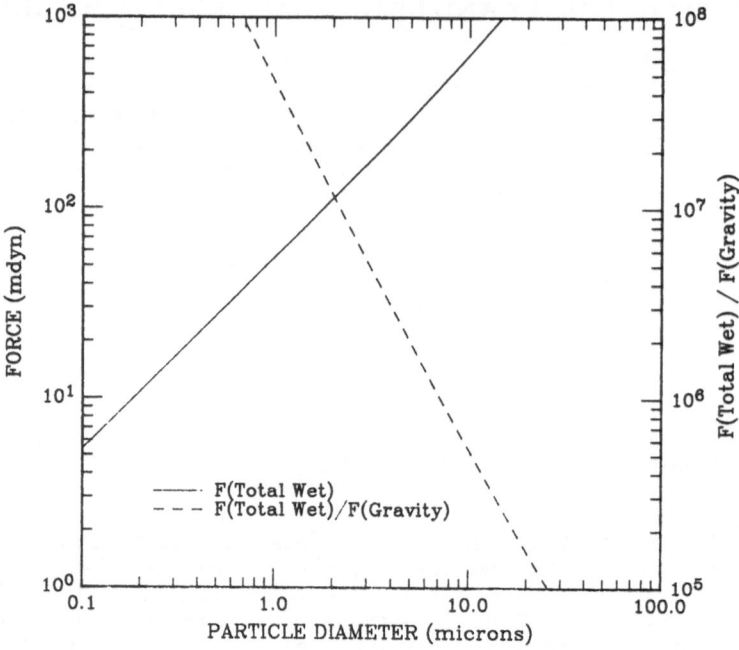

Figure 5. Comparison of the total wet force of adhesion to the gravitational force as a function of particle diameter.

By immersion of adherent particles in a liquid, the van der Waals force can be reduced by, in most cases, about a factor of two since the liquid partially shields the attraction. For calculation purposes, the van der Waals constant, h, must be replaced by an immersed van der Waals constant, h(im.). Table II shows some typical h(im.) values as compared to h values.[5] Electrostatic image forces are more or less eliminated by immersion and electrostatic contact potential forces are greatly reduced. The electrostatic forces generally tend to become negligible because of the enhanced static dielectric constant of the liquid, as compared to a gaseous or vacuum environment, and because of the sorption phenomena which tend to shield the charges.

TABLE II. Ccomparison of Immersed and
nonimmersed van der Waals Constants
(From reference 5)

PARTICLE-LIQUID-SURFACE	h (eV)	h(im.) (eV)
polystyrene-water-polystyrene	0.6	0.1
silicon-water-silicon	6.8-7.2	3.49
copper-water-copper	8.5	4.9
silver-water-silver	9.0	7.76

Immersion does not necessarily always cause a reduction in adhesion purely because of its shielding effect of charges. The liquid molecules and/or ionic impurities in the liquid can actually add to the total adhesion force. Likewise, in some cases, the reduction in total adhesion force by immersion is greater than predicted by shielding alone. To understand these possibilities, one must consider the interactive forces between the boundary layers present at all solid-liquid interfaces of the adherents (Note: these same discussions can apply to the humid and wet situations described above). Molecules of a liquid are often adsorbed at the solid surfaces and are polarized and/or oriented in a certain direction so that an electrical double layer is formed at each solid-liquid interface. If the processes occur in the same way on all surface, dipole layers can cause repulsive forces thus adding to the reduction in adhesion force and in some cases actually cause removal. If the liquid itself is dissociable into ions and/or if the liquid contains dissolved species which are present as ions, again dipolar alignments on surfaces can occur to cause double layer attractions or repulsions. These forces can be approximated by the formula for electrostatic contact potentials if the contact potential difference is determinable. For a particle on a surface, the repulsive or attractive force due to these charge double layers, for a certain charge density in solution, can be alternatively approximated by the formula:

$$F = \pm 64 \pi r N k T \delta$$

where r is the particle radius, T is the temperature in degrees Kelvin, N is the charge density in charges per cubic centimeter, and δ is the double layer thickness, which is typically about 50 angstroms. This reduces to approximately

$$F = \pm 7 \times 10^{-25} dTN$$

Whether the force is attractive or repulsive is completely system dependent. For particles which are the same materials as the surface, charges generally form double layers the same on both surfaces so that the double layer interactions are repulsive. Different ions may however collect differently at unlike surfaces so that either attractive or repulsive forces are possible depending on their alignment.

One final topic for consideration is aqueous and non-aqueous detergency. It is known that adhered colloidal particles below about 0.2 microns diameter are virtually impossible to remove and that removal is difficult even for particles even as large as 10 microns diameter or more. One paper has reported that significant detachment of small particles can occur in non-aqueous surfactant media by polymeric detergency.[19] The exact mechanism of this action is not known but it has been referred to an entropic repulsion mechanism where it has been hypothesized that the polymeric detergent may act as a "molecular wedge". Entropic action implies that the source of repulsion which counteracts particle adhesion is the reduction in configurational entropy of the adsorbed polymer chains when they are compressed in the space between particle and surface. The detergent molecule must not only be small enough to be able to be adsorbed in between the adhered particle and solid surface, but also must be large enough to exert an effective entropic repulsion. Therefore, the size of the detergent molecule is very critical.[19] These considerations imply that each particle size will be most effectively removed by a different detergent molecule size which is optimized for that particle size. Also, deformed systems in which the particle and/or surface have an extended contact area may not be so affected by this detergency.

REMOVAL OF PARTICLES

To gain some confirmation of these theoretical findings that micron sized particles adhere to surfaces with large forces of adhesion, experiments were conducted to attempt to remove small silicon particles from silicon wafers in a class 100 cleanroom. The calculations from above indicate that the forces of adhesion are great enough that noncontact methods of cleaning will be unable to remove micron size particles. This would be true for attempts at chemical removal as well as high pressure water and air sprays which cannot tranport the removal energy to the particle due to the fluid boundary layer at the surface. Some success has been noted in the past with ultrasonic and megasonic agitations of chemical solutions to effect small particle removal.[20] However, few results have been published which indicate that this works well for particles below about 2-5 microns diameter.[20]

To perform these experiments, silicon sawing dust was obtained from a wafer production area. The silicon dust was suspended in water solution. This silicon dust solution was filtered with a 10 micron Nuclepore filter membrane to filter out large particles. A liquid particle counter sampling of the remaining solution showed a large number of 0.5-2 micron diameter particles (>80% of the total particles) and about 20% 2-10 micron diameter particles. The particles were dispersed onto clean silicon wafers inside a class 100 cleanroom using a TSI aerosol generator equipped with an aerosol atomizer and particle diffusion dryer. The density and size of particles on the silicon wafer surfaces were measured with an Aeronca WIS-150 wafer surface particle counter. The Aeronca counts and sizes particles into two basic size categories, 0.2-2 and 2-20 micron diameters.

Twenty five silicon wafers with 98 to 2243 0.2-2 micron particles and 9 to 451 2-20 micron particles were cleaned chemically with the clean followed by a high pressure water spin scrub. The chemical clean consisted of an ammonium hydroxide/hydrogen peroxide cleanup with megasonic agitation. The scrub brush for the spin scrub was made of polyvinyl alcohol. Afterwards, wafers had 35 to 130 0.2-2 micron particles and 0 to 10 2-20 micron particles. The average percentages particles removed were 60% of the 0.2-2 micron particles and 95% of the 2-20 micron particles. Twenty four other wafers with 89 to 1160 0.2-2 micron particles and 24 to 1883 2-20 micon particles were cleaned chemically only. The clean did not include a scrub. Afterwards, wafers had 122 to 728 0.2-2 micron particles and 3 to 194 2-20 micron particles. The average percentages particles removed were 4% of the 0.2-2 micron particles and 85% of the 2-20 micron particles.

Clearly the cleanup involving only noncontact cleaning was much less effective at 0.2-2 micron particle removal, and was somewhat less effective at 2-20 micron particle removal. This gives some confirmation of the calculations from above in that the calculations correctly predict the size particles (about 2 microns) that form the transition region between easy and hard removal. Below this size physical contact must be used to effectively remove the particles. It should be noted, particularly for silicon wafer production, that physical contact with the surface also produces unwanted defects such as scratches in the surface. Thus, contact methods of cleaning are generally avoided in semiconductor fabrication.

One question frequently arises when considering the large forces of adhesion of micron size particles. What is the explanation for the observation of micron size particles with an air particle counter when the plastic tubing connected to the particle counter inlet is moved or flexed? It appears that particles on the inside surface of the tube are loosened from the plastic surface; they can be plastic particles breaking away from the surface or deposited particles breaking their adhesion forces. The observation that even rapid air flow through the tube does not loosen particles is a confirmation of the forces as calculated above. It is also not surprising that particles overcome their adhesion to the surface of the tube when the tube is moved or flexed, even very gently. The physical forces applied to the small area of tube surface by such movement or flexing is, in comparison to the forces of adhesion, quite significant. This is a mechanical transfer of energy much like the contact action of physical scrubbing of wafer surfaces described above. This phenomenon is further exemplified by the large particle generation of rubbing materials together at the inlet to a particle counter.

CONCLUSION

It has been shown that small particles on a dry surface are primarily held by strong van der Waals forces which can increase in magnitude with time due to deformation of the particle and/or surface which increases the total contact area. Immersion generally significantly reduces the total force of adhesion while the particle/surface system is still immersed. Upon removal from the liquid, however, predominating capillary forces of adhesion may be added due to the formation of liquid bridges between particle and surface. The same phenomenon of capillary action may also occur in high humidity. The total forces of adhesion for small particles are so large that they exceed the gravitational force on those particles by many orders of magnitude. This hopefully makes it clear that emphasis should be placed on prevention of particle deposition rather than counting on their subsequent removal. Experimental attempts of removal of small particles from silicon wafer surfaces support these findings. Additional references on particle adhesion, not already cited in this paper, appear as references 21 to 29.

ACKNOWLEDGEMENTS

The author would like to express special thanks to Olga Paradis for technical library assistance in the compilation of this information, to Graydon Larrabee for his helpful assistance in completing this work, to Keith Russell for help with the particle removal tests, and to Zack VanBlack for help with Aeronca tests.

REFERENCES

1. J. M. Duffalo, and J. R. Monkowski, Solid State Technol., 27 (3), 109 (1984).
2. M. Corn, J. Air Pollution Control Assoc., 11 (11), 523 (1961).
3. M. Corn, J. Air Pollution Control Assoc., 11 (12), 523 (1961).
4. J. Visser, Surf. Colloid. Sci., 8, 3 (1975).
5. H. Krupp, Adv. Colloid Interface Sci., 1 (2), 111 (1967).
6. S. Bhattacharya, and K. L. Mittal, Surf. Technol., 7, 413 (1978).
7. W. Kern and D. A. Puotinen, RCA Rev., 31, 187 (1970).
8. R. K. Iler, "The Colloid Chemistry of Silica and Silicates," p. 95, Cornell University Press, Ithaca, New York, 1955.
9. R. A. Bowling, J. Electrochem. Soc., 132, 2208 (1985).
10. M. Corn, in "Aerosol Science," C. N. Davies, Editor, p. 359, Academic Press, New York, 1966.
11. F. London, Z. Phys., 63, 245 (1930).
12. F. London, Trans. Faraday. Soc., 33, 8 (1937).
13. H. C. Hamaker, Physica, 4, 1058 (1937).
14. H. Rumpf, in "Agglomeration 77," Proc. of the Second Internat. Symp. on Agglomeration, Vol. 1, p. 97, K. V. S. Sastry, Editor, American Institute of Mining, Metallurgical, and Petroleum Engineers, Inc., New York, 1977.
15. J. A. Kitchener, J. Soc. Cosmet. Chem., 24, 709 (1973).
16. E. M. Lifshitz, Sov. Phys. JETP, 2, 73 (1956).
17. B. B. Morgan, The British Coal Utilization Research Assoc. Monthly Bulletin, 25 (4), 125 (1961).
18. B. V. Deryagin, and A. D. Zimon, Colloid. J. USSR, 23, 454 (1961).

19. E. J. Clayfield and E. C. Lamb, Disc. Faraday Soc., 42, 285 (1966).
20. C. Tipton, (1986) personal communication.
21. B. Y. H. Liu, Editor, "Fine Particles," Academic Press, London, 1976.
22. N. A. Fuchs, "The Mechanics of Aerosols," MacMillan Co., New York, 1964.
23. C. N. Davies, Editor, "Air Filtration," Academic Press, London, 1973.
24. R. G. Dorman, in "Aerosol Science," C. N. Davies, Editor, p. 195, Academic Press, New York, 1966.
25. S. N. Omenyi, J. Chappuis, and A. W. Neumann, J. Adhesion, 13, 131 (1981).
26. D. Bargema, and F. Van Voorst Vader, J. Electroanal. Chem. Interfacial Electrochem., 37, 45 (1972).
27. A. D. Zimon, "Adhesion of Dust and Powders," Second Edition, Consultants Bureau, New York, 1982.
28. B. V. Deryagin, N. A. Krotova, and V. P. Smilga, "Adhesion of Solids," Consultants Bureau, New York, 1978.
29. K. L. Mittal, Editor, "Surface Contamination: Genesis, Detection, and Control," Volumes 1 and 2, Plenum Press, New York, 1979.

THE ELECTROSTATIC FORCE ON A DIELECTRIC SPHERE RESTING ON

A CONDUCTING SUBSTRATE

Wm. Y. Fowlkes and K. S. Robinson

Copy Products Research and Development
Eastman Kodak Company
Rochester, NY 14650 USA

The electrostatic force of removal is calculated for a sphere in
contact with a grounded plane in an externally applied electric field
that is normal to the plane. The electrostatic force is given by
the sum of the Lorentz force QE_o, where Q is the free charge on the
sphere and E_o is the applied electric field, and the electrical force
between the sphere and the plane. The force between the sphere and
the plane can be described by the interaction between the bound and
free charges on the sphere, whose distribution is strongly influenced
by the polarization of the sphere, and their images in the plane.
The polarization charge distribution of the sphere is described by
a linear multipole expansion. The multipole terms are calculated
by a simple, iterative, self-consistent scheme, in which the externally applied field and the image charges induce the polarization of
the sphere. The net electrostatic force on the sphere is given by
the sum of the force on each linear multipole in the expansion. Two
novel results of this force computation are found. The force on the
higher order multipoles increases with the applied electric field
more rapidly than the Lorentz force. For a given charge level, a
field magnitude exists above which the net electric force is adhesional. Furthermore, an optimum charge level exists that minimizes
the field required for electrostatic removal.

INTRODUCTION

The electrostatic force on a charged insulating particle resting on a
plane conductor is important in a wide variety of applications including
the electrostatic transfer of toner from a photoconductor in the electrophotographic cycle.[1] High transfer efficiency is achieved when the electrostatic force of removal greatly exceeds the adhesion force. The adhesional
forces acting on an insulating particle resting on a substrate have been
broadly classified elsewhere.[2,3] In order of decreasing strength they
include chemical forces, the double layer force, the van der Waal/London
dispersive force, and the gravitational force. The image force is often
included as another source of adhesion, which is appropriate for charged
particles with no externally applied field. In the case of electrostatic
transfer or removal, the image force is a strong function of the applied
electric field, as shown in this paper. Therefore, the image force should
be considered as part of the net electrostatic force of removal.

Current flow determines the charge distribution[4] and the electrostatic force on particles with nonzero conductivity.[4,5] Electrostatic computations on a charged dielectric sphere with a fixed initial charge distribution and induced polarization is appropriate for insulating toner particles in the electrophotographic cycle because the charge relaxation time for toners is very long (hours or more) compared with the typical residence time on the photoconductor (seconds). Thus, the charge redistribution due to current flow within toner particles is negligible.

The image force on a point charge Q located at a distance R from an infinite plane conductor at zero potential is usually analyzed by use of the method of images and Coulomb's law. The resulting image force is given by

$$F_I = \frac{1}{4\pi\varepsilon_m} \frac{Q^2}{(2R)^2} \qquad (1)$$

Computing the force on a charged, dielectric sphere of radius R and permittivity ε_p in a medium of permittivity ε_m, resting in contact with an infinite grounded plane (Figure 1, s = 0) is greatly complicated by the induced polarization charge on the sphere. The sphere is polarized by the field produced by the image charge in the conducting plane as well as any externally applied electric field. For an isolated sphere in a uniform field, the polarization produces no net force. For a dielectric sphere near a grounded plane, however, the field due to the image charges is manifestly nonuniform. In that case the simple expression for the image force on the sphere must be modified to include the dielectrophoretic (DEP) force.

The dielectrophoretic effect has been modelled by representing the dielectric sphere by a simple dipole.[6] In the uniform field approximation,[6,7] the dipole moment is given by

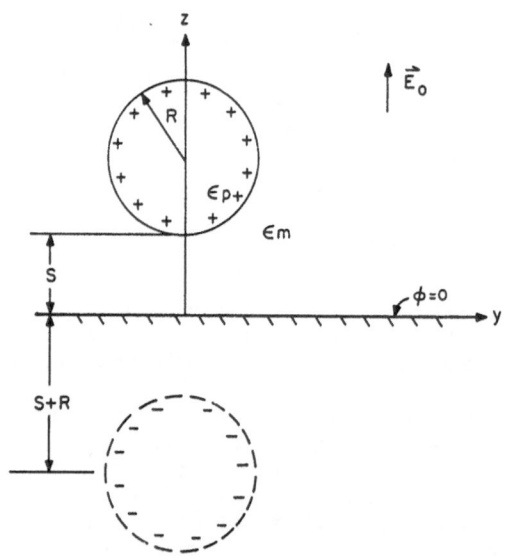

Figure 1. Dielectric sphere, of permittivity ε_p and radius R, a distance s from the conducting plane, located a z = 0, in a medium of permittivity ε_m. Also shown is an applied field E_o, normal to the plane.

$$\bar{p}_{eff} = 4\pi\varepsilon_m \frac{\varepsilon_p - \varepsilon_m}{\varepsilon_p + 2\varepsilon_m} R^3 \bar{E} \qquad (2)$$

where the value of $\bar{E}$ at the location of the center of the sphere, with the sphere removed, is used. The nonuniformity of the electric field is then used for the force expression

$$\bar{F}_{DEP} = (\bar{p}_{eff} \cdot \nabla) \bar{E} \qquad (3)$$

In the case of a sphere that is resting in contact with a conducting surface this approach is clearly inadequate. Because the image charge is located at $z = -R$, only 2R from the center of the particle, the variation of the field is large compared with the dimension of the particle. It has been shown that the simple approach to analyzing DEP can be corrected by the use of higher order multipoles.[8,9] In this case, the charge distribution for the sphere can be represented by a linear multipole expansion, located for convenience at the center of the sphere.

The solution set of multipoles and their images is found by simultaneously solving a set of three electrostatic equations. The first equation relates the source multipole, which describe the charge distribution of the dielectric sphere, to the generating field, which is comprised of the field due to the image multipoles and the externally applied field. The second equation describes the field in the half-space outside of the grounded plane due to the image multipoles by a multipole expansion of the potential. The third equation relates the image multipoles located at $z = -R$ to the source multipoles located at $z = +R$ by the method of images.

THEORY

Multipole Calculation

The potential outside of any localized charge distribution can be written as an expansion in spherical harmonics, located at the origin of the coordinate system,

$$\phi(\bar{r}) = \sum_{\ell=0}^{\infty} \sum_{m=-\ell}^{\ell} B_{\ell,m} \frac{Y_{\ell,m}(\theta,\phi)}{r^{\ell+1}} \qquad (4)$$

where the coefficients $B_{\ell,m}$ contain the multipole moments.[10] In this paper only the axially symmetric case is considered because the applied field is normal to the conducting plane (Figure 1). In the case of axial (azimuthal) symmetry, the spherical harmonics can be replaced by Legendre polynomials $P_n(\cos\theta)$

$$\phi(r,\theta) = \sum_{n=0}^{\infty} \frac{p^{(n)} P_n(\cos\theta)}{4\pi\varepsilon_m r^{n+1}} \qquad (5)$$

and the $p^{(n)}$ are defined to be linear multipoles.[11] Linear multipoles can be generated by combinations of point charges as shown in Figure 2. A dipole is generated by inverting the charge of the monopole and displacing it by a distance d from the positive monopole. A linear quadrupole is generated by the same operation -- inverting the charge of a dipole and displacing it from a positive dipole. An octupole is generated from two quadrupoles and so forth. Note that in each case the pole whose charge

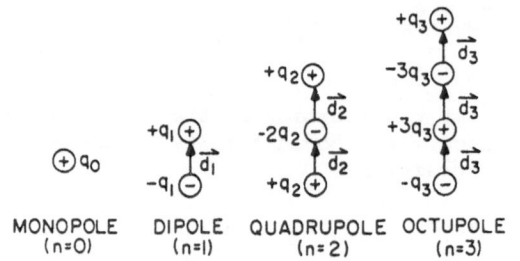

MONOPOLE DIPOLE QUADRUPOLE OCTUPOLE
(n=0)   (n=l)   (n=2)   (n=3)

Figure 2. Linear multipoles, constructed from point charges.

was reversed is displaced in the -z direction so that positive linear multipoles have the characteristic that the point charge furthest in the +z
direction is positive. In this point charge construction, the magnitude of
the q_n and the d_n determines the magnitude of the nth multipole. With the
appropriate linear multipoles in Equation 5, the potential in space due to
the dielectric sphere can be specified. A second expansion, closely related
to the first, is used for the image charges.

The multipole moments induced in a dielectric sphere by an externally
applied electric field have been calculated by Jones[8,9] in terms of the
axial field E and its derivatives, evaluated at the location of the center
of the sphere with the sphere removed.

$$p_i^{(n)} = \frac{4\pi\varepsilon_m(\varepsilon_p - \varepsilon_m) R^{2n+1}}{(n-1)! [n \varepsilon_p + (n+1)\varepsilon_m]} \frac{\delta^{n-1}E_z}{\delta z^{n-1}}, \quad n = 1,2,3,\ldots \qquad (6)$$

The monopole moment $p^{(0)} = Q$ is given by the net charge on the sphere. If
we use the definition

$$\frac{\delta^0 E_z}{\delta z^0} = E_z \qquad (7)$$

then Equation 2 is simply a special case of Equation 6.

Consider an initially unpolarized, dielectric sphere with charge Q
whose center is located at z = +R, as shown in Figure 1 with s=0. Associated with the ground plane is an image charge located at z = -R. The
axially symmetric field evaluated along the z-axis, with the sphere removed,
is given by

$$E_z = \frac{1}{4\pi\varepsilon_m} \frac{-Q}{(z+R)^2} \qquad (8)$$

Inserting Equation 8 into Equation 6 shows that the image monopole will
induce an infinite set of linear multipoles $p^{(n)}$ at z = +R in the dielectric sphere. This will, in turn, require an infinite set of image multipoles
$p_i^{(n)}$ located at z = -R to preserve the boundary condition along z = 0.
Thus, the potential due to the charge distribution on the grounded plane
is given, according to Equation 5, by

$$\phi(r,\theta) = \sum_{n=0}^{\infty} \frac{p_i^{(n)} P_n(\cos\theta)}{4\pi\varepsilon_m r^{n+1}} \qquad (9)$$

The axially symmetric field along the z axis is given by $E = -\nabla\phi$ evaluated at $\theta = 0$, $r = z + R$.

$$E_z = \sum_{n=0}^{\infty} \frac{(n+1)\ P_i^{(n)}}{4\pi\varepsilon_m\ (z+R)^{n+2}} + E_o \qquad (10)$$

Also included in Equation 10 is any externally applied field E_o that may be present.

The Method of Images

The image multipoles must be related to the source multipoles according to the method of images. The point charge representation illustrated in Figure 2 makes the construction of image multipoles straightfoward as shown in Figure 3. The method of images actually involves two operations, a spatial inversion about the $z = 0$ plane and charge inversion. Thus for a point source charge q located at z, the image charge is given by -q located at -z. The effect of these two operations on linear multipoles depends on the symmetry of the multipole with respect to the transformations. The charge inversion operator c produces an opposite polarity multipole.

$$c(p^{(n)}) = -p^{(n)} \qquad (11)$$

The effect of the spatial inversion operator m, however, depends upon the mirror symmetry of the linear multipole.

$$m(p^{(n)}) = (-1)^n p^{(n)} \qquad (12)$$

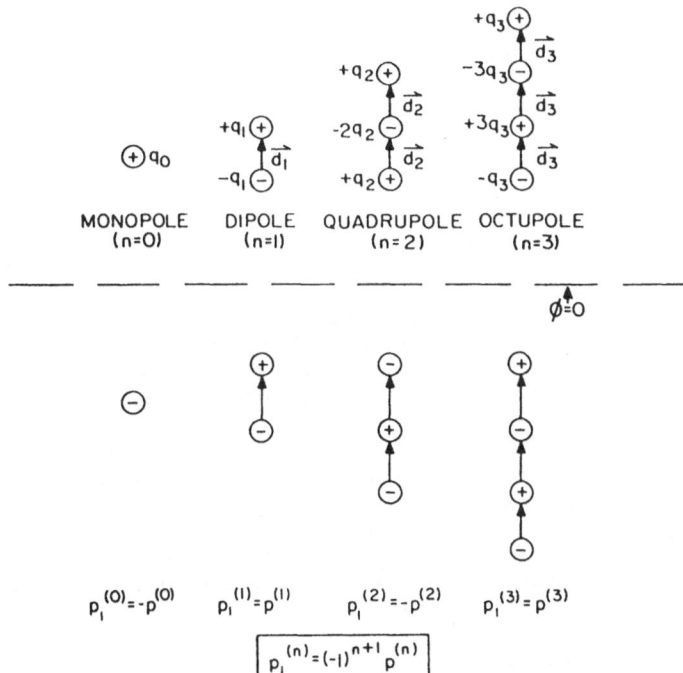

Figure 3. Linear multipoles and their images.

Combining the operators gives the complete method of images transformation

$$\mathcal{i} = \mathcal{c} * \mathcal{m} \tag{13}$$

$$p_i^{(n)} = \mathcal{i}(p^{(n)}) = (-1)^{n+1} p^{(n)} \tag{14}$$

Force Calculation

The electrical force on the dielectric sphere is given[9] by

$$F_e = \sum_{n=0}^{\infty} \frac{p^{(n)}}{n!} \frac{\delta^n E_z}{\delta z^n} \tag{15}$$

Inserting Equations 10 and 14 into Equation 15, and doing the differentiation term by term,

$$F_e = p^{(0)} E_o + \frac{1}{4\pi\varepsilon_m} \sum_{n=0}^{\infty} \sum_{k=0}^{\infty} \frac{(-1)^{n+k+1}(n+k+1)! p^{(n)} p^{(k)}}{n!\,k!\,(z+R)^{n+k+2}} \tag{16}$$

By solving Equations 6, 10, and 14 simultaneously for $z = R$, the solution set of multipoles necessary to evaluate Equation 16 are determined for the case of a dielectric sphere of radius R touching a conducting substrate. Equation 16 is evaluated using this set of multipoles and $z = R$. Note that the Lorentz force QE_o appears as the leading term of Equation 16. Similarly, the image force, Equation 1, appears as the $n = k = 0$ term.

RESULTS

The details for solving Equations 6, 10, and 14 are given elsewhere.[12] The solution is restricted to a finite number of multipoles, $N \leq 64$. Once the solution set $p^{(0)} \ldots p^{(N)}$ is found, Equation 16 is evaluated to find the net electrostatic force on the sphere.

Before analyzing the case of electrostatic removal, consider a charged, dielectric sphere a distance s from a conducting plane with no applied field, in a vacuum. Figure 4 is a logarithmic plot of the force attracting the sphere towards the plane as a function of the separation distance calculated using Equation 16. Note that for a sphere of relative permittivity $k_p = 1$ ($k_p = \varepsilon_p / \varepsilon_0$) no polarization is possible, this case represents Equation 1. At large separations, polarization effects are negligible and the force of attraction for all k_p is given by Equation 1. As the separation becomes small, the field produced by the image charge is sufficient to polarize the sphere for $k_p > 1$ and significantly alter the image force. Davis[13] analyzed this problem by numerically solving Laplace's equation in bispherical coordinates. His results are also given in Figure 4 for comparison.

A force of attraction between an uncharged sphere and a ground plane is induced by a strong external field because it polarizes the sphere. This case is shown in Figure 5 along with Davis's results for the identical problem. At large separations, only the dipole interactions are significant. As the separation becomes small, force contributions from higher order multipoles become important.

The problem of the electrostatic removal force of a dielectric sphere in contact with a conducting plane combines the features of Figures 4 and 5. That is, the attractive electrical force increases with the charge Q

148

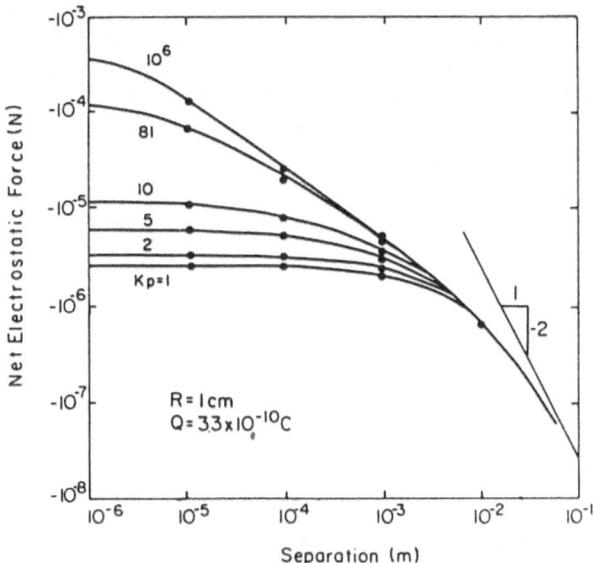

Figure 4. The net electrostatic force on a dielectric sphere, of charge $Q = 3.3 \times 10^{-10}$ C, in a medium of permittivity ε_0 vs. the separation between the sphere and a ground plane. Shown are the results from the multipole calculation for several relative permittivities of the sphere (solid lines). Also shown are the results from a solution to Laplace's equation in bispherical coordinates from Reference 13 (circles).

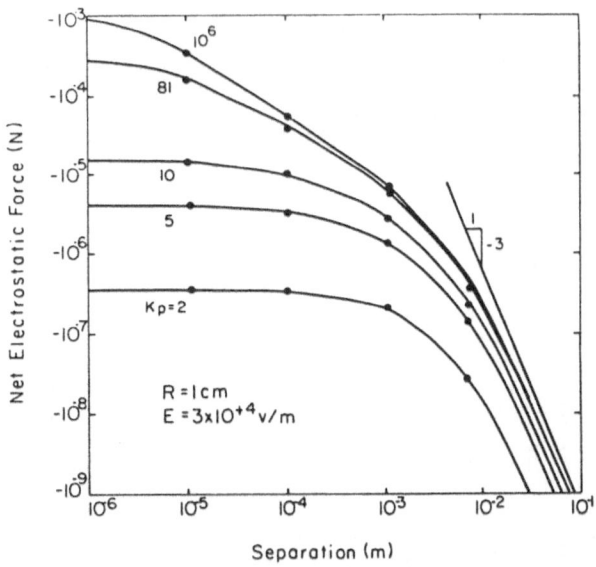

Figure 5. The net electrostatic force on an uncharged dielectric sphere in a medium of permittivity ε_0 with an applied field $E_0 = 3 \times 10^4$ V/m vs. the separation between the sphere and a ground plane. Shown are the results from the multipole calculation for several relative permittivities of the sphere (solid lines). Also shown are the results from a solution to Laplace's equation in bispherical coordinates from Reference 13 (circles).

149

(Equation 1) and with the dielectric constant ε_p (Figure 4) as well as with the applied field E_0 (Figure 5). On the other hand, the Lorentz force pulling the sphere away from the plane goes as QE_0 and is independent of ε_p. Thus, we can expect to find that the force of removal is a complicated function of Q, E_0, and ε_p.

The net electrical force may be plotted as a function of the particle charge and the applied electric field using equi-force contour plots as shown in Figure 6. Identified are the four regimes of importance.

Polarization Adhesion: $F_e < 0$
Image Adhesion: $F_e < 0$
Mechanical Removal: $0 < F_e < F_A$
Electrostatic Removal: $F_A < F_e$

where F_e is the net electrical force, Equation 16, and F_A is the adhesion force given by

$$F_A = F_V + F_G \qquad (17)$$

where F_V is the van der Waals force between a sphere and a plane.

$$F_V = \frac{h\,\omega_0}{8\pi z_0^2} R \qquad (18)$$

where $h\omega_0$ is the Lifshitz-van der Waals constant and z_0 is the distance of closest approach between the sphere and plane. The gravitational force is given by

$$F_G = \frac{4}{3} \pi R^3 \rho \qquad (19)$$

where ρ is the volume density of the sphere.

In the <u>Polarization Adhesion</u> regime, the applied electric field is large, but the particle charge is small. The net electrical force is dominated by the attraction of the induced polarization charge (multipoles)

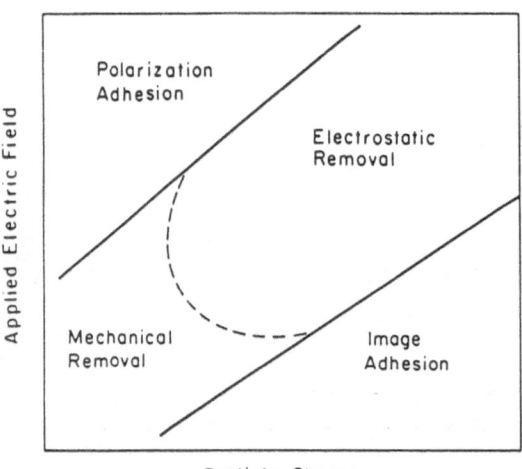

Figure 6. Sketch of the equi-force contours for the combinations of particle charge and applied electric field identifying the four regimes of interest.

to their images in the grounded plane. Thus, $F_e < 0$, or towards the ground plane. The existence of the Polarization Adhesion regime is the principal new result of this work. Since the magnitude of multipole interactions decreases rapidly with separation distance, removal by mechanical vibration or other means is possible if sufficient separation is achieved. Nevertheless, removal by mechanical means may not be very effective in the Polarization Adhesion regime because separation distances on the order of a particle radius must be achieved.

In the Image Adhesion regime, the applied electric field is small and the particle charge is high. The Coulombic force between the monopole ($p^{(0)} = Q$) and its image in the grounded plane dominates.

In the Mechanical Removal regime, both the applied field and the particle charge are small. The electrostatic force favors removal but is smaller than the force of adhesion (van der Waals and gravitational). Mechanical vibrations or other means of separating the particle from the plane momentarily are required for removal to occur.

In the Electrostatic Removal regime, both the particle charge and applied field are large. The Lorentz force dominates and a particle is removed electrostatically from the grounded plane. This condition is achieved in the electrophotographic cycle to accomplish high toner transfer efficiency. Of course, the location of the demarcation line between the Mechanical Removal regime and the Electrostatic Removal regime will depend upon the magnitude of F_A.

Computed results corresponding to Figure 6 are given in Figures 7 - 9 for a 10 μm radius sphere and a range of relative permittivities, as labelled on the figures. In each case the medium is vacuum, $\varepsilon_m = \varepsilon_0$ and the sphere is taken to be resting on the conducting substrate. The equiforce contours are given by the force ratio

$$\mathcal{f} = \frac{F_e}{(F_A + F_G)} \qquad (20)$$

The van der Waals force was calculated for Equation 18 using[3,14] $\hbar\omega_0 = 3$ eV and $z_0 = 0.6$ nm, the gravitational force was calculated from Equation[19] using $\rho = 2.5 \times 10^3$ kg/m^3 but is negligible for this size sphere.

The values for E_0 range from the upper limit of field induced emission at approximately $E_0 = 10^8$ V/m to $E_0 = 10^5$ V/m. Naturally, fields above 3×10^7 V/m or higher are possible for 10 μm spheres in air only if air gaps are kept small according to the Paschen curve[15].

Several observations may be made from Figures 7 - 9:

1. The monopole's attraction to its image in the ground plane (the familiar point charge image force, Equation 1) dominates the Lorentz force in the Image Adhesion regime. The transition into removal regimes is not strongly dependent on the relative permittivity of the sphere. As seen in Figure 4, the electrical force varies by less than an order of magnitude over a $2 < \varepsilon < 10$. The $\mathcal{f} = 0$ contour bordering the Image Adhesion regime has a slope of unity, and to first order is independent of the particle permittivity. This can be demonstrated analytically if we consider only the $n = k = 0$ term in Equation 16. Thus, along the $\mathcal{f} = 0$ line

$$\mathcal{f} = \frac{QE_0 - \frac{1}{4\pi\varepsilon_m}\frac{Q^2}{(2R)^2}}{F_A} = 0 \qquad (21)$$

151

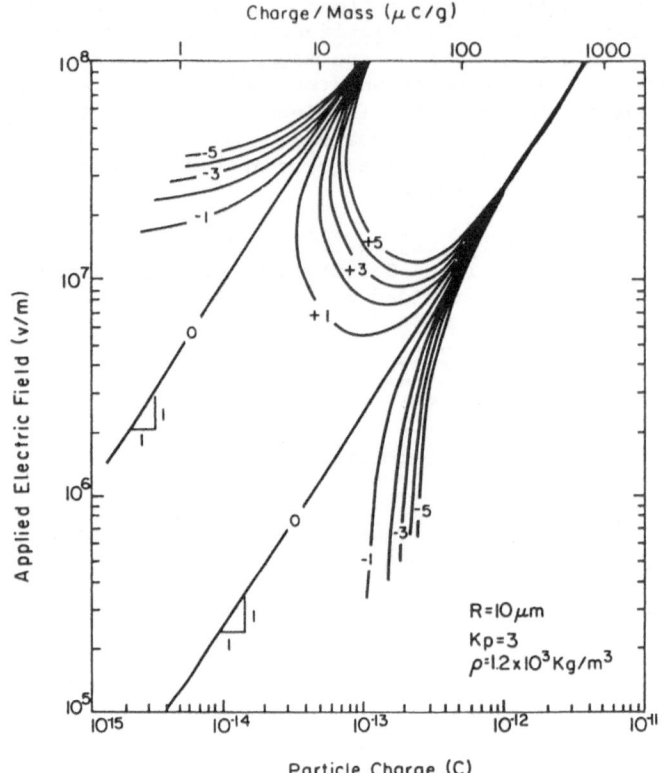

Figure 7. Equi-force contours for a 10 μm radius particle of relative permittivity k_p = 3, in units of adhesion force F_A = 5 x 10^{-7} N. The zero force lines indicate the boundary between the two adhesion regimes and the two removal regimes. The +1 force line is the boundary between the Mechanical Removal regime and the Electrostatic Removal regime. Also shown are charge/mass values calculated assuming a particle mass density of 1.2 x 10^3 kg/m, which is appropriate for Kodak Ektaprint[TM] toner.

therefore,

$$E_0 = \frac{1}{4\pi\varepsilon_m} \frac{Q}{(2R)^2} \qquad (22)$$

2. The attraction of the induced polarization charges to their images in the ground plane dominates both the Lorentz force and other adhesion forces in the Polarization Adhesion regime. The transition between Polarization Adhesion and the removal regimes is very sensitive to the relative permittivity of the sphere. As seen in Figure 5, the electrical force at close spacings varies by over an order or magnitude for 2 < ε_p < 10. The F = 0 contour bordering the Polarization Adhesion regime has a slope of unity indicating that the terms from Equation 16, along that contour, can be grouped into two factors, one proportional to QE_0 and the other proportional to Q^2, as in Equation 21 leading to a relationship similar to Equation 22.

3. The transition from the Image Adhesion regime to the Electrostatic Removal regime is very abrupt for a highly charged sphere as compared to a lower charged sphere's transition from Mechanical Removal to Electrostatic Removal.

152

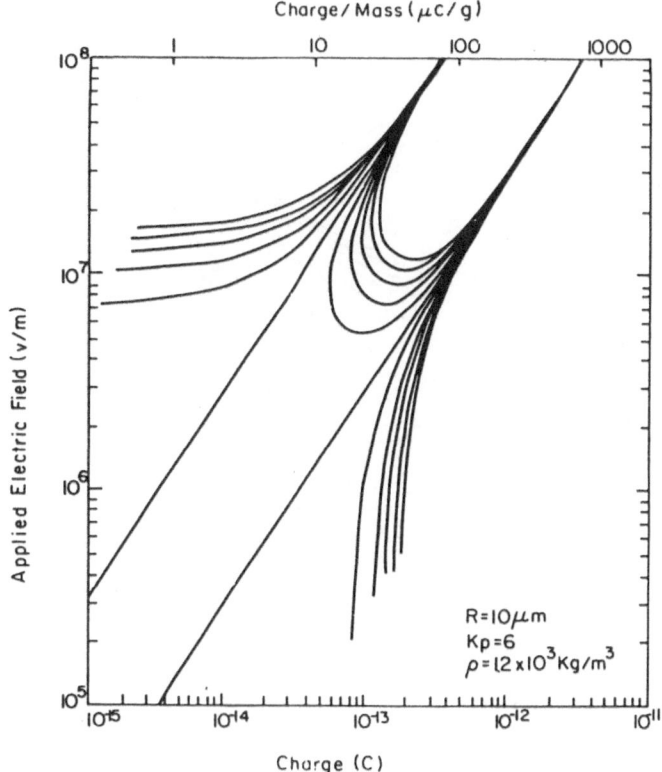

Figure 8. Equi-force contours for a 10 μm radius particle of relative permittivity k_p = 6, in units of adhesion force F_A = 5 x 10^{-7} N. The zero force lines indicate the boundary between the two adhesion regimes and the two removal regimes. The +1 force line (cf. Fig. 7) is the boundary between the <u>Mechanical Removal</u> regime and the <u>Electrostatic Removal</u> regime. Also shown are charge/mass values calculated assuming a particle mass density of 1.2 x 10^{3}kg/m^3, which is appropriate for Kodak Ektaprint^{TM} toner.

4. An optimum particle charge exists that minimizes the applied electric field required for electrostatic removal, corresponding to the minimum of the f = +1 contour. The optimum charge is insensitive to the relative permittivity of the sphere. For a 10 μm radius particle, the optimum charge is ∿1 x 10^{-13} C resulting in an optimum charge to mass ratio of ∿16 μC/g (mass density = 1.2 x 10^3 kg/m^3). In the vicinity of the optimum charge and minimum electric field, small changes in charge or applied field result in relatively small changes in force. The Lorentz force and the net electric polarization vary together. Of course, the optimum charge and field depend strongly upon the magnitude of F_A.

CONCLUSIONS

A new, heuristic method of calculating the net electrostatic force on a charged, dielectric sphere resting on a grounded plane in an externally applied electric field is presented. The polarization charge density, which can have a dramatic effect on the net force, is represented by a series of linear multipoles.

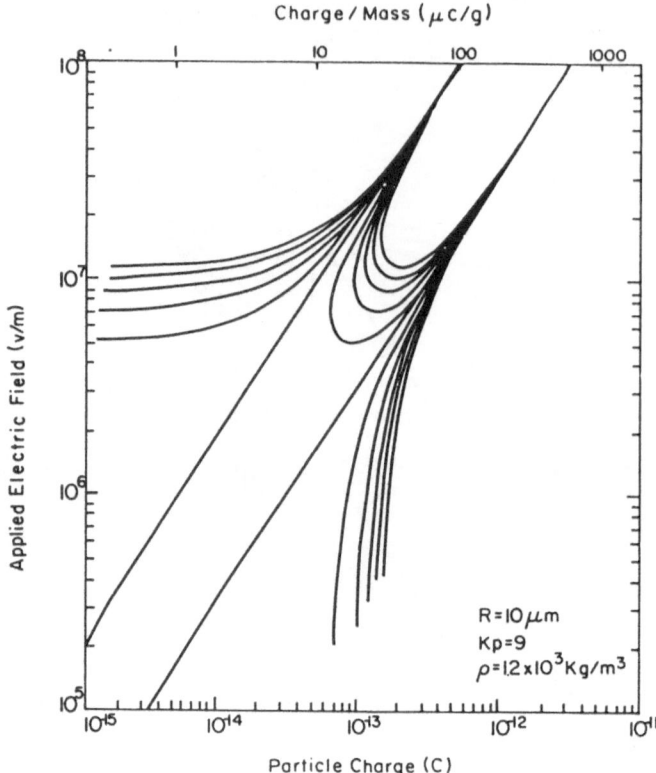

Figure 9. Equi-force contours for a 10 μm radius particle of relative per-mittivity k_p = 9, in units of adhesion force F_A = 5 x 10^{-7} N. The zero force lines indicate the boundary between the two adhesion regimes and the two removal regimes. The +1 force line (cf. Fig. 7) is the boundary between th Mechanical Removal regime and the Electrostatic Removal regime. Also shown are charge/mass values calculated assuming a particle mass density of 1.2 x 10^3 kg/m^3, which is appropriate for Kodak Ektaprint^{TM} toner.

Force computations made using the multipole expansion method[9] are in agreement with the results of a formal solution to Laplace's equation using eigenfunction expansions in bispherical coordinates.[13]

Using the multipole expansion method, the electrostatic removal of dielectric spheres is examined in detail. The two important results are:

1. A regime of Polarization Adhesion is identified in which the attraction between induced polarization charges and their images in the ground plane dominate. Mechanical vibration or other means of removing particles is inefficient in this regime because separation distances on the order of a particle radius must be achieved to overcome the electro-static adhesion.

2. An optimum particle charge exists that minimizes the electric field required for removal. For a 10 μm radius particle, assuming F_A = 5 x 10^{-7} N, typical[14] of the adhesion of Kodak Ektaprint^{TM} toner on an illuminated Ektaprint^{TM} film loop, the optimum charge to mass ratio is 16 μC/g.

REFERENCES

1. R. M. Schaffert, "Electrophotography", p. 52, Focal Press, London, 1980.
2. H. Krupp, Adv. Colloid Interface Sci., $\underline{1}$, 111 (1967).
3. N. S. Goel and P. R. Spencer, Polym. Sci. Technol., $\underline{9B}$ (Adhes. Sci. Technol.), pp. 763-829, L. H. Lee, Editor, Plenum Press, New York, 1975.
4. K. J. McLean, J. Air Pollution Control Assoc., $\underline{27}$, 1100 (1977).
5. P. W. Dietz and J. R. Melcher, in "Control and Dispersion of Air Pollutants: Emphasis on NO_x and Particulate Emissions", R. L. Dyers, D. W. Cooper, and W. Licht, Editors, p. 166, American Institute of Chemical Engineers, New York, 1978.
6. T. B. Jones and G. A. Kallio, J. Electrostatics, $\underline{6}$, 207 (1979).
7. T. B. Jones, J. Electrostatics, $\underline{6}$, 69 (1979).
8. T. B. Jones, in "Proc. IEEE-IAS 1984 Annual Meeting", p. 1136, IEEE, 1984.
9. T. B. Jones, J. Electrostatics, $\underline{18}$, 55 (1986).
10. J. D. Jackson, "Classical Electrodynamics", p. 136, John Wiley, New York, 1975.
11. J. A. Stratton, "Electromagnetic Theory", p. 172, McGraw Hill, New York, 1941.
12. K. S. Robinson and W. Y. Fowlkes, to be published, J. Electrostatics.
13. M. H. Davis, Am. J. Phys., $\underline{37}$, 26 (1969).
14. D. S. Rimai, (1986), unpublished data.
15. J. D. Cobine, "Gaseous Conductors", p. 164, Dover, New York, 1958.
16. L. Marks, (1987), personal communication.

ELECTROSTATIC CHARGE GENERATION ON WAFER SURFACES AND ITS EFFECT ON PARTICULATE DEPOSITION

Mark Blitshteyn* and Angel Martinez**

*The Simco Company, Inc., 2257 North Penn Road
Hatfield, Pa 19440
**Advanced Micro Devices 901 Thompson Place
Sunnyvale, Ca 94088

Semiconductor fabrication processes involving contact and separation of processing media with wafer surfaces can leave a wafer electrostatically charged. It was shown theoretically that the electrostatic charge may have an effect on the rate of particle deposition on exposed wafer surfaces. In this paper the results of the experimental studies of static charge generation on wafer surfaces and particle deposition on charged wafers in semiconductor manufacturing environment are presented. It was observed that in a very clean environment of the Class 10 cleanroom the effect of electrostatic potentials of up to 6000 volts applied to the wafers on the rate of > 1 μm particle deposition on the wafer surfaces could not be demonstrated, while the effect was statistically significant in a less clean environment of the Class 10,000 cleanroom.

INTRODUCTION

Semiconductor fabrication processes involving contact and separation of processing media with wafer surfaces can leave a wafer electrostatically charged. The role of these charges on the rate of particle deposition on exposed wafer surfaces in the manufacture of IC's has not been well understood. A number of theoretical articles on the role of electrostatic charges on particle transport and deposition in semiconductor manufacturing have been published.[1,2] However, the experimental data supporting any of the theoretical models suggested in these papers have been missing.

This paper presents experimental evaluation of the electrostatic charging of wafers in various process steps of photolithography and the effect of static charges on particle deposition on the wafer surface.

OUTLINE OF THE EXPERIMENTAL PROGRAM

Two photolithography cleanroom tunnels with full laminar
ceilings equipped with 0.12 μm high efficiency particulate
air (HEPA) filters and low-level side-wall air return were
chosen as the test sites. Additional tests were performed in
an aisleway of the diffusion area where 0.5 μm HEPA filters
provided coverage over workstations.

The evaluation consisted of two major parts;
1) measurements of static potentials on the wafers
generated in process steps at designated workstations in each
tunnel, and
2) particle collection experiments with unpatterned wafers
maintained at fixed electrostatic potentials and exposed to
the environment. Rates of particle deposition on wafer
surfaces were determined from surface particle counts. The
values of potentials applied to the wafers duplicated the
potentials measured in the first part of the evaluation.

The flow diagram of the experiments is shown in Fig. 1.

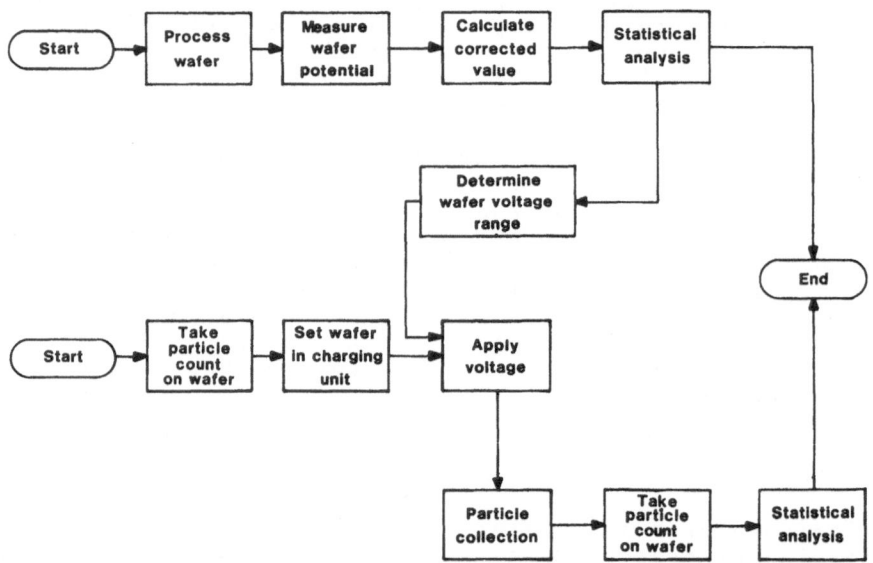

Figure 1. Chart showing main components of the test program.

ELECTROSTATIC CHARGE GENERATION ON WAFERS

Static potential measurements in Tunnel #1 were taken
after processing the 4-inch diameter wafers in the track spray
developer (manufacturer: Silicon Valley Group, Inc., San Jose,
CA) where a spray of developing solution dissolves exposed
photoresist to developed the image, oven vapor prime
(manufacturer: Lab-Line Instruments, Inc., Melrose Park, ILL)
where an adhesion promoter is deposited onto substrate prior
to photoresist application, and optical linewidth inspection
station (manufacturer: ITP, Inc., Sunnyvale, CA) where critical

dimensions on wafers are optically examined to verify
accurate image transfer; and the track spray developer
(manufacturer: Silicon Valley Group, Inc.), optical linewidth
inspection station (manufacturer: ITP Inc.), track photoresist
spinner (manufacturer: Silicon Valley Group, Inc.) where a
positive resist coating material is applied to a rotating
wafer and two rinser-dryers (manufacturer: Semitool,
Kalispell, MT) where wafers are rinsed in deionized water and
dried in heated and filtered nitrogen gas flow, in Tunnel #2.

The static potential measurements were conducted for a
period of one month. All measurements were made by Advanced
Micro Devices employees who received special training in
electrostatic measurements.

Two factors were taken into account in the development of
measurement procedures to assure accuracy of static potential
measurements. First, because of the low resistivity of silicon
(surface resistivity range between 10^6 and 10^7 ohms/square) it
is safe to consider a wafer as an isolated conductor when it
is placed on a non-conductive surface. Secondly, since a wafer
has a flat surface, surface charge distribution can be assumed
uniform.

Electrometer type electrostatic fieldmeters were found
to provide quick measurements with adequate accuracy. The
instruments of this type measure a charge induced on the
sensing electrode by the field of the charged object whose
potential is being measured.

When a fieldmeter is brought into a constant charge
system such as a charged wafer it influences the field of the
wafer. The meter then reads the value of the wafer potential
which differs from that in the absence of the meter. That
influence amounts to a change in the object's capacitance to
ground by the value of its capacitance to the fieldmeter and
can be significant. The solution is to determine surface
charge density, σ , on the wafer surface, and then recalculate

the potential, V_o , of the object without the meter in its
proximity.[3]

To determine surface charge density, σ , on the wafer,
the meter should be installed within a grounded metal plate
with an opening for the meter's sensor. The change in a
fieldmeter geometry is compensated for by a calibration
factor, K , for the specific arrangement and distance used in
the measurements. Because of low resistivity of silicon the
charge on the wafer surface is distributed uniformly and a
uniform electrical field will be provided between the meter
and the wafer (Fig. 2). An obtained fieldmeter reading, V_m ,
represents the wafer potential in the presence of the
fieldmeter. The value V_m then can be used to calculate charge
density, σ , as

$$\sigma = \frac{\varepsilon_o V_m}{d K} \qquad (1)$$

where ε_o is the permittivity of free space; d is distance.

As it was previously mentioned, capacitance of the wafer in a wafer cassette is different and, therefore, the potential of the wafer when it is in a cassette will be different from V_m.

This potential can be determined as

$$V = \frac{\varepsilon_o V_m A}{d\, C_o\, K} \tag{2}$$

where A is the surface area of the wafer, and C_o is wafer capacitance to ground in a Teflon wafer cassette placed on a grounded metal bench (4 picofarads).

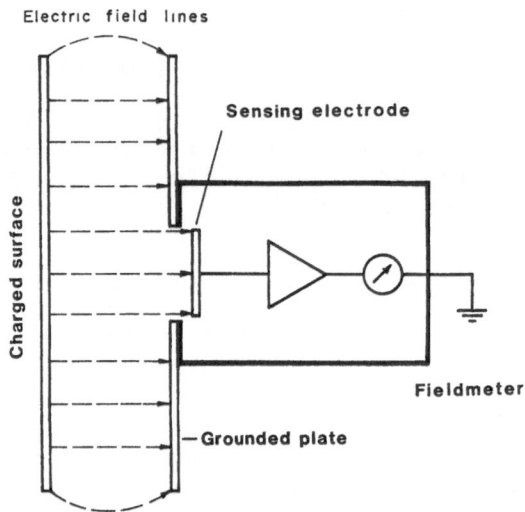

Figure 2. Method of providing a uniform electric field between the charged surface and the fieldmeter.

The operators were grounded via a conductive wrist strap while taking measurements. Immediately after completion of the process step, the operator would remove the top wafer from the cassette using an isolated vacuum pickup wand and make a measurement with a fieldmeter equipped with a 4-in. dia. metal disc. The reading was used for V_o calculations. The top wafers were chosen for measurements because, firstly, they are usually exposed to particle fallout significantly longer than any of the other wafers,[4] and secondly, the electrostatic field of the top wafer is least suppressed by the other wafers.

Statistical analysis was extensively used in processing the obtained results. The statistical method for analyzing outlying observations was used when a reading was suspect.[5] Average values of potentials and standard deviations for wafers were calculated for each station. Upper and lower limits of a 95% confidence interval of each potential mean were also determined. The average wafer potentials, V_o, along with the confidence ranges for wafers are plotted for

each station in Fig. 3. In the Figure the track spray
developer are designated as DEV, the oven vapor prime as
VAPOR, the optical linewidth inspection stations as INSP, the
track photoresist spinner as SPIN, and the rinser-dryers as
DRYER.

The potentials observed on wafers after processes
involving physical interaction of media with the wafer
(track spray developers, track photoresist spinner, rinser-
dryers) were significantly higher than the potentials observed
on wafers after processes where wafer surfaces were not
contacted (ITP inspection stations, oven vapor prime). In the
latter case the wafers were practically neutral. This
indicates that contact and separation may be a prime cause of
wafer charging.

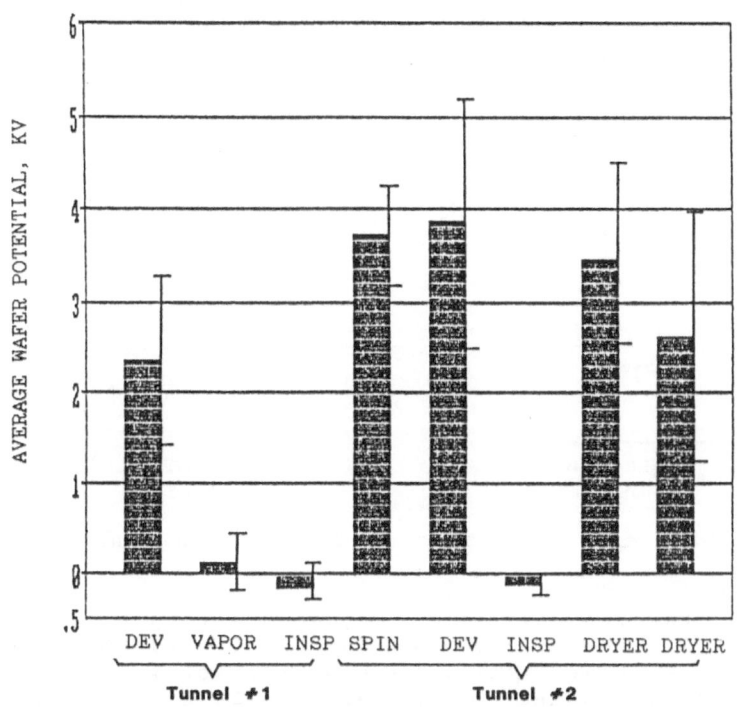

Figure 3. Average electrostatic potentials observed
on wafers after process steps.
Thin vertical lines are 95% confidence limits
about a mean value of wafer potential.

In this study, the highest observed electrostatic
potential of a wafer was +6,100 volts after processing in a
rinser-dryer. The measured potentials on the wafers were
significantly lower than previously reported values of 15-20
kV.[2,6]

Experiment design

The data collected in the wafer charging study was used to determine the range of wafer potentials for the particle collection experiment. The highest calculated upper confidence limit of all test locations was +5.2 kV for SVG Developer in Tunnel 2. Based on these results the range of wafer potentials for the particle collection experiment was chosen between 0 volts (neutral wafer) to +5 kV.

Bare silicon wafers were chosen for this experiment because of a convinience of surface particle counters for measuring particles on wafer surfaces, and, secondly, because low resistivity of the bare silicon wafers provides an excellent opportunity to control a potential on the wafer from an external power supply as is diagrammatically shown in Fig. 4.

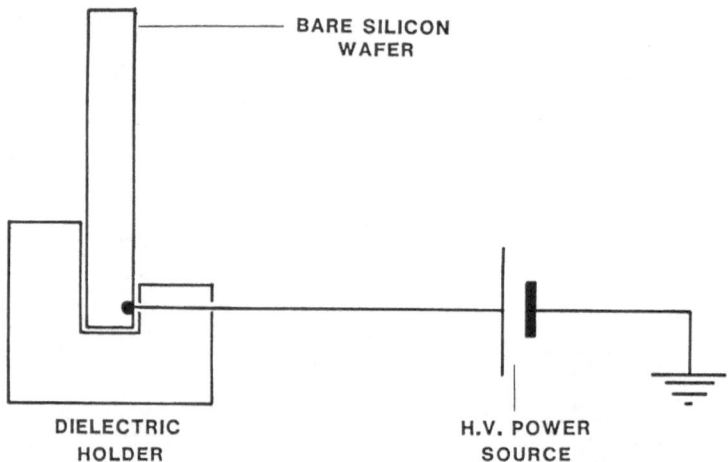

Figure 4. Schematic of a wafer charging unit.

Five charging units with adjustable output from 0 to +6,000 volts were fabricated for the experiment. Each unit had a Teflon block with a narrow slot for wafer insertion. A metal contact connected to a high voltage power supply was located on the bottom of the slot where it made contact with the wafer.

The objective was to determine whether significantly different numbers of particles will deposit on wafers maintained at different potentials. Five wafer charging units were set at one of the stations in a photolithography tunnel, as shown in Fig. 5.

In the experiment five wafers were under test at a time, and they were placed 12" to 15" apart from each other. In order to determine the effect of wafer charging and to account for the variation caused by the two restrictions of the experiment (number of wafers under test and their physical separation), a 5x5 latin square design shown in Table I was used.

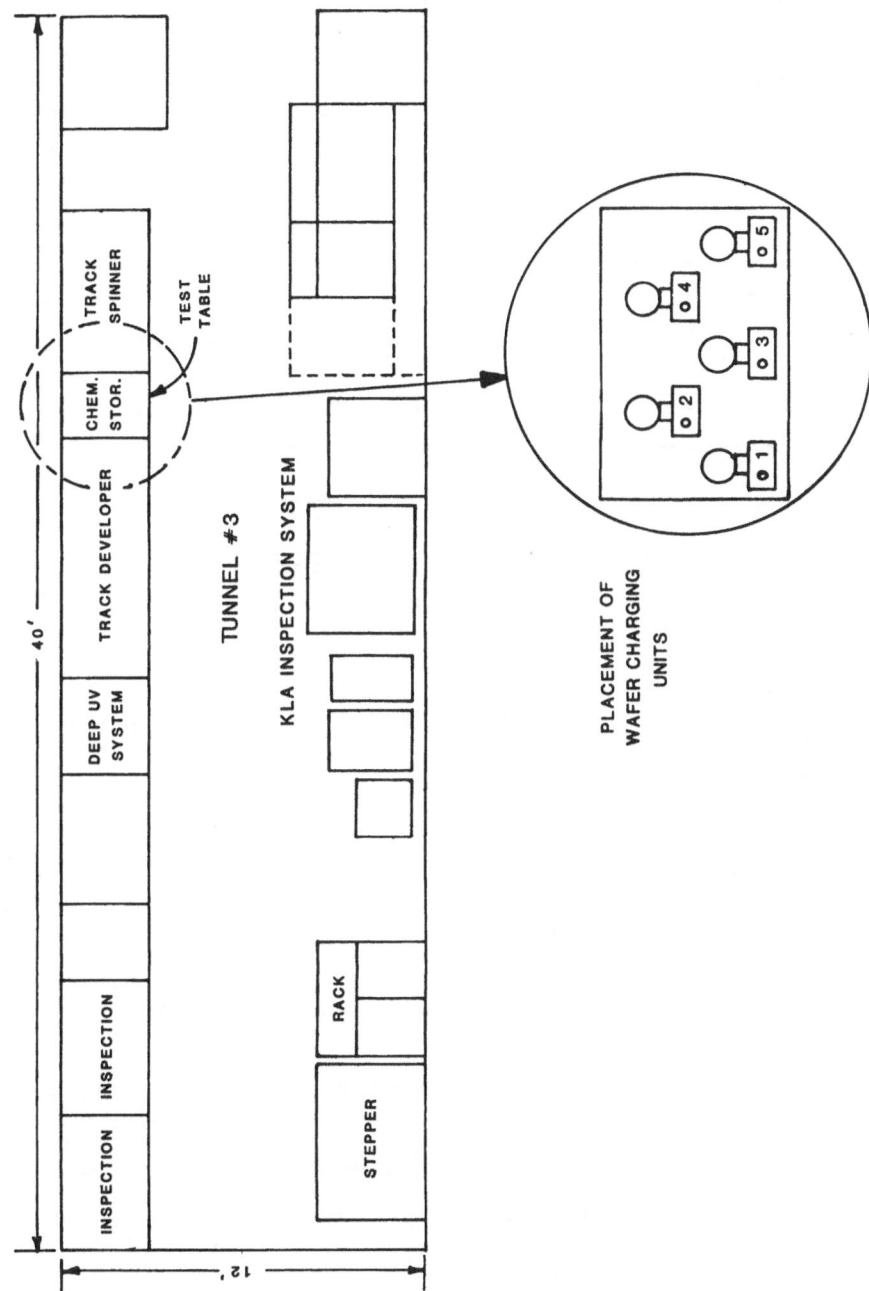

Figure 5. Placement of wafer charging units in the photolithography tunnel.

163

Wafers were subjected to five different voltage treatments V1 = 0, V2 = 1 kV, V3 = 2.5 kV, V4 = 4 kV and V5 = 5 kV. Each test (row) represented a single replication where a different voltage was applied to each wafer on the test. The voltage settings of the charging units were changed from test to test, so that over a 5-test sequence each voltage setting appeared once on every charging unit (column). This way consistent differences in particle accumulation which might exist between tests, conducted at different times, and charging units located in different positions were eliminated.

Table I. 5 x 5 latin square experiment design.

| | Wafer charging units | | | | |
	1	2	3	4	5
Test A	V1=0	V4=4	V2=1	V3=2.5	V5=5
Test B	V5=5	V1=0	V4=4	V2=1	V3=2.5
Test C	V3=2.5	V5=5	V1=0	V4=4	V2=1
Test D	V2=1	V3=2.5	V5=5	V1=0	V4=4
Test E	V4=4	V2=1	V3=2.5	V5=5	V1=0

Each data point was determined as a cumulative count of all added particles larger than 1 μm. Each of the clean unpatterned wafers was initially measured six times on a surface particle counter (Surfscan, manufacturer: Tencor Instruments, Mountain View, CA). After the initial baseline measurements, wafers were exposed to the ambient environment for 90 minutes while electrostatic potentials were applied to them. Finally, wafers were returned to the scanner and measured again six times. The difference between the averages of final and baseline measurements represented the average number of collected particles. The average number of particles contributed by the counter to the wafer was taken into account.

Airborne particle counts were taken to monitor changes in ambient conditions during particle collection tests. A laser based aerosol particle counter intake was set at the test location. The counter was programmed to produce a readout if the particle count exceeded the level of cleanliness required for a Class 10 cleanroom. Precautions were taken to avoid particle deposition during the transfer of wafers to and from the surface particle scanner.

Test results and statistical analyses

Testing yielded the results shown in Table II. It must be noted that very low numbers of particles accumulated in 90 minutes of exposure.

164

The underlying relationships or trends were tested in the
analysis of variance. Analysis of variance allows one to
judge quantitatively the contribution of each factor to the
variability of the observations. The procedure includes the
F-test of the hypothesis that the treatment means are equal.
The level of significance, which is the probability of
accepting a false hypothesis, was chosen at 5%. The F-test
amounts to a computation of the F-value and determining
whether it falls inside or outside the critical region. Based
on the outcome of that comparison the hypothesis is either
accepted or rejected.

Table II. Particles accumulated on wafers in laminar air flow.

| Test | Charging unit (position) | | | | | Row totals |
	1	2	3	4	5	
A	V1/(17.06)	V4/6.50	V2/7.84	V3/11.5	V5/5.67	48.57
B	V5/14.00	V1/3.00	V4/19.5	V2/8.0	V3/(12.21)	56.71
C	V3/ 7.87	V5/1.33	V1/8.00	V4/0.67	V2/1.37	19.24
D	V2/ 4.33	V3/0	V5/6.00	V1/6.83	V4/5.00	22.16
E	V4/12.00	V2/1.17	V3/0	V5/7.67	V1/14.50	35.34
Column totals	55.26	12.00	41.34	34.67	38.75	
Treatment	V1	V2	V3	V4	V5	
totals	49.39	22.71	31.58	43.67	34.67	
Averages	9.88	4.54	6.32	8.73	6.93	

Note: There were two missing values in the data which were
substituted in the table with the calculated values (shown
in parenthesis) using the formula suggested in Ref. 7.

The assumption is made that all samples are drawn at
random from normal populations with the same variance. The
randomness of samples, which is by far the most important
assumption, was achieved by the experiment design. Two other
assumptions were satisfied by a square root transformation
which is customary for proper analysis of counts data using
the analysis of variance. It was used in the presented
analysis in addition to the non-transformed version and
yielded the same result. The latter version is shown in the
text.

The analysis of variance for our experiment is summarized
in Table III. The sums of squares (S.S.) for total, rows,
columns and treatments were calculated. The error S.S. is
found by subtraction. Number of degrees of freedom (D.F.) for
error was reduced by two for each missing value. Mean squares
(M.S.) are determined by division of each sum of squares by
the corresponding number of degrees of freedom. F-value is
determined by division of error mean square into treatment
mean square. The critical region is where F>3.48 (the numbers
of degrees of freedom of F are 4 and 10).

Table III. Analysis of variance for a 5x5 latin square with two missing values.

Source of variation	D.F.	S.S.	M.S.	F	F(4,10) for 5%
Rows (tests)	4	211.79	59.95	2.55	3.48
Columns (charging units)	4	196.80	49.20	2.37	
Treatments (wafer voltages)	4	87.05	21.76	1.05	
Error	10	207.48	20.75		
Total	22	703.12			

Based on the results of the analysis of variance, F-value for treatments is outside the critical region, F = 1.05 <3.48, and therefore the hypothesis is not rejected. Hence, the conclusion is that the data (numbers of accumulated particles on wafers maintained at five different voltages) all came from the same population. In other words, it was not shown that wafer voltage affected accumulation of particles > 1 μm in size.

The F-tests for wafer position (charging units) and time of testing (test numbers) similarly indicate a lack of influence of these parameters on particle deposition.

SURFACE PARTICLE ACCUMULATION IN NON-LAMINAR AIR FLOW

Experiment design

In this test the charging units with wafers were set up in the non-laminar flow aisleway of the cleanroom. The area was rated Class 10,000, while individual workstations in the cleanroom equipped with vertical laminar flow hoods were rated Class 100. The cart with the test setup was located between two vertical laminar flow hoods. Four charging units were used so the experiment was designed as a 4x4 latin square. The test was slightly modified by using two low wafer voltages (V1 = 0 volts and V2 = 1 kV) and two high wafer voltages (V3 = 5 kV and V4 = 6 kV). Preliminary experiments showed that sufficient particle deposition could be obtained in less than 90 minutes. The exposure time was determined to be 15 minutes.

In a 4x4 Latin square sixteen wafers were divided into four groups, each consisting of 4 wafers for the number of wafers tested at one time. Four positions were designated for charging units.

The testing yielded the results shown in Table IV.

Table IV. Particles accumulated on wafers in non-laminar air flow.

| Test | Charging unit (position) | | | | Row totals |
	1	2	3	4	
A	V4/ 7.50	V1/ 5.25	V3/13.00	V2/ 2.25	28.00
B	V1/ 5.50	V4/12.25	V2/ 4.00	V3/ 8.25	30.00
C	V3/16.00	V2/10.75	V4/31.00	V1/17.00	74.75
D	V2/ 7.75	V3/16.25	V1/ 3.25	V4/26.75	54.00
Column totals	36.75	44.50	51.25	54.25	

Treatments	V1	V2	V3	V4
totals	31.00	24.75	53.50	77.50
averages	7.75	6.19	13.38	19.38

Noticeable increase in the rate of particle deposition was observed. In 15 minutes of exposure, on the average, nearly two times as many particles deposited on wafers compared to the 90 minute yield in the laminar flow environment. Inspection of the table also shows certain relationships which were tested in the analysis of variance. Again, the F-test was used to check the hypothesis that the treatment means are equal. The same assumptions were made. Level of significance was chosen at 5%. The critical region is where $F > 4.76$. (The numbers of degrees of freedom of F are 3 and 6).

Based on the results of the analysis of variance summarized in Table V, wafer position and time of testing (test numbers) did not have effect on particle deposition. However, F-value for voltage treatments is inside the critical region, $F = 5.04 > 4.76$. The statistical difference of the latter interaction is not overwhelming but yet indicative compared to the results of the previous test. Therefore, the hypothesis is rejected and the conclusion is that the data (numbers of accumulated particles on wafers maintained at four different voltages) did not all come from the same population. In actuality, there is a noticeable difference between means of the first and second treatment (low wafer voltages) and of the third and fourth treatments (high wafer voltages).

Table V. Analysis of variance for a 4x4 latin square.

Source of variation	D.F.	S.S.	M.S.	F	F(3,6) for 5%
Rows (tests)	3	367.17	122.39	4.29	4.76
Columns (charging units)	3	45.39	15.13	0.53	
Treatments (wafer voltages)	3	430.79	143.60	5.04	
Error	6	170.99	28.50		
Total	15	1014.34			

In order to establish if the differences were significant the estimated standard error of each mean was calculated as $\sqrt{(s^2/r)} = \sqrt{(28.5/4)} = 2.67$, where s^2 is the error M.S. and r is the latin square size. For testing the difference between a pair of means, the standard error is $\sqrt{2} \times 2.67 = 3.77$. Since the 5% t-value for 6 d.f. is 2.447, the difference between two means must be at least $(2.447)\times(3.77)$, or 9.24, in order to attain significance at this level. The differences between the mean of the last voltage treatment V4 = 6,000 volts and each of the two first voltage treatments V1 and V2 were higher than 9.24, and therefore, met the criterion.

CONCLUSIONS

The highest values of electrostatic potential were observed on wafers after processes involving physical interaction of a medium with the wafer (spray developer, photoresist spinner, rinser-dryer) while the lowest values, close to zero, were observed on wafers after processes where wafer surfaces were not contacted (ITP inspection station, oven vapor prime). This indicates that contact and separation is a prime cause of wafer charging. Polarity of charges on wafers after contact-intensive processes was always positive. The highest observed value was +6,100 volts.

An experiment in the photolithography area of the Class 10 cleanroom did not demonstrate that in laminar air flow a positive electrostatic potential of up to 6000 volts applied to the wafers had an effect on the rate of >1 μm particle deposition on the wafer surfaces. The lack of evidence of an increase in the particle attraction rate to the wafers maintained at a high potential demonstrates the effectiveness of the laminar flow in preventing particle deposition even on the electrostatically charged wafers.

An experiment in the Class 10,000 non-laminar flow aisleway of the operational cleanroom showed that electrostatic charges on the wafers had a measurable effect on the rate of > 1 μm particle deposition on the wafer surface.

REFERENCES

1. B.Y.H. Liu, presented to the 32nd Annual Meeting of the Institute of Environmental Sciences, Dallas, 1986.
2. M. Yost and A. Steinman, Microcontamination, June 1986, p.18.
3. M. Blitshteyn, Evaluation Engineering, November, 1984, p.70.
4. B. J. Tullis, Microcontamination, December 1985, p.14.
5. J.C.R. Li, "Statistical Inference", Vol.1, Edwards Brothers, Inc., Ann Arbor, Michigan, 1964.
6. B.A. Unger, R.G. Chemelli, and P.R. Bossard, presented at the 1984 EOS/ESD Symposium.
7. W.G. Cochran and G.M. Cox, "Experimental Design", Second Edition, John Wiley and Sons, New York, 1957, pp.117 - 127.

TONER ADHESION IN ELECTROPHOTOGRAPHY

M. H. Lee and A. B. Jaffe

IBM Almaden Research Center
San Jose, CA 95120-6099

The adhesion of electrophotographic toner to carrier and to photoconductor is commonly believed to be the result of electrostatic forces, van der Waals forces, or a combination of the two. We discuss the assumptions that are typically used and show that they are generally deficient. When the surface texture and geometry of the toner particles are considered, it appears that electrostatic forces should actually dominate both toner-carrier and toner-photoconductor adhesion. Data are presented to support this conclusion.

INTRODUCTION

Electrophotography is a multi-step technology which forms the basis of today's copiers and an increasing fraction of the printers. The marking technique involves the development of toner onto an imaged photoconductor (PC) surface followed by the transfer of the toner pattern to paper. The toner particles are charged so that they can be moved with electric fields supplied by coronas or voltages during each step of the process. Charged particles also bring to mind adhesion, which clearly plays an integral role in this technology.

In typical high-volume electrophotographic systems, the toner particles are charged by the triboelectric interaction with coated magnetic carrier beads in a sump. The toner-carrier mix is brought into the region of the imaged PC by one or more rollers, each with internal, stationary magnets which attract the mix. Adhesion is necessary to retain the toner on the bead until development, hold it against the PC after this step, and keep it on the paper following the transfer until it can be fused on. Adhesion is also a factor in the cleaning of the residual toner on the PC after transfer which affects the condition of the PC entering the next cycle. If the adhesion is too strong, development, transfer and cleaning may be impossible. If the adhesion is too weak, the toner can move to where it is not wanted, including background areas on the PC and vital surfaces in other parts of the system. Once there the toner may adhere sufficiently to cause even more problems.

Since electrical forces play such an important role in electrophotography, it is reasonable to assume that the adhesion is controlled by electrostatics. Historically, however, there was considerable opposition to this view. A number of studies of toner

169

particle adhesion to both carrier beads and to PC have been published.[1-9] Adhesion is generally attributed to electrostatic attraction, short-range van der Waals-type forces, or a combination of the two. More than a decade ago, Donald showed that the adhesion force increases with charge on the toner particles.[1-3] In general the force loosely follows Coulomb's Law (q^2/R^2 where q is the charge and R the particle radius), but there appears to be another force component. Nevertheless, he showed that van der Waals forces are not likely responsible.[2] In contrast, Krupp indicated that van der Waals forces could account for a significant portion of the adhesion.[4-6] Mastrangelo concluded that van der Waals forces also dominate toner-organic PC interaction.[7] Goel and Spencer suggested that electrostatic forces control toner-selenium PC adhesion for larger particles, but becomes less important relative to van der Waals forces as the particle size decreases.[8] On the other hand, Hays believed that any apparent non-electrostatic component that he observed could probably be the manifestation of a non-uniform charge distribution on the particle surface.[9]

In this paper we present the basic arguments for the two competing explanations for toner adhesion. In particular we examine the assumptions of each side and compare them with the actual conditions that exist. We show that both sets of assumptions are generally poor. When more realistic assumptions are made, the dominance of electrostatics becomes apparent. Some data are presented to back this assertion. The work complements our earlier studies which were restricted to toner-PC adhesion.[10,11]

ASSUMPTIONS IN TONER ADHESION

A typical toner particle is a dielectric material about 10 to 20 μm in size. Its adhesion characteristics can readily be calculated if certain assumptions are made. The usual first step is to assume that the toner particle is spherical since it has approximately the same dimension along each axis. The second step is to assume that a particle charges more or less uniformly on its surface by triboelectric interaction with the carrier beads with which it comes in contact. Since charge is transferred only where the toner and the carrier touch, the local density cannot be uniform. Nevertheless, the distribution should be even on a more global scale since the particle is believed to move from bead to bead randomly charging any part of its surface with equal probability.

With these two assumptions the van der Waals and electrostatic components of the adhesion force can be calculated. To a first order the van der Waals force between a sphere representing the toner particle and a plane representing the bead (or PC) is given by[5]

$$F_{vdW} = \frac{\hbar\overline{\omega}}{8\pi z_0^2} R \qquad (1)$$

where $\hbar\overline{\omega}$ is a property of the material of interest usually found experimentally to be a few eV and z_0 is the minimum separation between the particle and the surface typically assumed to be 0.35 nm. In comparison the electrostatic force between a sphere against a plane is given by[9,12]

$$F_e = \alpha \frac{q^2}{16\pi\varepsilon_0 R^2} \qquad (2)$$

where α is a constant close to 1. The appropriate values for the various parameters can be substituted into the two equations. For 10 μm particles one obtains van der Waals forces in the range of 0.40 μN, close to experimental values typically measured, while electrostatic forces is around 0.01 μN, more than an order of magnitude too low.[5,7]

The above results appears to show that electrostatic forces are not likely to be dominant.[1-3,9] But the finding can be challenged on the grounds that both assumptions on which the equations are based do not approximate the actual situation in question.[10,11] First, the toner particles cannot be considered spherical for estimating the value of a surface sensitive expression such as Equation (1). The particles are very rough locally. Regions that are expected to come close to a flat surface at any one time are only one or two μm across. Even these surfaces often have asperities. Thus the radius assumed in Equation (1) is probably far larger than the actual value. If these factors are taken into account, the van der Waals forces are likely to be only 0.3 that estimated above,[8] somewhere in the range of 0.1 μN. We show below that even this is an overestimate.

The assumption that toner particles are charged more or less uniformly is also not justified. We had found that toner particles tend to stay with the first carrier bead it touches.[10] This is based on on a series of similar experiments. For each one we placed a batch of toner followed by a batch of carrier into a jar and thoroughly mixed the contents by rolling over an extended period (30 minutes or more). The resulting carrier beads fall into two categories, those that are nearly covered with toner particles and those that are nearly bare, irrespective of the toner concentration (TC, the weight percent of toner in the mix). If toner in fact freely moves between beads, the distribution would have been much more uniform.

At TC's typically used in electrophotographic (EP) systems, an average particle sitting on one carrier bead is prevented from touching another carrier bead by particles on that bead. Since the particle probably charges only where it contacts the bead coating, its charge distribution can be highly asymmetric, assuming the toner is sufficiently insulating. Even where the TC is so low that more symmetric contact can occur, geometric constraints force the distribution to be highly localized.[10] If the toner is conductive enough to permit the charge to move, the situation is not too drastically different. The reason is that the surface to which a particle is attracted is either charged to the opposite sign (paper during transfer) or has metal underneath the surface (bead core or PC ground). The mobile charge on the particle must rearrange so that it faces the surface across the air gap.[12] Although the charge distribution on an insulative toner particle is not the same as that on a conductive one, both are likely to be very non-uniform on a global scale. This means that Equation (2) is not the proper expression to be used.

The correct calculation for the adhesion force due to electrostatics must take into account the highly localized charge distribution. Since the toner particles tend to rotate so that a highly charged surface is adjacent to the item to which it is attached,[13] a sizable portion of the toner charge is quite close to its counter charge. The adhesion force can then be approximated as in a capacitor by[10]

$$F_e = \alpha \, \frac{q^2}{2\varepsilon_0 A} \tag{3}$$

where A is the area over which the charge is distributed and α again around 1. We showed in Ref. 10 that Equation (3) in fact gives the correct magnitude for the adhesion of insulative toner using a reasonable estimate for A, typically a few percent of the particle surface area at the lowest average charge measured. Thus with the more appropriate assumptions we find that it is electrostatic forces which dominate and theoretically give the proper magnitude for the adhesion force. The van der Waals forces appear much smaller than expected from assuming spherical toner particles due to the lower total area between a particle and either a bead or the PC that can actually be in close proximity.

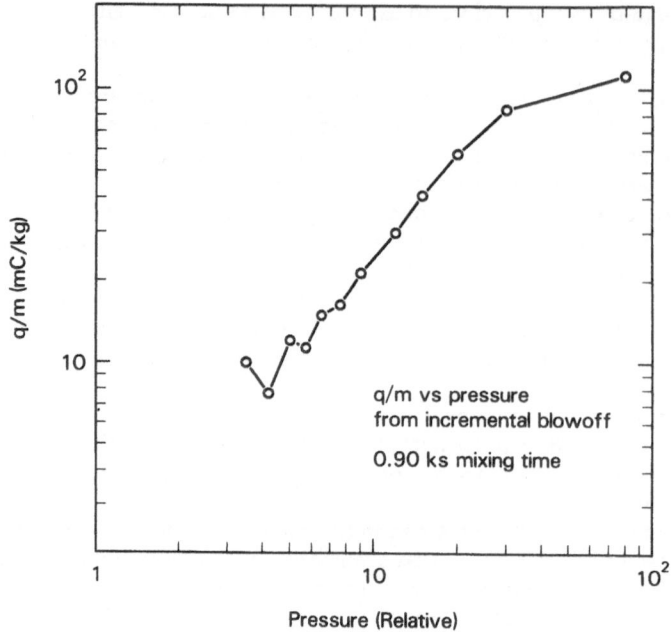

Figure 1. q/m of toner removed from cage versus air pressure for toner and carrier mixed for 0.9 ks. The increase is generally monotonic although there is some fluctuation due to weighing uncertainties, especially if the mass blown out is small.

EXPERIMENTAL RESULTS

The simplest test to see whether adhesion is due to van der Waals or electrostatic interactions is to observe its behavior as a function of the particle charge. A rapidly changing force would favor an electrostatic explanation while a lack of dependence on charge suggests van der Waals. Experimentally it means checking how the average adhesion force compares with the average charge-to-mass ratio q/m. In this work we examine both toner-carrier and toner-PC adhesion. Our toner-PC studies rely on centrifuge measurements which have been described in detail.[10,11] To determine the behavior of toner-carrier mixes, we used an incremental blowoff technique.[14] For this measurement the toner-carrier mix is placed in a metal cage with screens sufficiently fine to allow toner particles, but not beads, to pass through. The idea is to increase the air flow through the cage in discrete steps while monitoring the mass and charge on the cage.

As a first approximation we assume that the average force of the air impinging on a toner particle residing on a carrier bead is proportional to the air pressure applied. The incremental blowoff experiment starts at a low pressure so that particles removed at that point must have the lowest adhesion to the beads. As the pressure is increased in steps, particles with higher adhesion forces leave. The dependence of toner-carrier adhesion on toner charge can thus be inferred from the measurement of q/m as a function of the pressure. In general we find that the adhesion force increases with $|q/m|$. An example is shown in Figure 1 where the results are plotted for toner and carrier that had been placed individually in a jar and mixed for 0.9 ks. More pressure is clearly needed to remove the higher charged particles even though the blowoff technique is subject to some uncertainty if the mass involved in some steps in the series is low.

The increase of toner-carrier adhesion with the charge can also be discerned from a different perspective. For toner and carrier placed in a jar, the average charge

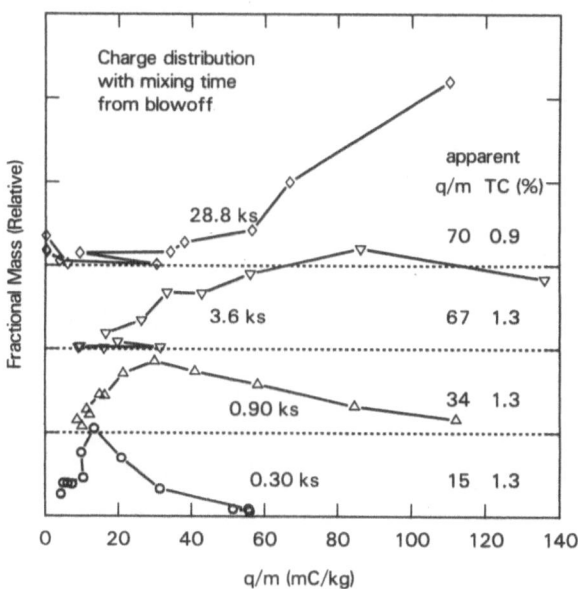

Figure 2.   Fraction of toner removed versus q/m for toner and carrier mixed over several different time periods. The various curves are staggered to minimize overlap. The dashed lines serve as the "zeros" for the respective curves. The average charge tends to increase with the mixing but this number does not give a complete picture. In general the distribution itself is more important.

typically increases with the mixing time of the materials. Figure 2 shows the fraction of toner mass removed as a function of q/m for four identical toner-carrier samples differing only in their mixing time. In order to put the data on the same sheet without extensive overlap, the curves are staggered from each other. The abscissa forms the "zero" for the 0.3 ks data. The lowest dashed line forms the "zero" for the 0.9 ks curve. The middle dashed line forms the "zero" for the 3.6 ks curve, etc. The points on each curve are connected to show the order in which the measurements were performed. In general q/m increases with the applied pressure (i.e., the order of the steps) as in Figure 1 although the points are again not completely monotonic due probably to the uncertainty of the masses involved. The columns on the right side of Figure 2 shows two measured values, the average q/m of the particles removed and the apparent TC of the mix (% of the mass removed from the starting mix mass). The average charge increases with mixing time but does not by itself give the complete picture. Even though q/m for 3.6 and 28.8 ks are quite comparable, the curves themselves look very different. The latter has a large fraction of particles removed at the highest pressure available suggesting that more would be extracted if the pressure could be increased. This is confirmed by the much lower apparent TC for 28.8 ks. In contrast the 0.3 and 0.9 ks curves give no indication of high charged particles.

Another indication of the importance of charge in toner-carrier adhesion can be found by looking at the carrier beads after blowoff. For the materials mixed at 0.3 and 0.9 ks, no toner can be seen after blowoff at the highest pressure. The same is not true at the longer mixing times. Figure 3 shows the surfaces of two carrier beads, one from the 3.6 ks mix (left) and the other from the 28.8 ks mix (right). Each of the beads is representative of toner-covered carrier beads in its respective mix although a significant fraction of the beads in both mixes have little toner coverage, as discussed in the last section. For the beads in Figure 3 the toner density is significantly higher on the bead from the 28.8 ks mix, just as expected from an electrostatic explanation of toner-carrier adhesion.

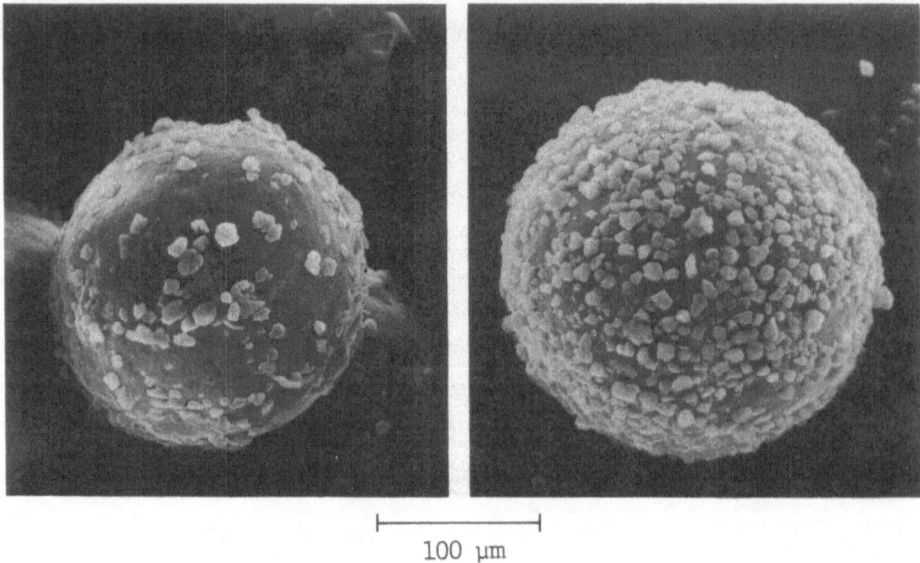

100 μm

Figure 3. Scanning electron micrographs of carrier beads after blowoff from a mix that had been rolled for 3.6 ks (left) and 28.8 ks (right). The increased mixing time caused a sizable fraction of the particles to remain on the carrier bead due to the high adhesion forces. The corresponding curve in Figure 2 also indicates a rising fraction of toner particles at high q/m leading to the conclusion that part of the toner is adhering too strongly to be removed. Some of the particles may be impacted onto the bead.

There are at least two other interpretations of our data. First, the toner stuck onto the carrier bead may have been impacted into the coating. The surface area in close contact can then be significant making van der Waals dominated adhesion more likely. We do not believe that this causes the increased adhesion because the compaction would concurrently increase the tribocharging. Nevertheless, the possible importance of van der Waals forces here cannot be ruled out since there appears to be no conclusive method of disproving either hypothesis. The second interpretation involves the act of removing the particles. The force of the air impinging on a particle increases with the size. Since the largest particles should be removed first, q/m would increase with each successive measurement if the average surface charge density is equal among particles of different size. We believe this argument is unlikely to be valid. The particle size distribution of the toner used here is not significantly different than that observed after blow off despite the more than order of magnitude range of pressures used. In addition the toner remaining maintains a large size spread as seen in Figure 3.

The dependence of toner-PC adhesion on particle charge is perhaps even more dramatic. We find that both the adhesion force and the toner charge increases with mixing time, just as with toner on carrier. Figure 4 shows a more general situation where the range of q/m is broadened by varying both the TC and the bead coating rather than mix time. The adhesion force as a function of q/m is obtained for size-classified 20 μm toner which has a relatively narrow size distribution in order to keep the spread of toner particle masses small. Both the adhesion force and q/m represent averages. The latter is obtained from a blowoff of toner developed onto PC from the mix under test.[10] The former is the force at which 50% of the particles are removed. The spread between the force necessary for 20% removal and 80% removal is typically a factor of 5.

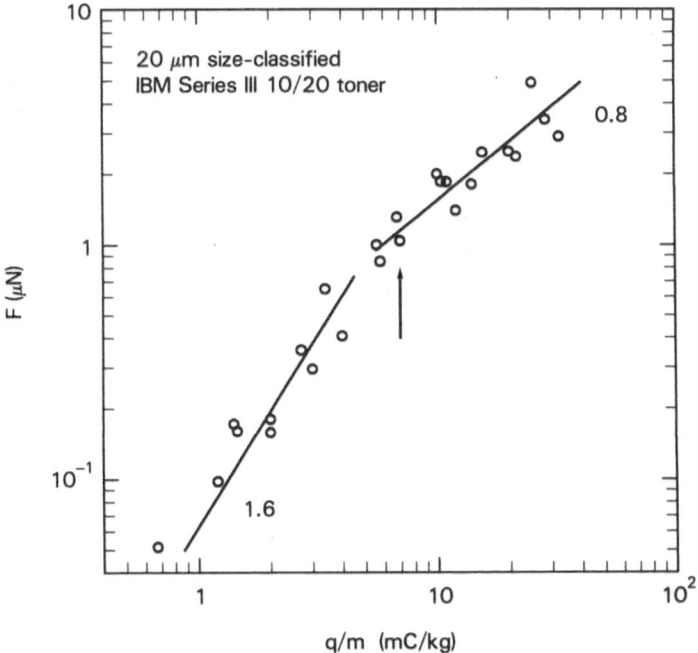

Figure 4.   Charge dependence of adhesion force for 20 μm toner. Each data point
represents the average adhesion force and average q/m of the toner deposited
onto PC from that mix. Toner charge is varied by using different carrier
coatings with varying thicknesses. The arrow points to charge corresponding
to half monolayer toner coverage on the beads. (q/m decreases as toner
coverage increases). The monotonic increase of the force with q/m suggests
that electrostatics dominate adhesion.

The adhesion force increases with about the 1.6 th power of charge at lower q/m
corresponding to high TC. Below about half-monolayer coverage on the carrier bead
prior to development (indicated by arrow), the adhesion increase with charge is closer to
the 0.8 th power. The decrease in the charge dependence is likely the result of increased
toner contact with other carrier beads at the lower TC. Since the additional charge is
restricted to the small region away from PC, its effect on adhesion is much smaller than
from the charge adjacent to the PC. If there is any effect of van der Waals forces, the
adhesion should have leveled off as the charge decreases. There is no indication of such
a trend down to the lowest average force that we measured. The same lack of leveling
was found with 10 μm size classified toner.[11] Note that at the smallest average adhesion
force indicated (0.06 μN), there is still the usual spread among the particles. Since a small
portion of the particles is removed at a force several times below the average, this makes
a possible upper bound on van der Waals forces to be around 0.02 μN, nearly an order
of magnitude less than the corrected value suggested in the previous discussion.

A schematic of the physical charge placement on a toner particle which would give
the functional dependence seen in Figure 4 is shown in Figure 5. The picture is suggested
by microscopic observations of toner sitting on carrier beads: if the mixing time between
the two is short, a sizable fraction of the toner particles is found touching the bead in a
single spot as at the top of Figure 5. This observation seems to confirm two assumptions
which together support an electrostatic model of toner adhesion. First, the particle charge

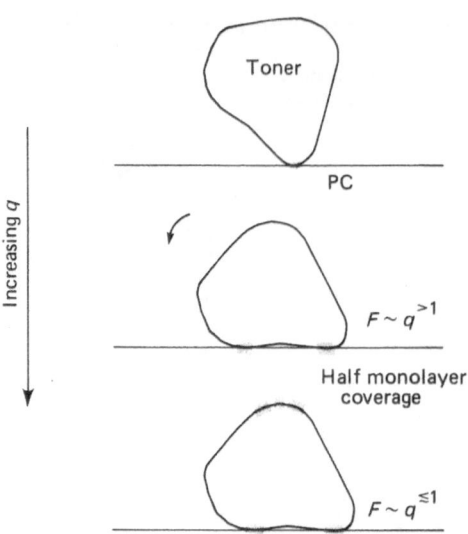

Figure 5. Location of charge on an insulative toner particle expected from geometric constraints and the restriction of toner movement between carrier beads during mixing. An average toner particle can contact other carrier beads below half-monolayer coverage. The charge distribution is nevertheless quite non-uniform. Note that m is assumed constant. Hence the dependence on q is the same as on q/m.

appears restricted to the small area where triboelectrification occurs if the toner is insulative. Second, the charge is sufficient to hold the particle in place even though the contact area is quite small. The isolated charge is also sufficient to hold the particle against the PC after development as shown in Figure 6. There the toner was deposited from two mixes, one where the TC is so high that a toner particle likely contacts the carrier bead to which it is attached in one spot (left) and one where the TC is sufficiently low that a toner particle can at least lie flat on the bead after some rough mixing (right). We believe the particle orientation on the PC reflects its relative position on the bead prior to development.

CONCLUSIONS

In this work we considered the adhesion of EP toner to carrier and to PC. We examined the usual assumptions on which the explanations for the adhesion forces are based and found them to be deficient. From a more reasonable set of assumptions we showed that electrostatic forces should dominate. The adhesion between toner and carrier is then explored using an incremental blowoff technique with the data generally confirming the importance of toner charge. If the q/m is sufficiently large, the toner cannot be removed from the bead under any condition that we can achieve. We believe the high charge is largely responsible even though toner impaction into the bead may increase van der Waals forces. Toner-PC adhesion is also discussed with respect to the centrifuge studies that we performed. The general behavior is quite similar to that between the toner and carrier. The adhesion force is also found to rapidly increase with the toner charge. Thus the data generally confirm our rough calculations which indicate that toner adhesion in EP is principally an electrostatic phenomenon.

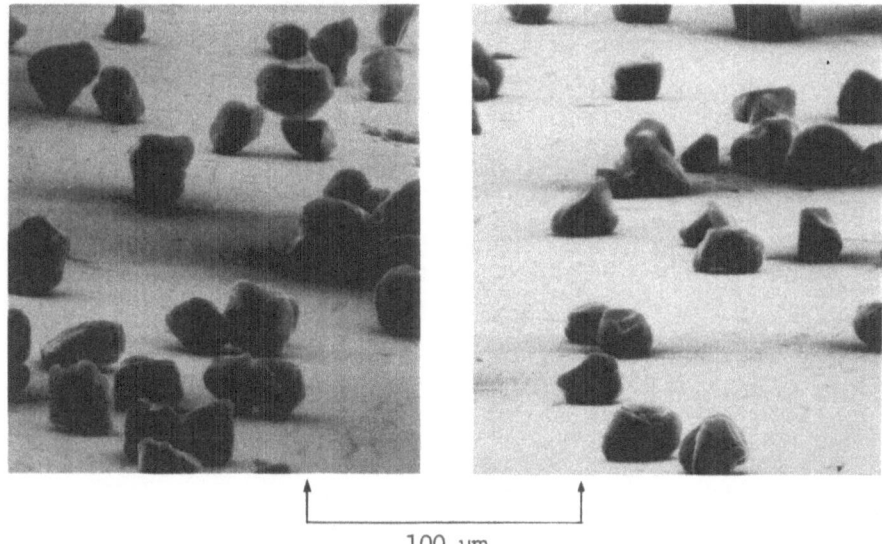

100 μm

Figure 6.   Scanning electron micrographs of 20 μm toner on PC deposited from a mix at 8% toner concentration (TC) (left) and 0.5% TC (right).   At high TC, the charge on a toner particle may be quite localized due to close packing of the particles on the carrier bead during the triboelectrification process.   Hence the toner particles stand in seemingly unstable positions in a sizable fraction of the cases.

ACKNOWLEDGMENTS

The authors thank L. B. Schein and D. M. Burland for their support.   They appreciate the SEM work performed by J. Duran.   They acknowledge R. T. Kerth, J. Ayala, F. Dodge, V. Ting, S. G. Steele, and G. P. Marshall for their various contributions to this work.

REFERENCES

1.   D. K. Donald, J. Appl. Phys. **40**, 3013 (1969).
2.   D. K. Donald and P. K. Watson, Photogr. Sci. Eng. **14**, 36 (1970).
3.   D. K. Donald and P. K. Watson, IEEE Trans. Electron Dev. **ED-19**, 458 (1972).
4.   H. Krupp and G. Perling, J. Appl. Phys. **37**, 4176 (1966).
5.   H. Krupp, Adv. Colloid Interface Sci. **1**, 111 (1967).
6.   W. Kottler, H. Krupp, and H. Rabenhorst, Z. Angew. Phys. **24**, 219 (1968).
7.   C. J. Mastrangelo, Photogr. Sci. Eng. **26**, 194 (1982).
8.   N. S. Goel and P. R. Spencer, in **Polymer Science and Technology**, Vol. 9B, p. 763, Plenum Press, New York, 1975.
9.   D. A. Hays, Photogr. Sci. Eng. **22**, 232 (1978).
10.   M. H. Lee and J. Ayala, J. Imag. Tech. **11**, 279 (1985).
11.   M. H. Lee, Proc. SID **27**, 9 (1986).
12.   M. H. Davis, Am. J. Phys. **37**, 26 (1969).
13.   D. K. Donald, J. Adhesion **4**, 233 (1972).
14.   L. B. Schein and J. Cranch, J. Appl. Phys. **46**, 5140 (1975).

ADHESION AND REMOVAL OF PARTICLES: EFFECT OF MEDIUM

M. B. (Arun) Ranade, V. B. Menon, M. E. Mullins, and V. L.
Debler

Research Triangle Institute
P. O. Box 12194
Research Triangle Park, NC 27709

Intermolecular forces of attraction between particles and
surfaces are generally lower in a liquid medium compared to
gases, and vacuum is lower by 1 to 2 orders of magnitude. Using liquids facilitates cleaning of particles from surfaces.
Although cleaning technology using liquids is available, no
systematic measurement of adhesion forces and cleaning effectiveness is reported in the literature. This paper presents
results of adhesion measurement using a new technique developed
for rapid and accurate data acquisition for liquid and gaseous
media.

INTRODUCTION

Forces between solids are predominantly attractive and cause
particles to adhere to each other and to surfaces. These forces become
increasingly significant for fine particles because the particle mass
varies to the third power of the particle size. A review of the literature in adhesion and removal of fine particles on surfaces is provided by
Ranade.[1]

Contamination of electronic components by submicrometer particles is
becoming a growing concern as the industry strives to reduce size and to
increase capacities. Particles are deposited on these surfaces by the
combined action of diffusion, sedimentation, inertia, and other factors
such as electrical charges and local electrostatic fields. The
probability of adhesion or the "sticking coefficient" subsequent to
deposition is the result of the adhesion forces and forces acting for
their removal. The following discussion assumes that particles are already in contact with the surface.

Silicon wafers themselves, and other environmental surfaces (such as
processing components, fluid filters, enclosures, and pipes) may also
come in contact with the particles and may be sources of contaminants if
these particles are released in processing operations. Adhesion and
reentrainment of particles from these surfaces must also be included in
the overall microcontamination control problem.

Particles in the microelectronics processing environment are deposited on surfaces from air and liquids, and from humans (e.g., skin flakes and hair). They range from about 0.1 μm to several micrometers in size and may be round or irregular grains, platelike, or cylindrical in shape. Composition of these particles ranges from metal oxides to soil and from sand to complex organics and polymers.

The environment, whether air or liquid, plays a role in the deposition and may influence the nature of the particle/surface contact zone.

THEORY

The principal interactions that are encountered in particle adhesion include molecular interactions, electrostatic interactions, liquid bridges, double-layer repulsion, and chemical bonds such as polar and metallic bonds. Of these, molecular interactions and double-layer forces are more dominant during the liquid cleaning process.

Molecular Interactions

The theories for molecular interactions are based on the van der Waals dispersion interactions. Atoms in the bodies are instantaneous dipoles, and the dispersion interaction between these dipoles and the induced dipoles in neighboring atoms is summed over all atoms. It is conveniently represented by a Hamaker constant, A.

The relations between the Hamaker constants of two dissimilar materials may be represented by:

$$A_{12} = \sqrt{A_{11} \cdot A_{22}} \quad ,$$

where A_{11} and A_{22} are the Hamaker constants for substances '1' and '2'. In the presence of a medium denoted by '3', the net interaction between substances 1 and 2 is given by:

$$A_{132} = A_{12} + A_{33} - A_{13} - A_{23}.$$

Visser[2] points out that it is possible to choose a medium '3' such that the net value of A_{132} is negative, resulting in repulsion! Examples of such combinations of the materials and media are rather rare and ambiguous.

Double-Layer Repulsion Forces

When particles are immersed in an electrolyte solution, any charge on the particles will attract a layer of ions of the opposite sign. These ions will, in turn, attract ions of opposite polarity in a diffuse layer. The suspension as a whole will be electrically neutral. This electrical phenomenon is characterized by zeta potential, ζ, which is measured by electrophoretic-mobility measurements and is the potential at the fluid shear plane.

MEASUREMENT OF ADHESION

Several methods that have been used to measure particle adhesion were devised for particles larger than 1 μm. Zimon[3] and Corn[4] have reviewed these extensively.

The following experimental methods have been used to measure the adhesion force of single particles on surfaces:

(a) slope variation of a surface

(b) microbalance technique

(c) pendulum method

(d) centrifuge method

(e) aerodynamic and hydrodynamic method

(f) vibration method.

Of these, only the latter three--d, e, and f--are useful for micrometer-sized particles.

Very limited data on the adhesion forces between micrometer-sized particles are available due to experimental difficulties. Because of the large surface area in relation to the mass of the particles, the removal force cannot be applied without pushing experimental techniques to the limit. In addition, examination of the fine particles on the substrates is difficult without electron microscopy.

Visser[5] used a hydrodynamic method to measure the force of adhesion between submicrometer carbon black (0.2 μm diameter) and cellophane substrates in rotating concentric cylinders. The hydrodynamic force, F_{HYD}, is given by:

$$F_{HYD} = 0.115 \ N^{3/2} \ r^2 \ \text{dynes} \quad ,$$

where N = revolutions per min (rpm), and

   r = particle radius (cm).

Speeds up to 8,000 rpm were achieved.

Deryagin and Zimon[6] used a vibrational method to produce accelerations of the order of 10^4 g. With the availability of frequencies in the megahertz range, accelerations of the order of 10^6 g are possible. Mullins and Ranade[7] used an ultrasonic horn to study adhesion of micrometer-sized metal flakes.

With the availability of high-frequency transducers, the vibrational technique holds promise for adhesion measurement of submicrometer sizes. However, violent cavitation in liquid media at high frequencies may cause problems due to material erosion. Use of high pressures may extend the range of applications as indicated by Davies et al.[8]

Particle removal from surfaces using a liquid rather than air or vacuum as the surrounding medium facilitates particle removal. Two features involving liquid media may be exploited. First, the van der Waals interactions are two to four times smaller. In addition, surface-active agents may be employed to take advantage of the double-layer repulsion forces.[9]

Kuo and Matijevic[10] and Kallay and Matijevic[11] studied removal of 0.17 μm hematite particles from stainless steel and showed that the ionic

strength and pH of the aqueous solution of sodium dodecyl sulfate and
ethylenediaminetetraacetic acid were important variables. The removal
was highest at a pH of around 11.5. Increasing temperature of the
solution also increased removal over the 25 °C to 80 °C range studied.

Brandreth and Johnson[12] describe particle removal from surfaces by
several common solvents such as alcohol mixtures with fluorocarbons. The
action was not by dissolution by solvents but by reduced molecular interaction as well as adsorption on the particle and substrate, changing the
original molecular interactions. Several types of Freons were shown to
be very effective in particle removal at relatively low accelerations.

The processes available for particle removal incorporate the same
mechanical actions indicated for adhesion-measurement techniques. The
acceleration may be provided by agitation, centrifugation, vibration, or
aerodynamic or hydrodynamic drag.

Zimon[3] discussed the role of drag on particle removal by air as well
as water flow. The force that is required for detachment of small
particles is expressed by:

$$F \geq \mu \, F_{ad},$$

where μ is the friction coefficient. The drag force, F, is given by:

$$F = C_d \, S \, \frac{\nu^2}{2} \, (\rho_p - \rho_f) \quad ,$$

where C_d is the drag coefficient, ν is fluid velocity, and ρ_p and ρ_f are
densities of particle and fluid, respectively. For a linear distribution
of velocity in the boundary layer on the substrate, "Stoke's Law" gives:

$$F = \frac{3\pi \, \eta \, \nu_o d_p^2}{2\delta}$$

where δ is the boundary-layer thickness, and ν_o is the average fluid
velocity. The boundary-layer thickness is usually sufficiently larger
than the particles.

Air or nitrogen blow-off guns are usually effective in removing
large particles (>10 μm) from the surface but ineffective in removing
smaller particles. Liquid jets are also employed for cleaning surfaces.
Stowers[13] applied a high-pressure 6.9-MPa liquid spraying technique using
a Freon-TF® solvent to remove Al_2O_3 particles larger than 5 μm in size
from glass and a metallic surface. Comparison was also made with other
removal methods such as compressed gas jets and ultrasonic cleaning also
using Freon-TF® as the medium. Close to 99 percent removal by the high-
pressure spray was reported, compared to about 60 percent for compressed
gas and about 90 percent for ultrasonic cleaning. The removal values
were based on the initial number of particles on the surface, which was
degreased and irradiated. The degreasing and irradiation removed about
33 percent of the particles before the ultrasonic or the high-pressure
jet treatments. The true removal efficiency numbers may be significantly
lower than reported. The paper also contains removal efficiency values
for ≳1 μm-sized particles. However, no discussion of how these values
were obtained is given in the paper. As expected, larger particles are
removed better, indicating that the technique may be effective for larger

particles only without the use of elevated temperature or use of surfactants.

Ultrasonic devices operating at around 20 kHz have been used for effective cleaning of surfaces in liquids. The mechanism of the action of ultrasound on a particle or agglomerate is not fully understood. In all types of particle processing systems, it has been reported that some form of cavitation is desirable to attain dispersion by ultrasonic means. Particle dispersion is thought to occur by the action of a collapsing cavitation bubble at the agglomerate interface.

Brodov et al.[14] reported that during precipitation of a particulate solid, the particle size decreased from >1 μm without ultrasound to <0.3 μm in an ultrasonic field under atmospheric pressure. At a pressure of 24 atm, the particle size further decreased to 0.08-0.1 μm. Holl[15] also noted that the dispersion of submicrometer particles was promoted at high pressures.

Kaiser[16] found that ultrasonic techniques were useful for separating submicrometer carbon, calcium fluoride, and other compounds and found that separation in fluorinated hydrocarbons with low surface tension was very efficient.

Reports of other uses of ultrasound for dispersion were given by Agabalyants et al.,[17] and by Lowe and Parasher,[18] who studied the dispersion of clay and soils, respectively.

Various types of ultrasonic devices were designed for the separation of submicrometer particles. Hislop[19] reported on two devices that used the ultrasonic energy most efficiently.

Davies et al.[8] designed and constructed a batch device for separating inorganic metal oxides and atmospheric aerosols from filters. This device could operate at various power, pressure, temperature, and time conditions. Operation in aqueous sodium pyrophosphate solutions was limited to low power and pressure due to erosion difficulties, but operation in fluorocarbons (e.g., 1 percent Krytox® 157 in Freon® E-3 at 100 psi and 100 W) was most effective in separating submicrometer particulates under conditions of negligible erosion.

Shwartzman et al.[20] described a "megasonic" cleaning system operating at frequencies from 850 to 900 kHz. It was claimed that in contrast to ultrasonics, the higher frequencies produce a cleaning action greater than the cavitation encountered in ultrasonic devices. A high-pressure wave is suggested as the mechanism for particle removal. The device was shown to be effective in removing 0.3 μm particles using a hydrogen peroxide solution. Water alone did not work as well. Although a surfactant solution (Triton-X®) worked better, it did not work as well as the peroxide solution did.

The qualitative hypothesis for the cleaning action does not explain the results reported by Shwartzman et al.[20] using different cleaning fluids. Further work is necessary to determine the best conditions for particle removal.

A proper choice of the cleaning system, the cleaning fluid, and the potential use of surfactants is not available at this stage, and the need for a systematic study of a cleaning mechanism is urgently needed.

EXPERIMENTAL

Recently Mullins and Ranade[7] have developed a vibrational technique for measuring particle adhesion in a gaseous atmosphere using an ultrasonic probe. Real-time observation of particles leaving the substrate leads to rapid data acquisition, and increasing acceleration in steps is possible without interrupting the experiment as with a centrifuge.

Adhesion Measurement in Different Media

For comparison, particle removal was studied for some particle/substrate combinations in air and in liquids. The apparatus described by Mullins and Ranade[7] and shown schematically in Figure 1 was used for experiments in air as a medium. Particles were deposited uniformly on the test substrate using a dispersion chamber also shown in Figure 1. A Climet 208 particle size analyzer was used as a real-time monitor. Particles deposited on the substrate were counted also with a microscope before ultrasonic treatment and after completion of the test to check for closure of the member balance.

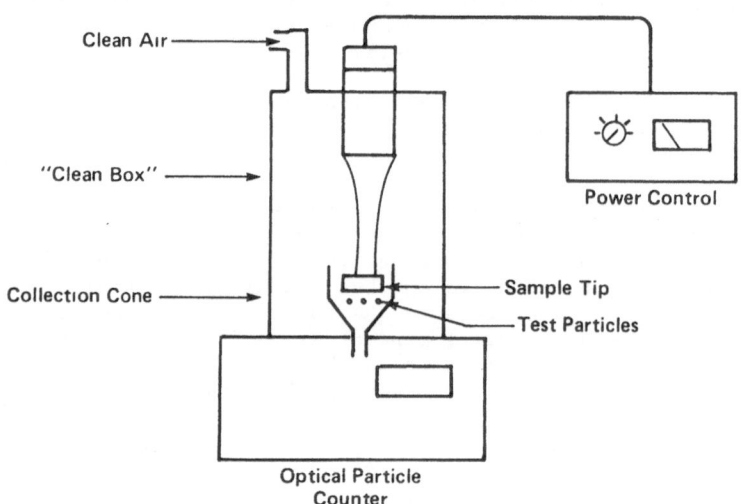

Figure 1. Schematic of ultrasonic adhesion test equipment.

For tests in liquid, a loudspeaker was used for low-frequency vibrations instead of the ultrasonic probe. Particle removal at each acceleration level was determined by counting particles through an optical microscope. In some of the tests at higher frequencies, the ultrasonic horn was used in a similar fashion.

Results for 2.5, 8, and 11.5 μm glass beads on glass substrates are shown in Figures 2, 3, and 4.

Particle Removal in Presence of a Second Parallel Substrate

Surface cleaning was also studied qualitatively in the presence of graphite. The principle involved in the experiments is as follows:

184

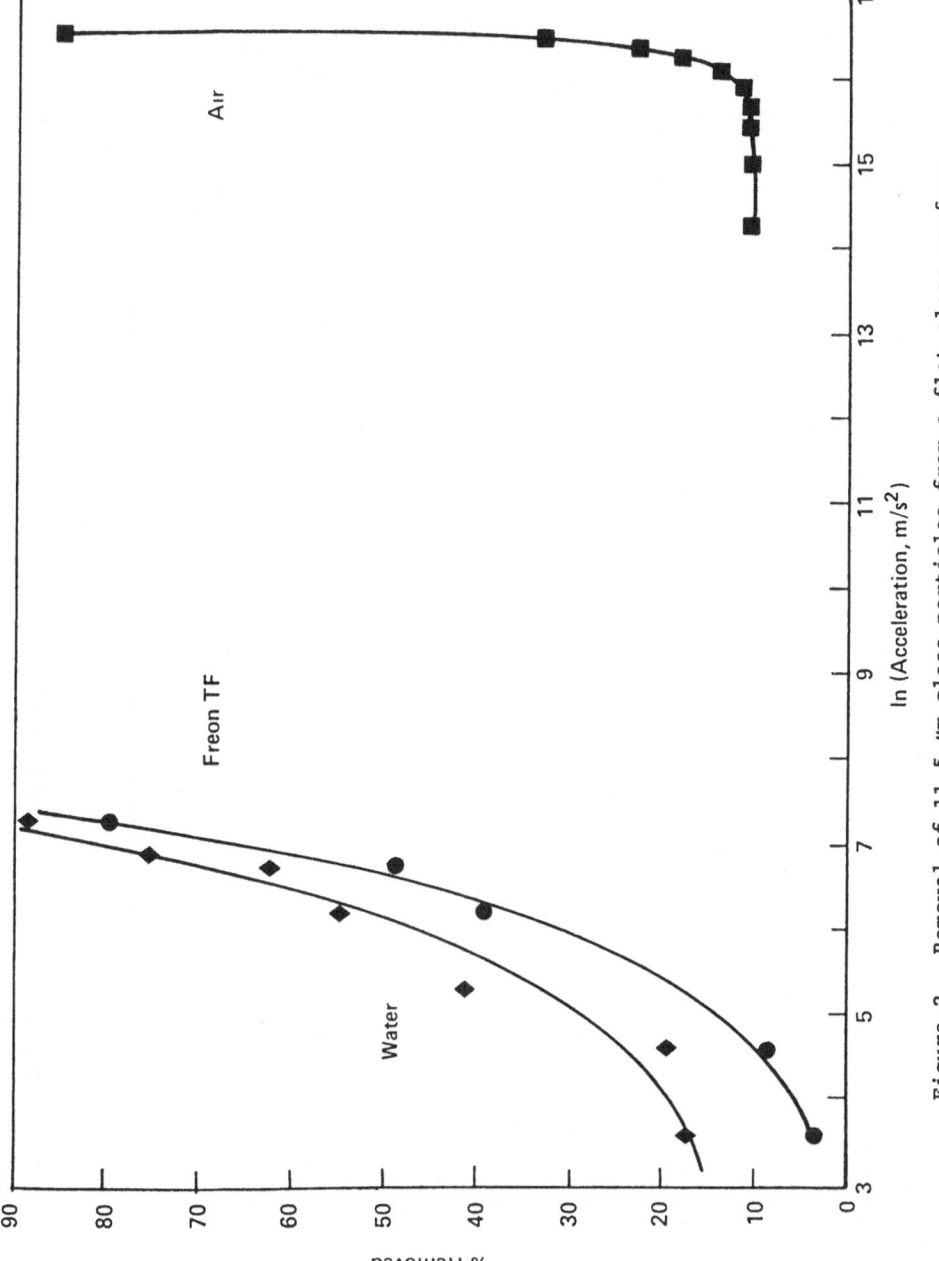

Figure 2. Removal of 11.5 μm glass particles from a flat glass surface by ultrasonic vibration.

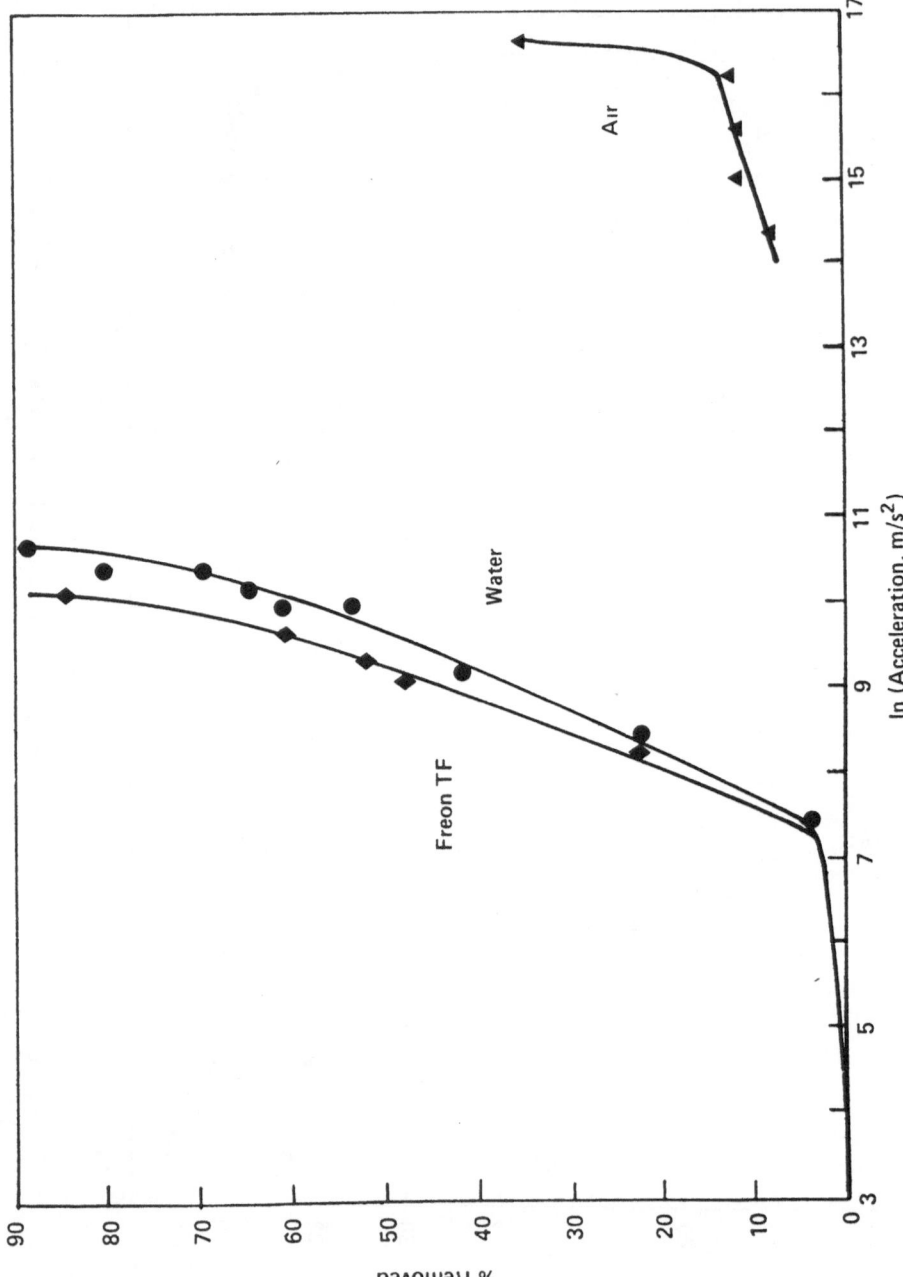

Figure 3. Particle removal in different media: 2.5 μm glass beads on glass substrate.

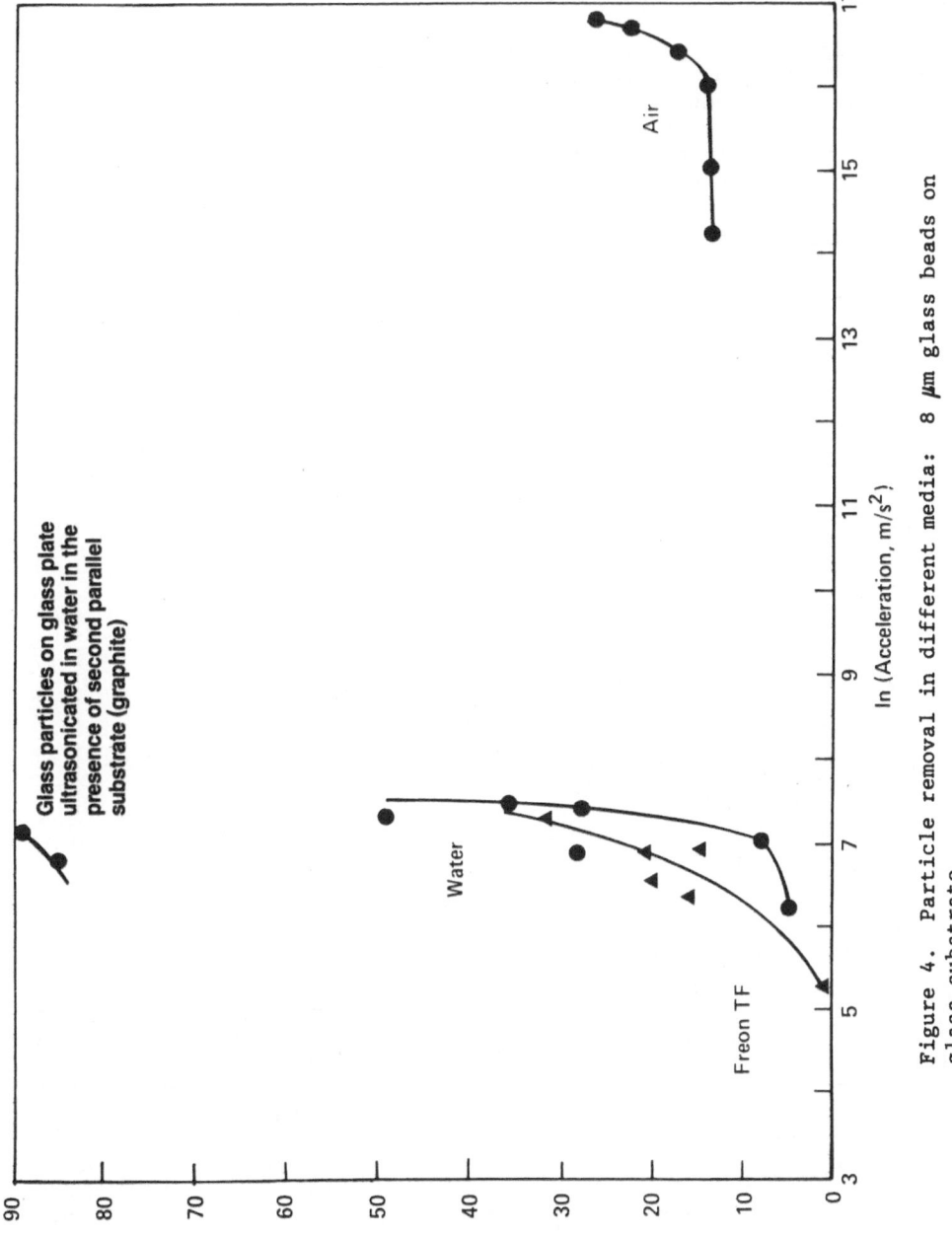

Figure 4. Particle removal in different media: 8 μm glass beads on glass substrate.

187

The van der Waals attractive potential for PSL on graphite is larger than the attractive potential for PSL on glass, when <u>water</u> is the continuous medium. So, if a graphite surface is brought very close to a glass plate covered with PSL and immersed in water, the PSL should transfer to the graphite surface. With Freon® as the medium this effect is not anticipated because the van der Waal's potentials are not favorable. The same applies for glass beads on a glass plate.

The experimental procedures involved comparing cleaning efficiencies by optical microscopy for two glass slides--one in the presence of a graphite surface about 30 to 50 μm from the glass slide surface, and the other in the presence of a graphite surface. These experimental arrangements are shown schematically in Figure 5. Two media were studied: water and Freon®. Results are summarized in Table I.

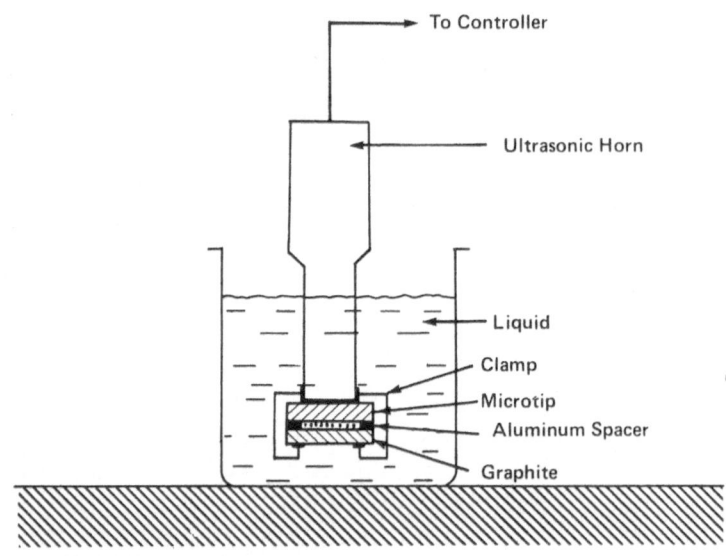

Figure 5. Experimental setup for the particle removal in presence of a second parallel substrate.

Table I. Particle Removal from a Glass Substrate in Presence of a Second Parallel Substrate

Medium	Particle Size (μm)	Particle Material	Second Substrate	Percent Removal
Water	8	glass	none	33.1
Water	8	glass	graphite	65
Water	4.8	PSL	none	9.4
Water	4.8	PSL	graphite	20.3
Freon-TF®	8	glass	none	44
(50% Freon-TF® Methylene Chloride)	8	glass	graphite	65

DISCUSSION OF RESULTS

Figures 2, 3, and 4 show that the acceleration required to remove particles was significantly greater in air as compared to the liquid media.

The value of the Hamaker constant for the glass was taken as 8.5×10^{-13} (from Visser[21]). The Hamaker constant for the glass/water/glass system was estimated as 6.76×10^{-14} using relationships between the values for individual components discussed earlier in this paper. Using these values the theoretical force of adhesion was calculated in air and water. The Hamaker constant for Freon-TF® was not available but estimates are in the same order of magnitude as water. A comparison between the theoretical and measured values is presented in Table II. Although the actual agreement is not very good, the trend is clearly indicated. The discrepancy in values for water may be caused by an actual separation other than the assumed 0.4 nm. The distance of separation may vary anywhere from 0.4 to 1 nm. If a 1.0 nm separation distance is used, the calculated force would be reduced by a factor of 0.16, i.e., the value of the force for 2.5 μm particles is 1.4×10^{-4} dyn.

Table II. Comparison Between Measured and Calculated Values of Force of Adhesion

Particle Size (μm)	Measured Force* (dyn)	Calculated Force** (dyn)
In Air		
2.5	2.7×10^{-2}	1.1×10^{-2}
8	1.2	3.7×10^{-2}
11.5	2.7	5.3×10^{-2}
In Water		
2.5	1.326×10^{-5}	8.8×10^{-4}
8	6.24×10^{-5}	2.8×10^{-3}
11.5	6.8×10^{-5}	4.0×10^{-3}

*Estimates from experimental data at 50 percent removal of particles.
**Assumed distance of separation of 4 Å.

The measured force varies linearly with particle size for water. In air the measured force is considerably larger than calculated. This may be caused by electrostatic charging of the glass beads.

The preliminary data in Table I show that cleaning is enhanced by the presence of a graphite surface parallel to the test substrate at close proximity. The data are at present too limited to draw conclusions, but further investigation is planned to systematically verify the observations so they could be at least qualitatively explained.

SUMMARY

Liquid media allow better cleaning of particles from surfaces. The forces of attraction are reduced and surface active agents may also be employed. Data for cleaning effectiveness are sparse and techniques for determining cleaning effectiveness for micrometer- and submicrometer-sized particles need to be developed. The authors plan to extend this work for submicrometer-sized particles on surfaces relevant to the electronics and semiconductor industries using well-characterized monodisperse particles.

REFERENCES

1. M. B. Ranade, Aerosol Sci. Technol., 7, 161-176 (1987).

2. J. Visser, Adv. Colloid Interface Sci., 15, 157-169 (1981).

3. A. D. Zimon, "Adhesion of Dust and Powder," Plenum Press, New York, 1969; also A. D. Zimon, "Adhesion of Dust and Powder," Consultants Bureau, New York, 2nd edition, 1980.

4. M. Corn, Chapter 11, in "Aerosol Science," C. N. Davies, Editor, Academic Press, London, 1966.

5. J. Visser, J. Colloid Interface Sci., 34, 26 (1970).

6. B. V. Deryagin and A. D. Zimon, Kolloidny Zhurnal, 23, 5, 544 (1961).

7. M. E. Mullins and M. B. Ranade, "Effects of Particle Geometry on Adhesion: Spherical and Platelike Particles," paper presented at the 16th Annual Meeting of the Fine Particle Society, Miami, Florida, April 1985.

8. R. Davies, M. B. Ranade, J. Puretz, and D. C. Freshwater, "Factors Affecting the Ultrasonic Dispersion of Submicron Particles for Size Analysis," Proceedings of Particle Technology, Nurnberg, 1977.

9. E. J. Clayfield and E. C. Lumb, J. Colloid Interface Sci., 36, 286 (1970).

10. R. J. Kuo and E. Matijevic, J. Colloid Interface Sci., 78, 407-421 (1980).

11. N. Kallay and E. Matijevic, J. Colloid Interface Sci., 83, 289-300 (1981).

12. D. A. Brandreth and R. E. Johnson, in "Surface Contamination: Genesis, Detection, and Control," K. L. Mittal, Editor, Vol. 1, p. 83, Plenum Press, New York, 1979.

13. I. F. Stowers, J. Vac. Sci. Technol., 15(2), 751 (1978).

14. Brodov et al., in "Poluch Svoistra Primen Tonkikh. Metal. Porosch Dokl., Vses Conf.," p. 113, Chemical Abstract 76, 116809 (1972).

15. P. Holl, in "Proc. Int. Symp. Particle Technol," IITRI, Chicago, August 1973.

16. R. Kaiser, "Particle Dispersion Studies," Final Report on USAF Contract F33657-72-C-0388, August 1973.

17. E. G. Agabalyants et al., Dopov, Akad. Nauk. Ukr. RSP Ser. B, 32 (9), 813 (1970).

18. L. E. Lowe and C. G. Parasher, Can. J. Soil Sci., 51 (1), 136 (1971).

19. T. Hislop, Ultrasonics, $\underline{8}$, 88 (1970).

20. S. Shwartzman, A. Mayer, and W. Kern, RCA Review, $\underline{46}$, 81 (1985).

21. J. Visser, in "Surface and Colloid Science," E. Matijevic, Editor, Vol. 8, pp. 3-84, Plenum Press, New York, 1976.

STRONG ADHESION OF DUST PARTICLES

Richard   Williams and Richard W. Nosker

David Sarnoff Research Center
Princeton, New Jersey   08543-5300

Dust particles lead to wear and damage in fine mechanical systems and to yield losses in the manufacture of electronic components. We have found experimentally that most of the damage is due to a relatively small fraction of the dust particles present.   Many cases involve composite particles that become much more strongly adherent after cyclic changes in relative humidity and deliquescence.   Water condenses around the water-soluble   part of a composite particle. The water-soluble particle then dissolves. The solution wets both the substrate and the undissolved part of the composite. Later, when the relative humidity decreases, the soluble material recrystallizes, forming a strong bond between   particle and substrate. A bond of this kind is much stronger than the original because the interfacial contact area has greatly increased. In many cases a particle cemented to the surface in this way cannot be removed without damage to the substrate. The right combination of soluble and insoluble components is found in only a small fraction of all dust particles. These cause most of the damage. We analyze   the physical chemistry of the   strong adhesion process and show the conditions   under which   it can take place.

INTRODUCTION

Adherent dust particles often cause wear and damage in fine   mechanical systems   Our own experience with the problem began with the RCA Video Disc System. The   video disc has the overall size of a 12" standard   33-1/3   audio record, but the grooves are much finer.   The video signal is present as a relief pattern in a spiral groove that has a pitch of   2.5 µm. The signal is read out with a capacitive probe on the tip of a fine diamond stylus. The stylus tip rides along in the groove and makes contact to the surface. It is very sensitive to the mechanical disturbance caused by an encounter with a small dust particle on the surface.[2]   This property gave us a unique tool for studying the adhesion of small particles to the surface.   We could apply known quantities of standard dusts, vary external conditions, such as temperature and humidity, then scan over the surface with the stylus. In a   period of an hour, the stylus scans over the surface, systematically covering every part of an area of about 500 cm^2 with a resolution   of 2.5 µm. A characteristic disturbance in the video signal indicates whenever there has been an encounter with an adherent dust particle in the groove.   These events could be counted and compared with the total number of dust particles present on the surface. Strongly adherent dust particles could be distinguished from ordinary particles. By counting the number of strongly adherent particles before and after a change in the ambient conditions, we got information on the mechanism of strong adhesion.[2,3,4] We will review the general results and give some additional information.

EXPERIMENTAL METHODS AND RESULTS

We used dust samples from several sources, including the industry standard, *"Arizona Road Dust"*, and material collected in various parts of the country. We got similar results with all the dust samples used. Using a suitably dimensioned box with a timed shutter, (Figure 1), we sedimented known amounts of dust onto test surfaces in such a way that we knew the total particle count and size distribution.

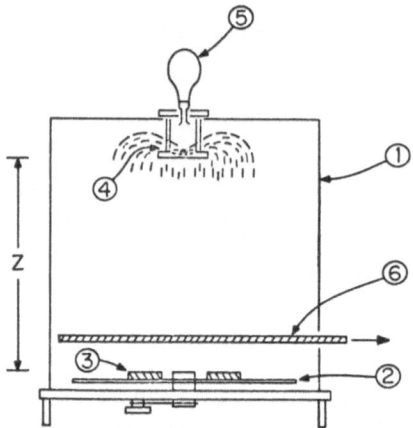

Figure 1.   Apparatus for the controlled sedimentation of dust onto experimental substrates.   (1) Transparent cover.   (2) Turntable rotating at 1 rpm.   (3) Sample to be dusted.   Dust sample, (4), is dispersed by a pulse of air from the   syringe, (5).   Shutter, (6), can   be opened for a   pre-determined time interval to allow only particles within a given   size range to reach the substrate.

We then scanned the surface with the diamond stylus, using suitable circuitry to count the number of events in which the stylus was kicked out of the groove by an encounter with a strongly adherent particle.

If the relative humidity (RH) never goes above 80% after dusting the surface, almost none of the dust particles are strongly enough adherent to kick the stylus out of the groove and give rise to an electronic signal that is counted. When the stylus encounters a particle that is not strongly adherent, the particle is simply moved out of the way by the stylus without   event. If, after dusting the surface, the relative humidity is cycled above and below 80%, the behavior is quite different. The humidity cycling was done by taking the dusted test surface from the usual room air environment (RH = 50%) ,   storing it in a controlled humidity chamber, (RH = 90%), for several hours, finally returning it to the room environment.   After this treatment we found by microscopic examination that about 0.1 to 1% of the particles were different from the others. Each
of these particles was   surrounded by a stain in the form of a visible halo.[3] Stained particles adhere to the substrate much more strongly than their unstained neighbors, often by several orders of magnitude. Counting the probe mistracking events gave us the total number of such strongly adherent particles.

A variety of evidence leads us to conclude that the strongly adherent particles, often surrounded by a stain, are actually composite particles, of which one component is a crystal of a water-soluble material. The key part played by water-soluble material can be seen from the following experiment. Before us, a dust sample was subjected to prolonged washing with distilled water in a Soxhlet extractor. to remove all water-soluble material. Using this washed dust, with or without the humidity cycling treatment, we found no stains, no probe mistracking, and no particles with exceptionally strong adhesion. We conclude that a water-soluble material is responsible for the strong adhesion.

The strongly adherent particles are so firmly bound to the substrate that it is often not possible to remove them without damaging the substrate. The importance of composite particles and water-soluble materials has not been treated in earlier discussions of particle adhesion. Reviews by Zimon[5] and Corn[6] have examined the role of van der Waals forces, capillary condensation, and electrostatic forces, among others. More recently, Bhattacharya and Mittal [7] have shown the effect of capillary condensation on the particle removal force, and Bowling [8] has treated the van der Waals adhesion mechanism as it applies to semiconductor problems. Whitfield[9] has suggested a possible effect of hygroscopic material, but without reference to composite particles.

SEQUENCE OF EVENTS LEADING TO STRONG ADHESION

The process begins when, at low relative humidity, a suitable composite particle settles on the substrate. At first, it is bound to the surface only by the relatively weak van der Waals forces. There may be an additional electrostatic force if charges are present. With increasing relative humidity, deliquescence begins at some threshold RH value that depends on what water-soluble material is present. [10] Water condenses on the surface of the water-soluble crystal, dissolving part or all of it. If the resulting solution wets both the substrate and the insoluble component of the composite particle, the contact area between particle and substrate will be greatly increased. When the RH later decreases again to a lower value, the condensed water evaporates. The dissolved salt crystallizes and can form a bond between particle and substrate. This adhesive bond can be very strong because the adhesive, the salt, makes good contact to both surfaces to be joined.This is the basic requirement of any good adhesive: the glue must flow, wet both surfaces to be joined, and then solidify. In our case, it is the cyclic change in relative humidity that makes this possible.

MECHANISM OF STRONG PARTICLE ADHESION

Deliquescence is the condensation of atmospheric moisture on the surface of a water-soluble material. This happens when the vapor pressure of water in the air is higher than that over a saturated solution of the water-soluble material. Table 1 shows the equilibrium relative humidity over their saturated solutions for several inorganic salts that may be present as airborne dust particles.[4]

Where do such particles come from? Sodium carbonate particles are emitted in great numbers by the cooling stacks of power stations and industrial plants. Calcium chloride is widely used as a de-icing salt for highways and coal storage facilities. The "Water Atlas" shows de-icer usage in various states.[11] Under car wheels, the de-icer solution forms a fine spray that then evaporates, leaving airborne salt crystals behind. Ammonium chloride and other soluble salts are widely used as fertilizers. Sodium chloride particles are produced in great abundance by the spray from ocean waves. The formation of airborne salt particles over the ocean has been studied in detail by meteorologists[12] because of the part such crystals play in the nucleation of cloud droplets. A single bubble from a breaking wave rises to the surface,

Table 1.   Atmospheric Relative Humidity   Above Which Salts Deliquesce

Soluble Salt	Critical Relative Humidity %
LiCl	15
$CaCl_2$	32
$MgCl_2$	33
K_2CO_3	44
NaCl	76
Na_2CO_3	78
NH_4Cl	79
Na_2SO_4	83
KCl	86
K_2SO_4	97

where it bursts, giving rise to hundreds of tiny droplets. These then evaporate, leaving behind airborne salt crystals. These particles, along with other natural dust particles, can be transported great distances over land.[13]   There is even a significant transport of dust from the Sahara Desert all the way across the Atlantic Ocean.[14, 15]   Thus, there are many opportunities for the formation of composite dust particles of the kind that we have found to be important for strong   adhesion.

Figure 2   illustrates the sequence of events that we have described.

Several glass spheres, 0.1 mm in diameter, rest on a glass microscope slide. We are looking at them from below, through the slide, using a metallurgical microscope.   One of the spheres is in contact with a small particle of sodium carbonate.   In the top photo the RH is 50%. Under these conditions, nothing changes with time.   The particles do not adhere strongly to the substrate.   A gentle stream of air or even tipping the slide a little will cause them to roll away.   In the center photo, the RH is 90%. The slide is covered by a humidistat chamber.   After a minute or two, water begins to condense on the sodium carbonate crystal, dissolving most of it.   Note that unbalanced surface tension has pulled the two spheres apart.   In the bottom photo, the RH is again 50%. The solution has now evaporated, leaving the sodium carbonate as a dry crystalline material.   Compared with the top photo, the salt now makes contact to a much larger area of both sphere and substrate.   As a result, the sphere now adheres very strongly to the substrate.   It is not blown away by a strong jet of air. In a similar experiment, using larger spheres[4], we measured the force needed to move a sphere along the surface, before and after cycling the relative humidity. The force needed to move the particle increased by a factor of 10^5 after the humidity cycling.   This experiment illustrates the origin of most of the strongly adherent particles that we found to cause mistracking, wear, and damage in our system.   The essential ingredients are: a composite particle with one water-soluble component, cycling of the RH above and below the threshold value for deliquescence,   and the ability of the   intermediary solution   to wet both the substrate and the insoluble particle.

CLIMATIC DATA AND THE FREQUENCY OF OCCURRENCE OF HUMIDITY CYCLES

For several of the most commonly found soluble particles, the threshold   RH value is around 80%. For other salts, such as calcium chloride, the threshold RH value is 32%.   To be specific, we will consider the 80% value. How often does a cyclic change   around 80% take place? Indoors, it all depends on air conditioning and humidity control. It is easier to look at what happens outdoors.   The weather bureau provides the needed information in the form of climatic data summaries. These are available for many locations and list hourly

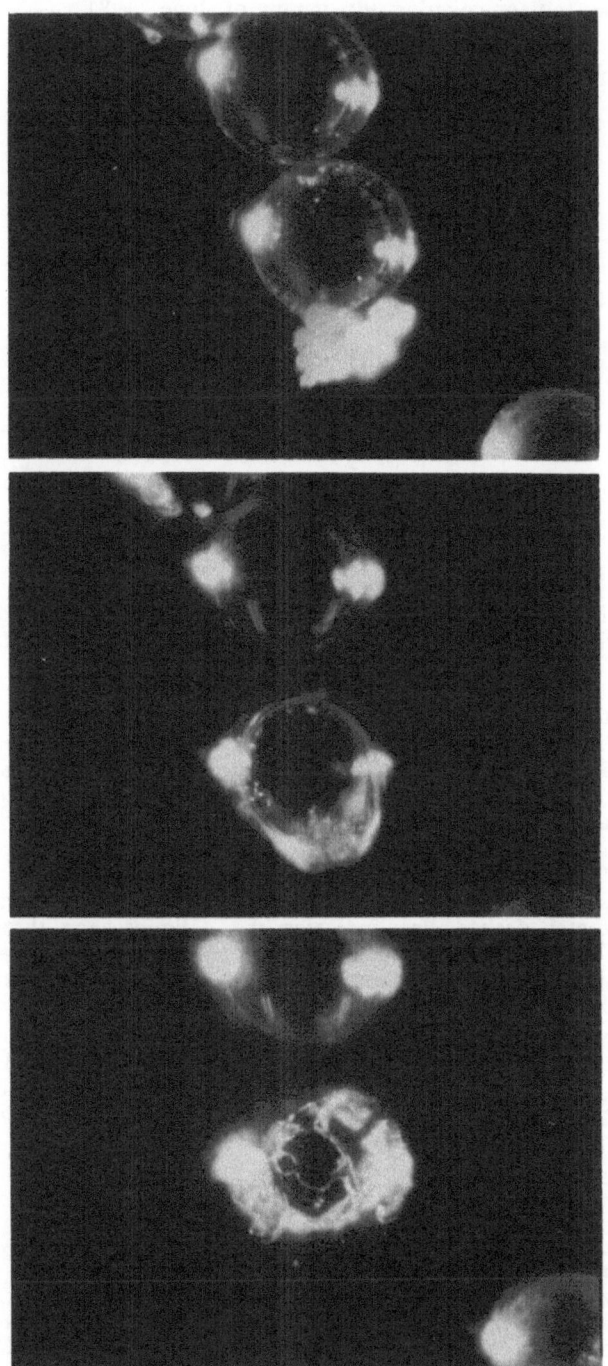

Figure 2. Glass spheres, 0.1 mm diameter, resting on a glass microscope slide, viewed from below, using a metallurgical microscope. In the top photo, alongside one of the spheres is a particle of sodium carbonate. <u>Relative humidity values:</u> Top photo, RH = 50%; Middle photo, RH = 90%; Bottom photo, RH = 50%.

values of relative humidity, together with other climatic information.   Figure 3 shows data for two cities, covering the month of June, 1981.

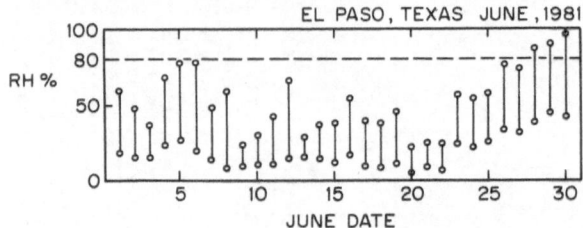

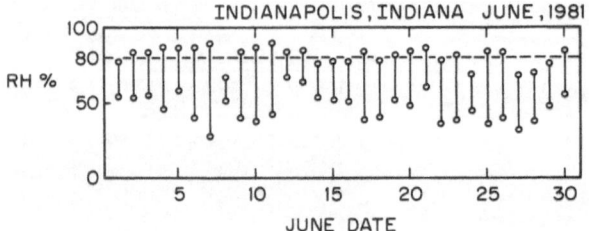

Figure 3.   Maximum   and   minimum   values   of   relative   humidity   for   June, 1981,   in   El   Paso, Texas   and Indianapolis, Indiana.   Data are taken from local

climatic data summaries,   available from U.S. Dep't of Commerce.   The dashed horizontal line in each drawing represents the threshold for deliquescence of soluble salts typically present in the atmospheric aerosol.,   The strong particle-adhesion   process that we have described can take place on those days for which the dashed line goes between maximum and minimum RH values.

The plots show the maximum and minimum RH values for each day of the month.   A horizontal dashed line indicates the deliquescence threshold value of 80%.   According to our ideas, strong adhesion can develop on those days on which the maximum RH values is above the 80% line and the minimum RH value is below the line.

In the relatively   dry climate of El Paso,   this happens on only   3   days out of the month, though several other days come close.   In the more humid climate of Indianapolis, the critical requirement is met on 19 days out of 30, with several other days that are close. This gives us an idea of the relative exposure for the strong adhesion process at these two locations.

SUMMARY AND CONCLUSIONS

We have looked at strongly adherent particles experimentally, and find that most of the damaging particles are   composite particles, cemented to the substrate by a water-soluble particle. The process is activated by natural cycling of the relative humidity. Early   on   in our work , we were surprised to find out how prevalent composite particles are, in which one component is a water-soluble salt particle.   The combination   seems   rather improbable.   More recently,   it has been   found [16] that such composite particle are quite common. It is believed that the composite particles are generated   by the action of fair weather cumulus clouds.   Cloud droplets are nucleated by salt particles,   and then scavenge insoluble dust particles, such as silicates.   Later,   the cloud evaporates,   leaving behind composite particles consisting of a   silicate particle and a salt crystal.   In dust samples collected over the ocean,   a large fraction of the particles were of this kind.[16]

REFERENCES

1. J.R. Clemens, RCA Review, 39, 33 (1978)
2. D.L. Ross, RCA Review, 39, 136 (1978)
3. R.W. Nosker, L.A. DiMarco, and R. Williams, Appl.Phys.Lett., 38, 1023 (1985)
4. R.Williams and R.W. Nosker, CHEMTECH, 15, 434 (1985)
5. A.D. Zimon,"Adhesion of Dust and Powder", Plenum Press, New York, 1969
6. M. Corn, J.Air Pollution Control Assoc., 11, 523 1961)
    M/ Corn, J. Air Pollution Control Assoc. 11, 563 (1961)
7. S. Bhattacharya and K.L. Mittal, Surf. Technol., 7, 413 (1978)
8. R.A. Bowling, J. Electrochem. Soc., 132, 2208 (1985)
9. W.J. Whitfield in "Surface Contamination; Genesis, Detection, and Control" K.L. Mittal, Editor, Vol. 1, pp73-81, Plenum Press, New York (1979)
10. H.R. Pruppacher and J.D. Klett, "Microphysics of Clouds and Precipitation" pp 83-85, D. Reidel Publishing Co., Boston (1980)
11. J.J. Geraghty, D.W. Miller, F. Nan der Leeden, and F.L. Troise, "Water Atlas of the United States", plate 66, Water Information Center Publiocation, Port Washington, New York (1973)
12. Reference 10, pp 194-197
13. W, Lockeretz, American Scientist 66 560 (1978)
14. J.M. Prospero, R.A. Claccum, and R.T. Nees, Nature, 289, 570 (1981)
15. In-Young Lee, J. Climate Appl. Meteorol., 22 632 (1983)
16. M.O. Andreae, R.C. Charlson, F. Bruyseels, H. Storms, R. Van Grieken, and W. Maenhut, Science, 232, 1620 (1986)

PREVENTION OF STRONG ADHESION OF DUST PARTICLES

Richard W. Nosker and Richard Williams

David Sarnoff Research Center
Subsidiary of SRI International
Princeton, New Jersey 08543

Strong adhesion of dust particles results when water
soluble components in dust particles are first
solubilized, and then recrystallized, after a successive
rise and fall of the relative humidity. The recrystallized
salt greatly increases the contact area with the
substrate, thereby increasing the adhesion. We have
investigated the use of thin film lubricants and
surfactants to avoid strong adhesion after the humidity
cycle. We found that, while some surfactants can prevent
the strong attachment of particles by causing secondary
nucleation in some precipitating salts, the most effective
means of preventing strong adhesion is to use a thin film
lubricant that cannot be displaced from the substrate by a
forming water droplet. We analyze the requirements for a
successful system in terms of the polar and dispersion
components of surface energy.

I. INTRODUCTION

In the development of RCA's capacitive stylus-in-groove VideoDisc
system[1], we encountered a problem that had to be solved before the
product could be marketed. The problem was that locked grooves and
signal loss were encountered altogether too frequently, especially
after environmental stress of the disc. We found that strongly adherent
dust particles were the primary cause of these problems. In the case of
locked grooves, the playing stylus was knocked into a previous groove
by the adherent particle. In the case of signal loss, the particle
collided with the stylus and stuck, causing the stylus to ride far above
the disc and lose the signal. The VideoDisc groove pitch was 2.5µm, some
fifty times smaller than a standard LP audio disc, and the stylus
system had an extremely low mass, well below 0.1 gram. Therefore,
average-size dust particles that would not bother a stylus-in-groove
audio disc, could, in our system, cover several grooves and cause the
problems mentioned above.

II. THE PRIMARY CAUSE OF STRONG DUST ADHESION

Early in our investigation we found that nearly all cases of strong dust adhesion are caused by deliquescence and subsequent recrystallization[2]. Above a certain relative humidity, a water-soluble salt that is present as a component of a small fraction of dust particles takes up moisture and forms a liquid droplet that surrounds the entire dust particle. When the relative humidity drops, the salt recrystallizes and may form a large-area contact with the underlying substrate. As long as the relative humidity stays below the critical level for the salt in question, the particle is strongly adherent. Humidity variations follow weather patterns, and also a natural 24-hour cycle.

We observed, in scaled-up experiments, the crystallization of salt around an inert particle as a drop dried. Figure 1 shows the evolution of salt (KH phthalate) deposits around a 2mm glass ball, for a humidity that begins to drop from well above the critical level. When about 65% of the water has evaporated the salt begins to form at the liquid-air interface.

When about 80% of the water has evaporated, "flying buttresses" begin to form between the ball and the substrate. There are azimuthal variations around the ball, but there is substantial ball-substrate contact after drying is complete with several flying buttresses typically formed. The force necessary to dislodge these glass balls is discussed in the next section. For some salts, flying buttresses form, while for others the precipitation occurs almost entirely upon the substrate or the ball and strong adhesion does not occur.

III. PREVENTION OF STRONG ADHESION

The RCA VideoDisc substrate is a carbon-loaded polyvinylchloride. It is coated with 200 Å of a purified version of GE's SF1147, a silicone-hydrocarbon oil having the structure shown in Figure 2, to prevent rapid wear of the diamond stylus. This low viscosity (20 cs) oil was chosen because it is nonvolatile, is inert, and wets the disc sufficiently well. However, whether this oil was present or not, even in appropriately scaled thickness, the process shown in Figure 1

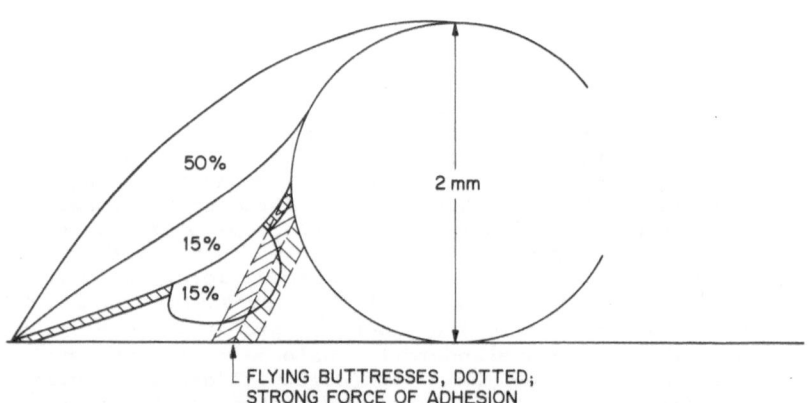

Figure 1. Depiction of large area contact forming between particle and substrate as salt precipitation occurs in a scaled-up experiment. Percentages indicate the evaporating portion of water within the region shown.

occurred and produced strong adhesion. To prevent this, we tried two approaches, both involving lubricant additives. First, we tried a variety of surfactants of varying generic description: cationic, anionic, amphoteric, and nonionic. These surfactants were first tested in scaled-up glass ball experiments on unoiled discs. A pure water drop on these unoiled discs formed a contact angle of about 75°. We found that when a small quantity of surfactant was added to the water the contact angle of the drop was lower, in the range from 50° to under 10°. When various salts (e.g. NaCl, Na bitartrate, KH phthalate, oxalic acid) in addition were added to the drop, the contact angle varied, sometimes rising and sometimes falling, in accordance with the resulting concentration of surfactant molecules at the liquid-air interface. We found that lower contact angles forced the precipitation to take place away from the insoluble particle and thus the area of ball-substrate contact was reduced. We also found that, for some combinations of surfactants and salts, the surfactant apparently facilitated the nucleation of very many particles so the resulting precipitate had no mechanical strength. Table I shows the force necessary to dislodge balls for two relatively successful cationic surfactants. The best of the surfactants were tested as lubricant additives on environmentally stressed discs, and while they were found to improve the problems, they did not offer a satisfactory solution. We believe the reason is that the nucleation mechanism is specific to certain classes of salts and does not extend universally to the vast variety of salts found in airborne dust. The contact angle has a rather small effect when the humidity gets only a little above the critical value, or when the insoluble part of the particle has a non-ideal shape.

The second approach was successful. Dr. C.C. Wang at our laboratory varied the end groups and chain lengths of lubricant molecules and tested their performance on environmentally stressed discs. His studies led him to have a lubricant additive custom formulated. It is now called Wang dopant and its formula is shown in Figure 2. This oil is nonvolatile, low in viscosity, and soluble in SF1147 in the 15 to 20% concentrations used. The doped SF1147 oil almost totally eliminated the

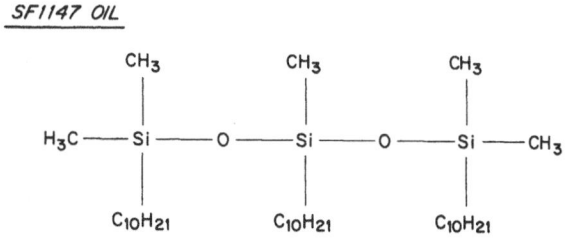

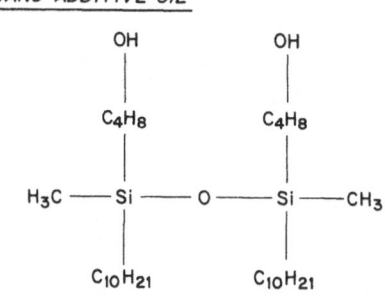

Figure 2. Structure of GE SF1147 and Wang additive oil.

Table I. Force required to dislodge 2mm glass balls
from the surface of a PVC substrate after
precipitation from 40 µl droplets formed over the
balls with the liquids indicated.

Solution	Median Force to dislodge Ball (grams)	Range (grams)
7g KH phthalate/0.1g Cyastat 609*/100 ml H_2O	0.03	0-0.1
7g KH phthalate/0.1g Cyastat SN*/100ml H_2O	0.30	0.05-2.2
7g KH phthalate/100 ml H_2O	6.0	0.3-12.5

* Cationic surfactants manufactured by American Cyanamid Corp., Bound Brook, N.J.

problems of locked grooves and signal loss. Table II shows the results of ball-pushing experiments for several salts with no lubricant, SF1147 alone, and SF1147 with 20% Wang dopant. In each case the experimental observations were the same: the crystalline deposit formed as usual but there was no adhesion to the substrate when Wang dopant was used.

IV.   MECHANISM OF ACTION OF WANG DOPANT

The experiments and calculations presented below show that Wang dopant prevents adhesion to the substrate by reducing the oil-water interfacial tension so much that a forming water drop is unable to displace the lubricant when Wang dopant is present. Therefore, even though the large area substrate contact appears to form as the salt precipitates, the contact remains lubricated and there is no strong adhesion. When SF1147 or any other hydrocarbon-like oil is present, the oil is displaced as the water droplet forms, allowing strong particle-substrate adhesion.

We were able to demonstrate experimentally in several ways that SF1147 oil is displaced by water while Wang doped oil is not. First, a large water drop was formed on a lubricated disc, then the disc was slightly tilted so that the drop moved across the disc at 0.3 inches/minute (Figure 3). The oil thickness was measured by x-ray fluorescence, before and after the passage of the water drop. A 297Å SF1147 film became 18Å thick, while a 314Å Wang doped film measured 311Å after passage. The second experiment is shown in Figure 4. A beaker was

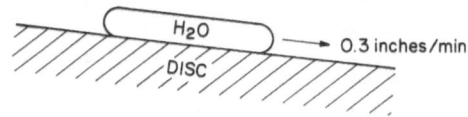

297Å SF1147 OIL → ROLL DROP → 18Å REMAIN

314Å WANG DOPED OIL → ROLL DROP → 311Å REMAIN

Figure 3. The rolling drop experiment, in which a moving drop of water removes SF1147 oil, but does not remove Wang doped SF1147.

Table II. Force required to dislodge 2mm glass balls from the surface of a PVC substrate after precipitation from 40 µl droplets formed over the balls with the liquids indicated (concentration 7g salt/100 ml H_2O). Lubricant thickness approximately 10 µm.

Salt	Lubricant	Median Force to Dislodge Ball (grams)	Range (grams)
KH phthalate	None SF1147 20% Wang/80% SF1147	6.0 2.35 0	2.3-37 0.6-35 0
Oxalic acid	None SF1147 20% Wang/80% SF1147	4.4 2.3 0	2.8-6 0-3.0 0
Na bitartrate	None SF1147 20% Wang/80% SF1147	10.0 0.35 0	4-18 0-1.4 0
NaCl	None SF1147 20% Wang/80% SF1147	1.35 0 0	0.7-2.0 0 0

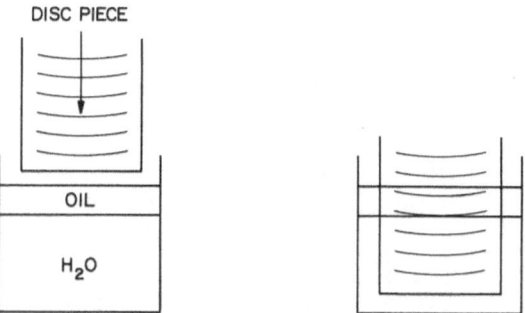

DISC PIECE

OIL

H_2O

TIME (min.) FOR OIL TO RETRACT COMPLETELY FROM H_2O

SF1147/H_2O
2 min.

15% WANG DOPED SF1147/H_2O
∞

Figure 4. As a disc piece moves downward it first contacts the oil. This oil is carried on the disc to the bottom of the container. The time for the oil to retract back up above the water is measured.

205

filled to a certain height with pure water. A layer of oil was floated on top of the water. Then an unlubricated disc piece was slowly inserted through the oil to the bottom of the container. The disc picked up a heavy coating of oil, which was dragged to the bottom of the container. With SF1147 oil it took about 2 minutes for the oil to retract completely from the portion of the disc under water (several inches in length). With Wang doped oil there was no evidence of retraction, even after several months. In the third experiment, a flat piece of unlubricated disc material was put on the bottom of a shallow glass dish (Figure 5). A volume of water was then poured into the dish, covering the disc. Then, an oil-filled syringe with a flat-tipped needle was inserted almost to the surface of the disc and a large oil drop was formed on the disc and then retracted, allowing measurement of the advancing and receding oil-to-disc contact angles. For SF1147 oil the advancing angle was 72°, and the receding angle was 34.5°. For Wang doped oil the advancing angle was 42°, and the receding angle 0°. The Wang doped oil therefore did not dewet the disc even when water had contacted the disc first.

To gain a better understanding of why Wang doped SF1147 is not displaced by water we measured and calculated the surface energetics of the oil-water-disc system. As water condenses on the oiled disc it will not displace the oil as long as the interfacial energy terms, $\gamma_{so} + \gamma_{ow}$, represent a lower energy situation[3] than the case in which the oil has been displaced, γ_{sw}. The subscripts s, o, and w are for solid, oil, and water, respectively. The oil will not be displaced, therefore, when the differential energy

$$\Delta = \gamma_{sw} - (\gamma_{so} + \gamma_{ow}) \qquad\qquad (1)$$

is positive. We observe experimentally that Δ is positive when Wang doped oil is used, and negative when SF1147 alone is used. For the negative case, the oil retracts into droplets which form a 34.5° receding contact angle allowing Young's equation to be used to recast Equation (1) as

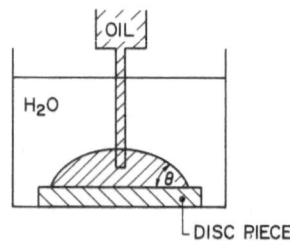

	SF1147/H₂O	15% WANG DOPED SF1147/H₂O
ADVANCING	72°	42°
RECEDING	34.5°	0°

Figure 5. A disc piece is first covered with water. A large drop of oil is then formed on the disc and the contact angle at the oil/water/disc interface is measured as the drop advances and recedes as controlled with the syringe.

$$\Delta = \gamma_{ow} (\cos \theta_r - 1) \qquad\qquad (2)$$

We will use Eq. (2) for the SF1147 case, since both γ_{ow} and θ_r can be experimentally measured. In Equation (1), γ_{sw} and γ_{so} must be estimated. We believe the most self-consistent values for solid-liquid interfacial tension are generated using the reciprocal mean formula due to Wu[4]

$$\gamma_{sl} = \gamma_s + \gamma_1 - \left[\frac{4\gamma_s^d \gamma_1^d}{\gamma_s^d + \gamma_1^d}\right] - \left[\frac{4\gamma_s^p \gamma_1^p}{\gamma_s^p + \gamma_1^p}\right]$$

$$\qquad\qquad (3)$$

in which the superscripts refer to dispersion and polar components of surface energy. In practice, several liquids with known parameters ($\gamma_1 = \gamma_1^d + \gamma_1^p$) are used to make contact angle measurements on the solid substrate in question. The term γ_{sl} is eliminated by using Young's equation, and γ_s^d and γ_s^p are iterated to give a best fit for the liquids used. A complication is that liquids typically show substantially different contact angles depending on whether the drops reach their static measured condition by advancing or receding. We believe this effect can be attributed to surface chemical heterogeneity in our case. As a result of this complication, we restricted ourselves in most instances to computing mean surface energy terms based on average contact angle $\bar{\theta} = (\theta_a + \theta_r)/2$. The liquids we used were water, formamide, tri-o-tolyl phosphate, glycerol, and decalin, which have surface energy values given in Table III. We measured total surface energy (γ_1) very close to these literature values. We also got excellent agreement between predicted and measured mean contact angles for these liquids on Teflon, using a literature value for Teflon[4] ($20.5^d + 2.0^p$). This technique gave us $\gamma_s = 36.2^d + 5.5^p = 41.7$ erg/cm^2 for the disc.

The measurements of liquid-air and liquid-liquid interfacial energy were made using the drop-weight method [6],[7]. This method gives much better than 1% accuracy if the edges of the tip are sharp and if the drops are formed slowly so that the forces remain quasi-static.

Table III. Literature values for dispersion and polar components of surface energy for liquids used in our experiments.

Liquid	SURFACE ENERGY (ERG/CM^2)		
	γ_1^d	γ_1^p	γ_1
Water[3]	21.8	51.0	72.8
Formamide[3]	39.5	18.7	58.2
Tri-o-tolyl phosphate[3]	39.2	1.7	40.9
Glycerol[3]	37.0	26.4	63.4
Decalin[5]	29.9	0	29.9

The interfacial surface energy is

$$\gamma_{ab} = \frac{(\Delta \rho_{ab}) VgF}{r}$$

where $\Delta \rho_{ab}$ is the difference in density between a and b, V is the volume of a single drop, g is 980 cm/sec^2, r is the tip radius from which the drop breaks, and F is a compiled empirical correction factor. For oil in air this method gives (in erg/cm^2) 27.0 for SF1147, and 30.4 for pure Wang dopant. For oil in water the measured values of γ_{ow} are 41.5 for SF1147 and 9.8 for Wang doped oil or for pure Wang dopant.

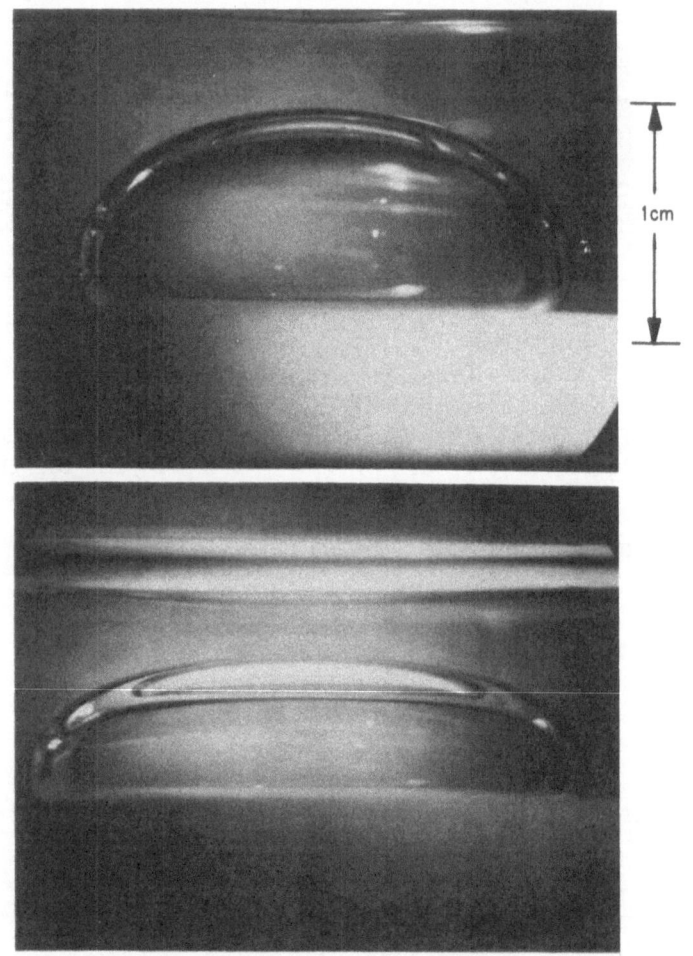

Figure 6. Sessile Drop Experiment. Each photograph shows a drop of water, under oil, resting on a teflon substrate. In the top photograph the oil is pure SF1147 oil. In the bottom photograph the oil is SF1147 containing 15% of the Wang oil additive. The smaller height of the lower drop indicates that the interfacial tension is less. The density of the pure SF1147 oil is 0.878. The density of the mixture containing 15% Wang oil is 0.881.

Then by measuring the contact angles of pure SF1147 and pure Wang dopant on Teflon and on polyvinylidene fluoride[4] (20.6^d+12.6^P) we found using Equation (3) that

$$\gamma(SF1147) = 24.6^d + 2.4^P = 27.0$$

$$\text{and } \gamma(\text{Wang doped SF1147}) = 25.1^d + 5.3^P = 30.4$$

The solid-water interfacial tension γ_{sw} is 40.3 erg/cm^2 from Equation (3). For Wang doped SF1147, γ_{so} is computed to be 2.0 erg/cm^2, giving Δ = 40.3 - (2.0 + 9.85) = 28.45 erg/cm^2. Wang doped oil is therefore extremely stable under a water droplet. For pure SF1147, Equation (2) gives Δ = 41.5 (cos 34.5^o-1) = -7.3 erg/cm^2. Using Equations (1) and (3), a less accurate means of determining Δ for pure SF1147, we get Δ = -4.6 erg/cm^2. It is the great reduction in γ_{ow} with the addition of Wang dopant that provides the solution to the adhesion problem.

To give a direct illustration of the reduction in γ_{ow} caused by addition of the Wang oil we show a sessile drop experiment in Figure 6. Each photograph shows a drop of water, under SF1147 oil, resting on a teflon substrate. In the top photograph we used pure SF1147 oil. In the bottom photograph the oil contains 15% Wang oil.

For a drop large enough to have a flat top, the height, z, of the drop above the equatorial plane depends on γ_{ow}[7]. Specifically:

$$z = \sqrt{2\gamma_{ow}/\Delta\rho g}$$

$$(5)$$

Measuring z from the photographs we find, for pure SF1147 oil, γ_{ow} = 33 ergs/cm^2. For Wang doped SF1147, γ_{ow} = 15 ergs/cm^2. The reduction in interfacial tension is very clearly seen from the difference in drop heights. The quantitative energy values are somewhat different from those we got by the drop weight measurement. This is due in part to the fact that we used different oil sources, but may reflect other experimental uncertainties as well.

V. CONCLUSIONS

We have shown that strong adhesion due to deliquescence can be prevented by lubricating a substrate with a thin film of oil that has a low interfacial tension with both the substrate (γ_{so}) and with water (γ_{ow}). On the RCA VideoDisc the polyvinylchloride substrate was primarily dispersive in character, so an oil was chosen that had a predominantly hydrocarbon character and was therefore able to wet the substrate well and give low γ_{so}. A low γ_{ow} was achieved by using an additive made of molecules with a hydrocarbon-like tail and a hydroxyl group on the other end that could present itself at the oil-water interface and greatly reduce γ_{ow}. In this way, condensing water was unable to displace the lubricant, and so strong adhesion could not develop between deliquescent dust particles and the substrate. In general, strong adhesion due to deliquescence is prevented when a lubricant is chosen such that Δ in Equation (1) is positive. Following the same procedures we have described here, it should be possible, for almost any substrate of interest, to choose or design a lubricant that is capable of preventing strong adhesion.

REFERENCES

1.   D.L. Ross, "Coatings for VideoDiscs," RCA Review, 39,136 (1978).

2.   R. Williams and R.W. Nosker,   these proceedings.

3.   D.H. Kaelble, "Physical Chemistry of Adhesion," Wiley, New York (1971).

4.   S. Wu, J. Polymer Sci. C, 34, 19 (1971).

5.   F.M. Fowkes, editor, "Chemistry and Physics of Interfaces," American Chemical Society Publications, Washington, D.C.(1965).

6.   W.D. Harkins, in "Physical Methods of Organic Chemistry", Vol. 1, A. Weissberger, editor, Interscience, New York (1971).

7.   N.K. Adam, "The Physics and Chemistry of Surfaces," 3rd Ed., Oxford University Press, London (1941).

DYNAMIC ADHESION OF PARTICLES IMPACTING A CYLINDER

Hwa-Chi Wang and Walter John

Air and Industrial Hygiene Laboratory
California Department of Health Services
2151 Berkeley Way
Berkeley, CA 94704

Adhesion efficiencies for ammonium fluorescein particles impacting a stainless steel cylinder were measured by a fluorometric technique. Below the critical velocity, the adhesion efficiency reached only 80%. When the efficiencies were divided by 0.8, they agreed with the theory of Wang[1]. Counting of particles around the cylinder by a microscope showed rebound to increase rapidly with angles beyond 15° from the stagnation point, accounting for the missing 20% in the overall adhesion efficiency. The data imply that the tangential velocity component can cause particle bounce even when the normal component is below the critical velocity. The critical velocity is found to depend on particle diameter to the power -1.29, in agreement with the data of Cheng and Yeh[2] but in disagreement with the theory of Dahneke[3]. Comparison of the magnitudes of the critical velocity from various experiments suggests that the observed critical velocity is lower when the angles of impaction and rebound are not restricted to 90°.

INTRODUCTION

The removal of particles from the carrier gas is accomplished in two steps, impaction with the collection surface and adhesion to it. Impaction of particles with collectors has been extensively studied and in most cases the impaction efficiency can be accurately predicted by theoretical models. Prediction of adhesion efficiency, defined as the fraction of particles captured after impaction, is more complicated because it involves unknown surface properties of both particles and collectors. Based on the equation of energy balance, Dahneke[4] demonstrated that the dynamic surface properties could be characterized by the critical velocity and the coefficient of restitution, where the critical velocity was defined as the largest impact velocity where no rebound occurred. Since these two quantities can be directly determined from experiments[5], they can be treated as independent variables, and thus enable the prediction of the dynamic response after impaction. Recently, Wang[1] studied theoretically the adhesion of particles impacting a cylinder at high Reynolds number. As a first approximation, only the normal component of the particle impact velocity was used in the rebound criteria. With the appropriate normalization techniques, the author was able to obtain a set of predictive equations to calculate the adhesion efficiency.

Some measurements of critical velocity taken from the literature are listed in Table I. For a given particle size, the results can differ by more than an order of magnitude. Comparisons are complicated by the variety of particle-surface materials used and by the different experimental configurations. The latter differences imply a variation between experiments in the angle of incidence and rebound. Aylor and Ferrandino[6] suggested that particle bounce might be facilitated at oblique angles of incidence and that the tangential velocity and not just the normal velocity component might be involved in determining bounce.

In the present study, adhesion efficiencies were measured for particles impacting a cylinder. Total adhesion efficiencies were measured as well as the dependence on the angle of incidence. Critical velocities were obtained through analysis of the data and compared to theory. The dependence of adhesion on the angle of incidence and the tangential velocity are discussed.

THEORETICAL BACKGROUND

A brief review of Wang's model is given here for the convenience of discussion, while detailed treatment can be found in Wang[1]. The equation of particle motion around a cylinder was solved numerically to obtain the particle trajectory for a given initial Y coordinate (Figure 1), assuming a potential flow field around the cylinder. As a first approximation, the rebound criteria were assumed to be:

(1)   if the normal component of the impact velocity is less than the critical velocity, no rebound occurs, and the tangential component is dissipated through friction.

(2)   if the normal component of the impact velocity is larger than the critical velocity, the particle rebounds, and the tangential component is conserved.

Multiple impacts were considered and a family of significant Y coordinates, corresponding to sticking (S_n) or rebounding (R_n) trajectories at nth impact (Figure 1), were identified to determine the adhesion efficiency. Numerical results indicated that the adhesion contribution of three or more impacts was less than 1% and could be neglected. With proper normalization, the adhesion contribution of the first two impacts (H_1 and H_2) can be calculated from

$$H_1 = 1 \text{ for } V_{cr} \geq V_s \tag{1}$$

$$H_1 = 1 - \left[1 - (\frac{V_{cr}}{V_s})^2 \right]^{\frac{1}{2}} \text{ for } V_{cr} < V_s \tag{2}$$

$$H_2 = 0 \text{ for } V_{cr} \geq V_s \tag{3}$$

$$H_2 = 0.054\sqrt{e} \left[1 - (\frac{V_{cr}}{V_s})^2 \right]^{\frac{1}{4}} \text{ for } V_s > V_{cr} \geq 0.27e^{0.85}V_s \tag{4}$$

$$H_2 = 0.054\sqrt{e} \left\{ \left[1 - (\frac{V_{cr}}{V_s})^2 \right]^{\frac{1}{4}} - \left[1 - (\frac{V_{cr}}{0.27e^{0.85}V_s})^{\frac{1}{2}} \right] \right\} \text{ for } V_{cr} < 0.27e^{0.85}V_s \tag{5}$$

where V_{cr}, V_s, and e are the critical velocity, the impact velocity at the stagnation point, and the coefficient of restitution, respectively. The adhesion efficiency (H) is the sum of H_1 and H_2. For a given particle-surface system, V_{cr} and e are fixed, and V_s can be varied by

Table I. Particle Impact Velocity or Flow Velocity at the Onset of Rebound for Various Particle-Surface Systems.

Particle Material	Particle Size (μm)	Target Material	Impact Velocity (cm/s)	Reference
Lycopodium	24.9	Glass rod	40	(6)
Ragweed	18.3		70	
Lycopodium	24.9	Glass rod	62	(7)
Glass beads	20-40		25	
Glass beads	10	polyamide fibre	3*	(8)
	5.0	20 μm	6*	
KHP	2.0	Nuclepore 12μm	19**	(9)
	1.3	Nuclepore 8μm	44**	
	0.7	Nuclepore 5μm	24**	
PSL	5.7	Stainless	43.3	(2)
	3.4	steel	88.1	
	2.02		184	
	1.01		355	
	0.82		551	
PSL	1.27	Quartz	120	(5)
Uranine	15.5	Brass coated	6.7	(10)
	12.4	with carbon	7.1	
	11.6	black	7.3	
	8.8		8.0	
	6.2		9.2	
	4.4		11	

*   Flow velocity
**   Estimated from the face velocity and porosity

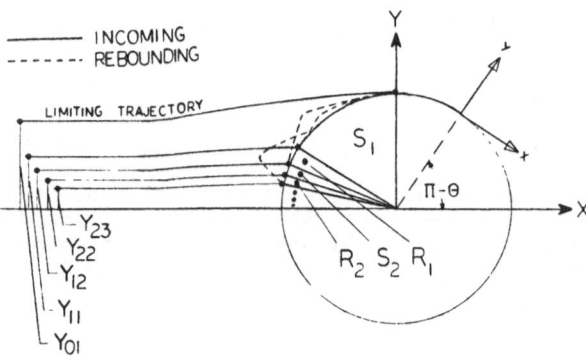

Figure 1. Coordinate system used by Wang[1] for the calculation of Y coordinates corresponding to significant trajectories.

213

varying the particle size and velocity, which allows an easy test of the theory using common aerosol instruments.

EXPERIMENTAL DESIGN

To obtain the adhesion efficiency requires two quantities: (1) the number of particles impacting the cylinder, and (2) the number of particles that are captured by the cylinder. Simultaneous measurements of these two quantities are preferred to minimize the error from variation of aerosol concentration. Monodisperse aerosol was generated and directed to impact a cylinder having half of its length greased. The greased surface gave the first quantity, while the ungreased surface gave the other. Furthermore, this technique also provided a means to obtain the particle adhesion at a specific location by counting the number of deposited particles under a microscope.

The schematic diagram of the experimental setup is shown in Figure 2. Monodisperse aerosol of ammonium fluorescein was generated by a vibrating orifice aerosol generator[11], neutralized by a Kr-85 radiation source, and transported to a manifold. The aerosol was driven through a rectangular slit 0.66 cm wide by a vacuum source at a regulated flow rate. The monodispersity and size of the generated particles were continuously monitored by an optical counter (Climet 201) and a pulse height analyzer. Polished stainless steel cylinders, located at the exit of the rectangular slit, were used to collect particles. Two cylinder diameters, 0.32 and 0.16 cm, were used. The cylinder was separable; one half was coated with a fluorocarbon grease (Halocarbon) to provide a bounce-free surface, while the other half remained ungreased. The particle mass collected on greased or ungreased surfaces was dissolved in ammonium hydroxide solution and determined by a fluorometer (Aminco, J4-7439). The ratio of the collected mass on these two surfaces gave the adhesion efficiency.

The incidence angle for particles impacting a cylindrical surface can vary from 0° to 90°, and therefore the overall adhesion efficiency is the integrated measurement of all possible incidence angles. To understand the angle dependence, we designed a cylinder holder that allowed us to rotate the cylinder and examine it under a microscope. The orientation of the cylinder was determined by the alignment mark on the Plexiglas wheel attached to the end of the cylinder (Figure 2). This mark, aligned with the centerline of the slit during experiments, gave the stagnation point ($\theta = 0$, Figure 1). The numbers of particles deposited on greased and ungreased surfaces were counted through a rectangular window within the eyepiece of the microscope. The width of the counting window subtended 5° at the center of the cylinder.

RESULTS AND DISCUSSION

Impaction Efficiency

To test the experimental techniques employed, the impaction efficiency of the cylinder was determined using liquid aerosol of oleic acid, which should have negligible bounce, and compared with the well documented theory[1,12,13]. The impaction efficiency is calculated from

$$E = \frac{(N_0 - N_1)/N_0}{R \int_0^c UdY / \int_0^b UdY} \qquad (6)$$

where N_1 and N_0 are the number concentration with and without the presence of the cylinder, R_c and b are the cylinder radius and the half width of the slit, and U is the flow velocity. A parabolic velocity profile is assumed because the Reynolds number based on the width of the slit is within the laminar range for most cases. The mean flow velocity is determined from the measured flow rate. Assuming a parabolic velocity profile, the maximum (centerline) velocity is 50% greater than the mean velocity. The particle Stokes number, defined as the ratio of particle

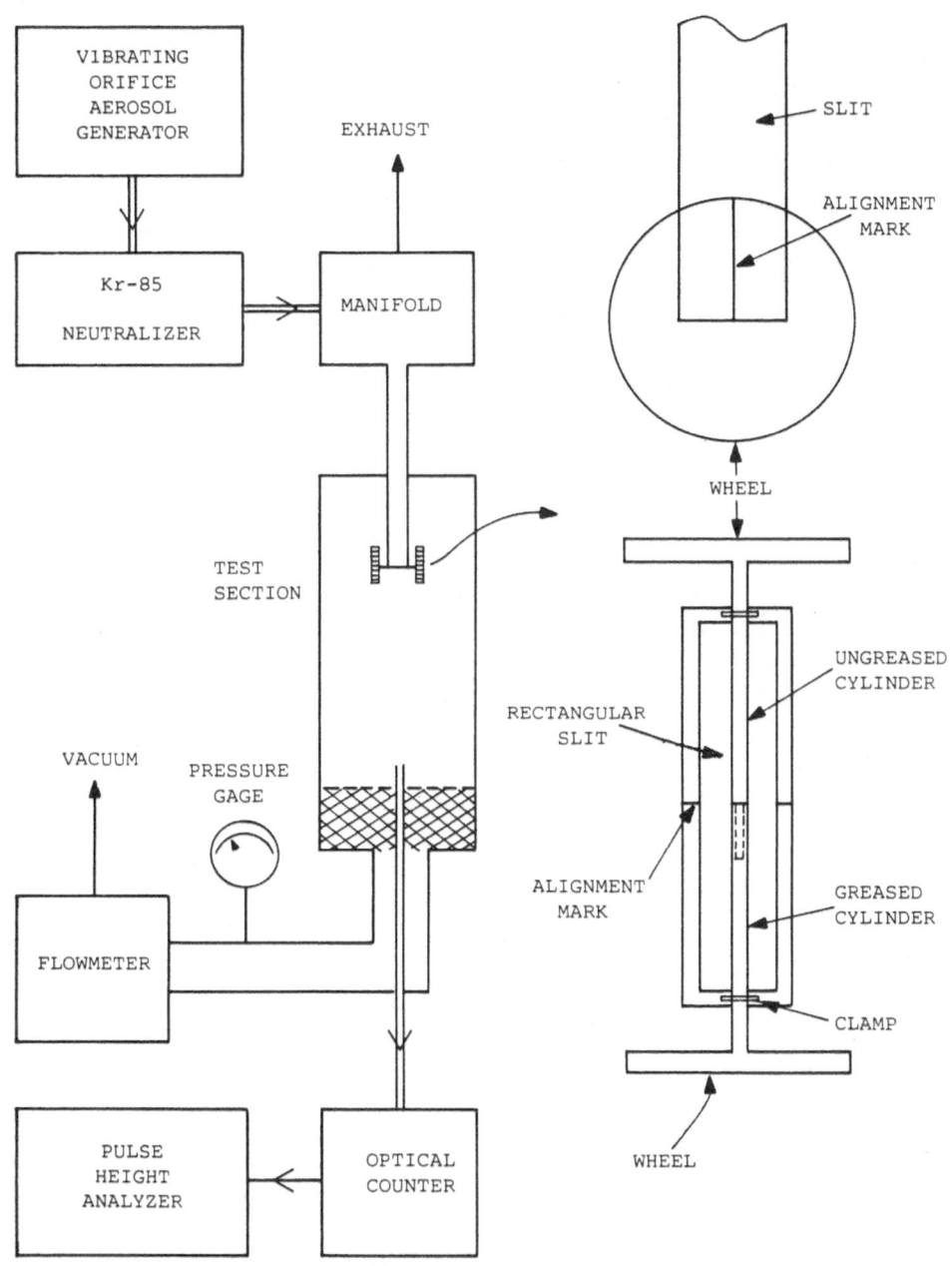

Figure 2. Schematic diagram of the experimental design for the measurements of impaction efficiency and adhesion efficiency.

stopping distance to the cylinder radius, is based on the velocity averaged over the slit area subtended by impacting particles, i.e., from centerline to ER_c, or $\int_0^{ER_c} UdY/ER_c$. For E=50%, the averaged velocity is within 2.5% of the centerline velocity. The difference is even smaller for lower impaction efficiency. For all the bouncing experiments performed, the impaction efficiency rarely exceeded 50%, and the averaged velocity corresponding to E=50% was used to represent the flow velocity.

The impaction efficiency as a function of the Stokes number is shown in Figure 3 for liquid particles. Also included is the theoretical prediction based on potential flow. The Reynolds number based on the cylinder diameter ranges from 150 to 1200, and therefore potential flow is a good assumption. Good agreement between theory and experiment is obtained, confirming the choice for the flow velocity.

Adhesion Efficiency

The adhesion efficiency (H) of solid ammonium fluorescein particles as a function of flow velocity is plotted in Figure 4, where H is simply the ratio of masses detected on ungreased and greased surfaces, and the flow velocity is the averaged value corresponding to 50% impaction efficiency. The adhesion efficiency decreases rapidly as velocity increases. For example, H decreases by one order of magnitude for about 40% increase in velocity. Larger particles start to rebound at lower velocity as expected. For each particle size H decreases almost linearly with increasing velocity in a log-log plot; the results for different particle sizes are parallel.

The particle impact velocity at the stagnation point can be calculated by solving the equation of motion[1] for each set of data. It is noteworthy that the impact velocity at the stagnation point is almost one order of magnitude smaller than the flow velocity. Figure 5 shows the same adhesion results as Figure 4 but as a function of the particle stagnation velocity (V_s). The slope is less steep as compared to Figure 4, but the linear dependence (log-log plot) in the medium range of H and the near-parallelism of the lines are preserved. The experimental results within 5-80% adhesion efficiency were fitted by

$$\log H = a + b \log V_s \tag{7}$$

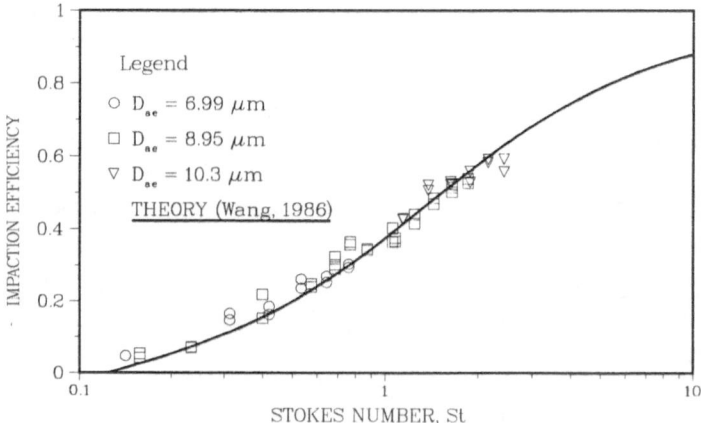

Figure 3.   Impaction efficiency as a function of Stokes number for liquid aerosol. The impaction efficiency is derived from the measurements of particle number concentration with and without the presence of the cylinder.

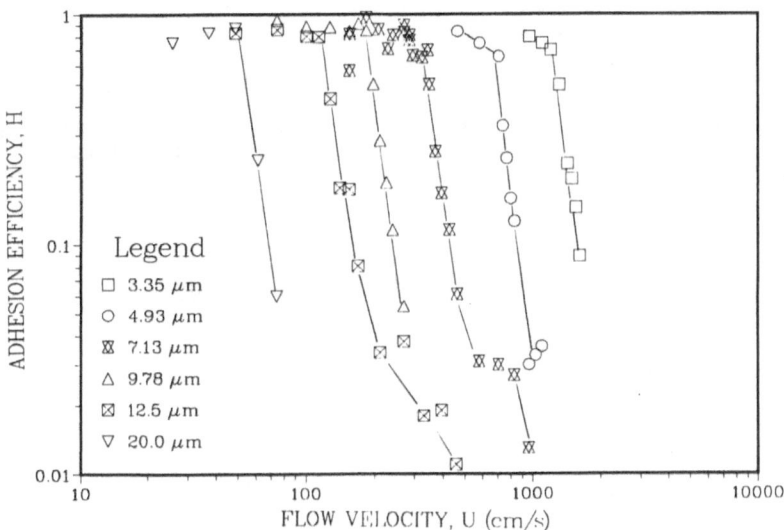

Figure 4. Adhesion efficiency as a function of flow velocity for ammonium fluorescein particles impacting a polished stainless steel cylinder.

The fitting parameters are given in Table II, where n, a, b, and r are sample number, intercept, slope, and the coefficient of correlation, respectively. The excellent coefficient of correlation (>0.98) confirms the linearity. The mean slope is 2.27 with a relative standard deviation of 5.6%. Considering the estimated experimental error of ±5%, the results for different particle sizes can be assumed parallel.

To compare with theory, the particle stagnation velocity has to be normalized by the critical velocity. For each particle size, the adhesion efficiency shows a sharp drop beyond a certain impact velocity as expected (Figure 5); however, below that point the efficiency is less than 100%. The cause for this discrepancy will be discussed later. By taking the critical velocity for each size to be the stagnation velocity at 50% adhesion efficiency, we can normalize the results and compare with theory as shown in Figure 6A. The results from different sizes converge to a single curve and the change of slope occurs at $V_s = V_{cr}$. This suggests that the selection of V_{cr} may be correct. V_{cr} would have been about 20% smaller if it had been chosen as the break point in the curves in Figure 5. In Figure 6A, theoretical curves based on Equations (1)-(5) are plotted for three coefficients of restitution. The experimental results are consistently smaller than the theoretical predictions. In Figure 6B, the theoretical results are multiplied by 0.8 which significantly improves the agreement between theory and experiment.

The cause for the 20% difference between theory and experiment may be: (1) systematic error introduced by the experimental techniques, or (2) rebounding mechanisms unaccounted for in the theory. However, we were not able to find any systematic error of experiments that can contribute as much as 20%. Reexamination of the assumptions used in the theory is necessary. In particular, the assumption that the tangential velocity component can be neglected is suspect. To investigate how the adhesion efficiency depends on the angle of incidence, we performed a differential measurement by comparing the number of deposited particles on greased and ungreased surfaces for each 5° around the cylinder.

217

Table II. Regression Parameters for the Correlation between Adhesion
Efficiency (H) and Particle Stagnation Velocity (V_s) for
Various Aerodynamic Diameters (D_{ae})

Data Set D_{ae}, μm	n*	Fitting Equation Log(H) = a + b Log(V_s) a	b	r**
3.35	5	4.27	-2.01	0.9880
4.93	6	3.96	-2.21	0.9879
7.13	8	3.43	-2.23	0.9821
9.78	7	3.14	-2.32	0.9820
12.5	5	2.52	-2.14	0.9950
20.0	3	2.30	-2.53	0.9936

*   Number of data
**   Coefficient of correlation

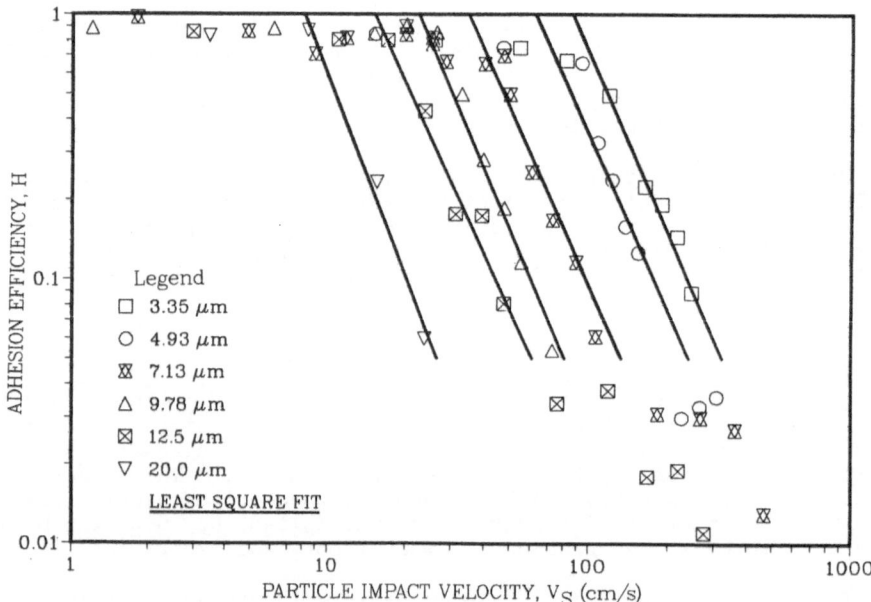

Figure 5. Adhesion efficiency as a function of the particle impact
velocity at the stagnation point. The stagnation velocity is
calculated assuming a potential flow field.

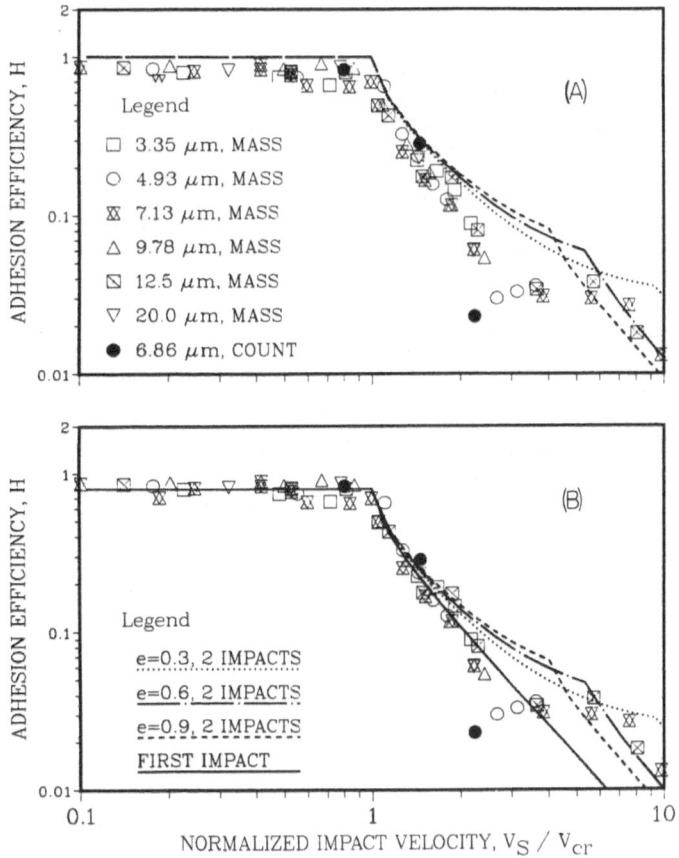

Figure 6. (A) Comparison of experimental and theoretical adhesion efficiencies as a function of the normalized impact velocity.
The experimental results for different particle sizes converge
to a single curve but are smaller than the theoretical predictions. (B) same as (A) except a correction factor of 0.8 is
applied to the theoretical values. The theoretical results at
the first impact agree reasonably well with the data.

Angle Dependence

Figures 7-9 show the counting results for three Stokes numbers (St)
of 0.402, 0.511, and 0.625. The same particle aerodynamic diameter of
6.86 μm was used and the Stokes number was increased by increasing flow
velocity. The upper half of these figures indicated by (A) gives the
number of particles in each counting window for greased and ungreased
surfaces as a function of the position angle θ, while the lower half
indicated by (B) gives the cumulative percentage of deposited particles
on greased surface and the incidence angle α as a function of θ. The
theoretical curves in (B) for the cumulative deposition and incidence
angle were obtained from a trajectory calculation for the given Stokes
number. The good agreement for the cumulative deposition between theory
and experiment confirms our counting technique.

Interestingly, the agreement is even better when only the first impact is
considered (using Equations (1) and (2)). It should be noted that the
results at less than 5% efficiency may not be reliable because of end
effects on the cylinder.

For St=0.402, the stagnation velocity of 42 cm/s is smaller than the critical velocity and, in Figure 7A, essentially the same counts are obtained near the stagnation point for both surfaces. The theory would predict no particle rebound; however, rebound was observed starting at θ=15° and increased rapidly with θ. In fact, less than 10% of impacting particles stick for θ greater than 25°. This indicates that the tangential component of the impact velocity, which is smallest at the stagnation point and increases with θ, can overcome the adhesion energy, and thus needs to be taken into account in the theory. The measured overall adhesion efficiency, 0.83, reflects the bounce at large θ, accounting for the 20% discrepancy discussed above.

For the higher Stokes number of 0.511, the stagnation velocity increases to 76 cm/s, exceeding the critical velocity. Figure 8a indicates greater than 90% rebound near the stagnation point. As θ increases, the fraction of bounce for each counting interval (5°) decreases as expected by the theory since the normal velocity component decreases. However, the fraction of bounce increases again for θ>30°, which once again demonstrates that the tangential component of the impact velocity plays an important role for large θ. The overall adhesion efficiency is 0.29.

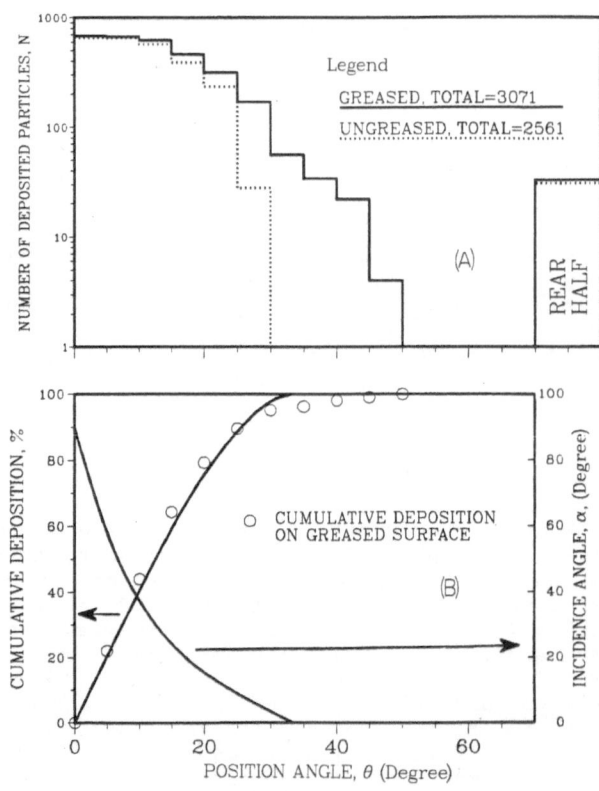

Figure 7.   (A) The number of deposited particles as a function of position angle on the cylinder for greased and ungreased surfaces at a Stokes number of 0.402.   (B) Comparison of the cumulative percentage of deposited particles on the greased surface with the theoretical prediction.   Also included is the theoretical curve of incidence angle as a function of position angle.

Further increase in flow velocity causes most particles to rebound after impact as shown in Figure 9A. Now, the overall adhesion efficiency is 0.023 and the stagnation velocity is 116 cm/s. Under this extreme condition, we would expect that most particles rebound after the first impact and make a second impact with smaller impact velocity. Therefore, there should be more particles at large θ than at small θ. However, on the ungreased surface there are fewer particles at large θ. This implies that particles rebound at the second impact, again probably due to the tangential velocity component. This also explains why the theoretical curve for the first impact agrees better with the experimental results in Figure 6B.

For particles deposited on the rear half of the cylinder (90°-180°), no significant difference was found for greased or ungreased surfaces, i.e., most of the particles stick. Presumably, the impact velocity is low, but no velocity calculations are available.

Figures 7-9(A) show that particles rebound at large position angle θ, even though the normal component of the impact velocity is smaller than the critical velocity. The incidence angle α (from the surface) as a function of the position angle for these three Stokes numbers is shown in Figures 7-9(B). For St=0.402 and θ=25° (Figure 7), corresponding to the point where significant rebound begins, the incidence angle is about 10°. Similarly, for St=0.511 and θ=30° (Figure 8), the incidence angle is about 13°. At such small incidence angles, there are apparently mechanisms by which the tangential velocity can cause particle bounce. One may involve surface roughness. At grazing incidence, a particle "sees" the front side of a surface protrusion. If such a surface feature has a slope with angle β to the surface and the particle approaches at incidence angle α, the actual particle-surface angle is $\alpha+\beta$, i.e., the

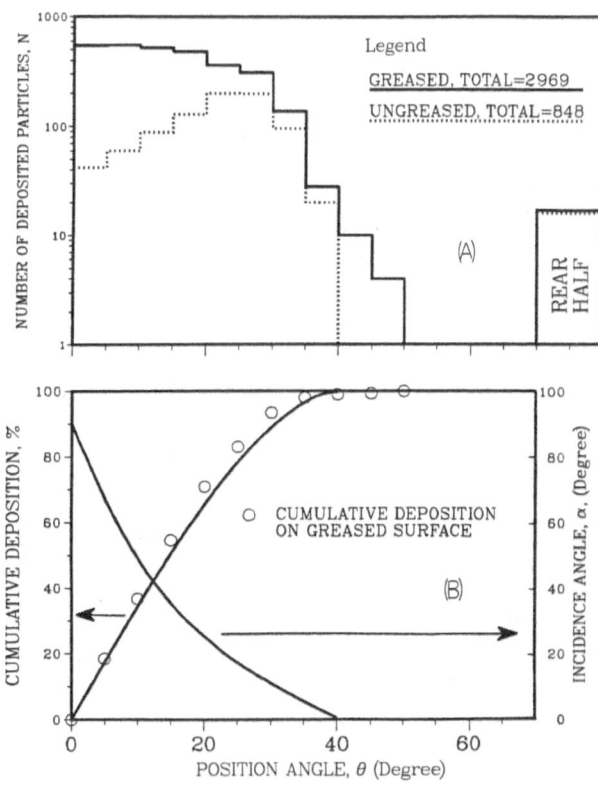

Figure 8. Same as Figure 7 except for a Stokes number of 0.511.

normal component of velocity to the surface is increased. A collision with a small surface feature can also impart angular momentum to the particle. The resulting rotation energy, which is derived from the tangential motion, can help to overcome the adhesion energy.

Critical Velocity

From the lines fitted to the data in Figure 5 the critical velocities corresponding to 100%, 50%, and 5% adhesion efficiency can be determined for each particle size. The results are plotted in Figure 10. The three lines drawn through the points are parallel, indicating a common dependence on particle diameter. For comparison, data from two other experiments are shown. Cheng and Yeh[2] investigated the bounce of latex particles from stainless steel substrates in the Sierra radial slit jet impactor. They took the critical velocity from the point where the efficiency for solid particles started to deviate from that for liquid particles. Data are also shown which derived from an ongoing experiment in our laboratory[14] involving laser Doppler velocimetry (LDV) measurements on ammonium fluorescein particles bouncing from a molybdenum surface.

From Figure 10 it is evident that all three experiments show quite closely the same dependence of the critical velocity on particle diameter, from which the average slope is

$$V_{cr} = C D^{-1.29} \qquad (8)$$

where C is a constant. Based on this size dependence and the energy balance equation for critical velocity[4]

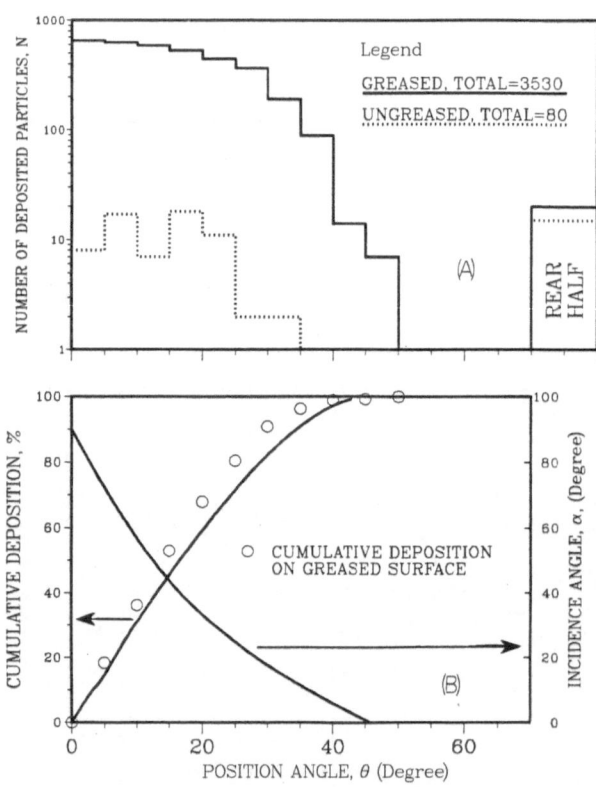

Figure 9. Same as Figure 7 except for a Stokes number of 0.625.

$$V_{cr} = (\frac{1-e^2}{e^2} \frac{2E}{m})^{\frac{1}{2}} \qquad\qquad (9)$$

where m is particle mass, we find that the adhesion energy E depends on particle diameter to the power of 0.42. The coefficient of restitution, e, has been taken to be independent of particle size. The classical theory of Bradley[15] and Hamaker[16] indicated that E increases linearly with particle size. Dahneke's flattening theory[3] predicts an even stronger size dependence than the classical theory. Therefore, both theories disagree with the experimental size dependence.

Evaluation of the adhesion energy from Equation (9) gives order-of-magnitude results consistent with Dahneke's flattening theory but 1,000 times larger than classical theory. If other physical mechanisms were involved such as plastic deformation or electrostatic interaction, E in Equation (9) would include more than the adhesion energy, affecting this discussion.

The other aspect of Figure 10 is the absolute value of the critical velocity for a given particle size. The present results for 100% adhesion are in good agreement with the data of Cheng and Yeh[2]. The present results for 100% adhesion have an estimated uncertainty of 20%. The LDV results, however, are higher by a factor of 4, and lie on the 5% adhesion line. The geometry of the LDV experiment is such that both the angles of incidence and reflection are constrained to be near 90°. By contrast, in the present experiment the angle of incidence varies from 90° to quite small angles and the angle of reflection is not determined. Similarly, in Cheng and Yeh's impactor measurements the angle of incidence varies considerably from 90° and the angle of reflection is not determined. Therefore, it appears that the critical velocity is highest for 90° incidence and reflection angles and lower for angles deviating from 90°. To further investigate this effect, measurements are needed on the critical velocity as a function of the angles of incidence and reflection.

CONCLUSIONS

The impaction efficiency of ammonium fluorescein particles on a stainless steel rod is in good agreement with a theoretical calculation by Wang[1]. The measured adhesion efficiency, however, reaches only 80% instead of 100% below the critical velocity. When the adhesion efficiency is divided by 0.8 and plotted vs. the normalized impact velocity,

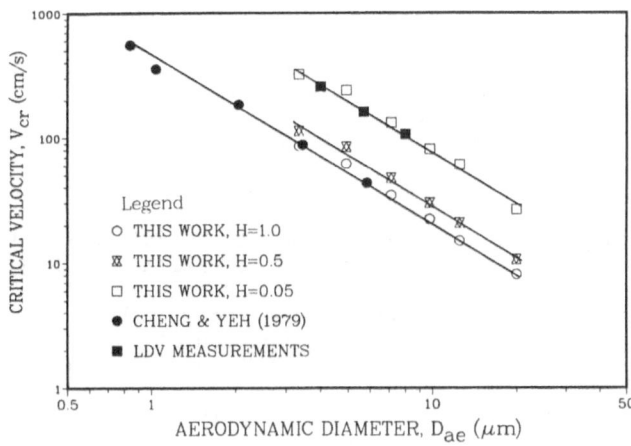

Figure 10. Comparison of the particle size dependence of the critical velocity with available data in the literature.

it agrees well with the theory of Wang. Measurements of particle rebound as a function of the position angle on the cylinder show that rebound increases rapidly with angle. Even when the velocity at the stagnation point (θ=0) was below the critical velocity, rebound started at θ=15° and exceeded 90% for θ>25°. The rebound of particles for θ>15° accounted for the 20% discrepancy in overall adhesion efficiency. The data imply that the tangential velocity component can cause particle bounce, which possibly occurs through interaction with surface roughness.

The critical velocity is found to depend on particle diameter to the power of -1.29. This is in close agreement with the results of Cheng and Yeh[2] and with unpublished LDV measurements from our laboratory. Such a dependence is not in agreement with classical adhesion theory or the flattening theory of Dahneke[3]. In magnitude, the present results for critical velocity agree with Cheng and Yeh's impactor data, but are considerably smaller than the LDV data. The LDV measurements are restricted to 90° angles of incidence and rebound whereas the present experiment and that of Cheng and Yeh[2] admit small angles of incidence and any angle of rebound. It appears likely that the lower critical velocities observed in these two experiments arise from angles differing from 90°.

ACKNOWLEDGEMENTS

We thank Professor S.L. Goren for technical review and discussions. This work was supported by the National Science Foundation under grant CPE-8442795.

DISCLAIMER

The mention of commercial products, their source or their use in connection with material reported herein is not to be construed as an actual or implied endorsement of such products.

REFERENCES

1.   H. C. Wang, J. Aerosol Sci., 17, 827. (1986)
2.   Y. S. Cheng and H.C. Yeh, Environ. Sci. Technol., 13, 1392. (1979)
3.   B. Dahneke, J. Colloid Interface Sci., 40, 1. (1972)
4.   B. Dahneke, J. Colloid Interface Sci., 37, 342. (1971)
5.   B. Dahneke, J. Colloid Interface Sci., 51, 58. (1975)
6.   D. E. Aylor and F. J. Ferrandino, Atmos. Environ., 19, 803. (1985)
7.   K. T. Paw U, J. Colloid Interface Sci., 93, 442. (1983)
8.   R. Hiller and F. Loffler, Ger. Chem. Eng., 3, 327. (1980)
9.   W. John, S. Hering, G. Reischl, G. Sasaki, and S. Goren, Atmos. Environ., 17, 373. (1983)
10. N. A. Esmen, P. Ziegler, and R. Whitfield, J. Aerosol Sci., 9, 547. (1978)
11. R. N. Berglund and B. Y. H. Liu, Environ. Sci. Technol. 7, 147. (1973)
12. J. B. Wong and H. F. Johnstone, Technical Report 11, Engineering Experiment Station, University of Illinois, Urbana, Illinois. (1953)
13. R. Israel and D. E. Rosner, Aerosol Sci. Technol., 2, 45. (1983)
14. W. John and S. M. Wall, "Fundamental and Applied Studies of Particle-Surface Interaction." NSF Annual Report under Award CPE-8442795. (1986)
15. R. S. Bradley, Phil. Mag., 13, 853. (1932)
16. H. C. Hamaker, Physica, 4, 1058. (1937)

CROSSED FIBER MODELS OF THE PARTICLE-SURFACE INTERACTION

William S. Bickel and Thomas M. Wentzel

Department of Physics
University of Arizona
Tucson, Arizona 85721

Crossed fibers are a useful model for studying the
adhesion forces that exist between sphere-sphere and sphere-
surface systems. Crossed fibers whose radii, composition and
surface structure are known can easily be micromanipulated to
establish various contact geometries. In addition, the same
two fibers can be used repeatedly to acquire good statistics
of the adhesion force. We describe a specially built
apparatus which exploits the crossed fiber geometry to
generate large numbers of contact and separation events
between the same two fibers. For each event, the apparatus
records a time which is monotonically related to the adhesion
force. These data show several interesting effects,
including "quantization," i.e., discrete values of the
adhesion force, which we describe in the paper.

INTRODUCTION

The four fundamental forces of nature are the gravitational,
electrical, strong, and weak forces. (Strong and weak forces are very
short-ranged forces which act within atomic nuclei.) The adhesion force
between a particle and a surface is not one of these fundamental forces,
but rather is a complicated mixture of several types of forces which can
act concurrently or successively to varying degrees depending on the
details of the particle-surface separation. These forces are dependent
on the large-scale curvature of the materials,[1] as well as the fine
structure of the geometry involving bumps and roughness on the surfaces.[2]
Bulk solid constants and surface energies,[2,3] both of which are material-
dependent properties, also affect the forces, as well as the presence of
liquids or gases in the region of contact.[3,4,5] The entire adhesion
process is time dependent with different effects having characteristic
times ranging from nanoseconds to years.[4,5,6,7]

Real world particles are rough and irregular. A logical place to
begin learning about the adhesion force is by dealing with perfect
surfaces or close approximations to them which can be theoretically
described. Often-modeled adhesion geometries include those of a perfect
sphere adhering to a perfect plane surface or to another perfect sphere.
Deryagin[8] has shown in his thermodynamic theory of adhesion at "convex"

contact that the force of adhesion between two convex surfaces is dependent on the curvatures and relative orientation of the surfaces. It follows from this that the contact region between two cylinders at right angles to each other, which are just touching, is geometrically equivalent to those of sphere-sphere contact and sphere-plane contact. (In addition, if the two cylinders contact each other at an angle other than a right angle, their contact region resembles that of ellipsoids in contact with each other or with planes.) This approximation breaks down if the slopes of the contacting surfaces become appreciable within the range of the forces involved. Several researchers have made use of a crossed cylinder geometry in their experiments, including Tomlinson[9] in 1928 and more recently, Gane et al.,[10] Israelachvili,[11] and Briscoe and Kremnitzer,[12] among others.

One of the aims of our research was to study the adhesion force between small particles in a statistical fashion. Taking a statistically significant number of trials is often a difficult chore. We have designed an apparatus which makes this easier. We can take hundreds or thousands of trials with the same two particles, each as small as 0.1 μm, under constant experimental conditions. We can vary these experimental conditions during or between runs. Our results will be strengthened by the statistics of many experimental trials, and perhaps we can use information from these statistics as a probe of the adhesion process.

Small spherical particles are difficult to hold on to; so it is hard to recover and reuse a given particle from trial to trial. We avoid the problem of how to hold on to such a particle by using long, small diameter cylindrical fibers. Such fibers are easy both to mount and micromanipulate to establish various contact geometries. Because of the geometrical equivalence found by Deryagin,[8] adhesion force measurements we make on our crossed fibers should be extendable to systems of spheres (or ellipsoids) in contact.

Our research group in the University of Arizona Department of Physics has already had several years of experience in making cylindrical quartz (SiO_2) fibers as small as 0.1 μm, and working with small particles in general. By using our light scattering nephelometer system and computer calculations of exact light scattering theory, we can determine the diameter of our cylindrical quartz fibers to within 0.005 μm. It is for all these reasons -- ease of fabrication, ease of diameter measurement, ease of handling, need for repeatability, and geometrical equivalence -- that we decided to use a crossed cylinder geometry in our experiments.

THE CROSSED FIBER APPARATUS

Our experimental apparatus is essentially an environmental chamber to study the adhesion force between two crossed quartz fibers of various diameters under various environmental conditions. The experimental apparatus is described with the help of Fig. 1.

A fiber of radius r_1 is mounted vertically inside an environmental chamber. Another fiber of radius r_2 is mounted 90 degrees to the first on a moveable arm, pivoted at the base of the first fiber. A motorized cam, driven at constant angular velocity, moves the arm so that the crossed, horizontal fiber comes into contact with the vertical fiber and pulls it away from its vertical equilibrium position. As the arm moves back, the force between the crossed fibers increases until at some deflection distance, x, contact is broken when the restoring force becomes greater than the force of adhesion. The fiber then swings back

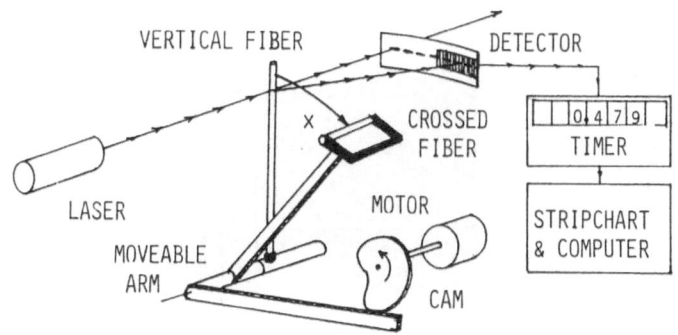

Figure 1. Schematic of the crossed fiber apparatus.

to its equilibrium position and the arm returns to make contact again. The process is continuously repeated with an adjustable cycle time depending on the motor speed.

The cam was a linear spiral, so that x varied approximately linearly with time. The adhesion force F is related to the deflection distance, x, of the fiber by a special timing circuit. A laser beam (HeNe 6328A) was directed through the chamber past the vertical fiber and on to a screen about 50 cm from the fiber. A detector was placed several degrees from the incident beam in the plane of scatter by the fiber. With the motor in operation, rotating the cam at constant angular velocity, the timing proceeded as follows:

1. A mechanical trigger from the cam set up the timing circuit to ready it for an optical pulse.

2. The moveable crossed fiber on the arm moved forward to make contact with the vertical fiber. In doing so it pushed it slightly forward (~1 mm) into the laser beam.

3. In this position the fiber scattered the laser beam into the scattering plane and into the detector creating an optical pulse which started the timer. This is the t=0 pulse.

4. As the arm moved back, the fiber was pulled out of the laser beam and away from its vertical equilibrium position. This motion continued until the fiber restoring force exceeded the adhesion force and the fiber was released.

5. After release, the fiber swung back through the vertical position into the laser beam creating a second optical pulse. This is the stop pulse for the timer. The time between the start and stop pulses showed on a digital display and was recorded in the timer's memory. For convenience this is called the "contact time," but it really is the sum of the true contact time from the t=0 pulse to the moment of fiber separation at displacement x, plus the time for the fiber to swing back through its vertical equilibrium position.

6. The arm continued to move away from the fiber to its maximum displacement and then returned to repeat the process again.

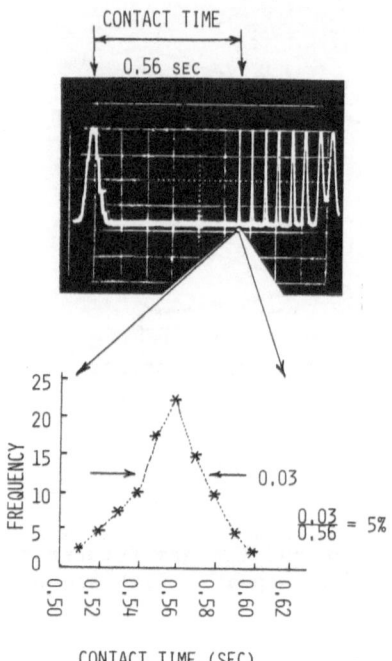

Figure 2. Top: Signal from light detector. Bottom: Histogram showing expected statistical fluctuations in contact time.

Figure 2 (top) shows an oscilloscope trace of the pulses seen by the detector and timer. The first pulse is the start pulse, the next pulse (first of a train of pulses) is the stop pulse occurring 0.55 seconds later. The extra pulses are caused by the fiber oscillating through the laser beam as it comes to rest at equilibrium. These pulses are ignored by the timer circuit.

There are several features of this experimental setup that we are using to learn about environmental effects on contact forces between crossed fibers. The chamber can be flushed with any gas. We can use dry nitrogen to reduce humidity to close to zero. The humidity can be increased by bleeding in humid air at controlled rates. The humidity and temperature in the chamber are measured by a Vaisala humidity probe.

Although experiments so far have been done at room temperature, provisions exist to flush the chamber with vapors from liquid nitrogen for cooling and to heat with small internal heaters. Therefore, we can carry out experiments in a range of controlled environmental conditions -- humidity, temperature, pressure, partial pressures of mixtures, etc.

In addition, the fiber contact geometry can be varied by changing the radii r_1 and r_2 of the fibers, the angle between the crossed fibers, and the location of the contact point by adjusting the position of the moveable fiber along the arm. Various cams can be constructed to vary the time for initial fiber contact, and rate of separation. Various speeds can be used to study other time dependent effects. The timer can measure times from 0.001 sec to 20 sec, which embraces the range of contact times we use.

The vertical fiber acts as its own force gauge. The farther from equilibrium it is pulled, the larger the force and the longer the contact time. The contact time can be related to the adhesion force at the

228

separation position when the fibers and the apparatus are calibrated.
Calibration of the apparatus involves making measurements to determine
the functional relationship between the fiber displacement x and the time
elapsed since the t=0 pulse. Calibration of the fibers involves
measuring the diameters of both, and for the vertical fiber, measuring
the length from the fixed end to the point of contact, and the time for
the fiber to fall back to equilibrium (which is independent of x for
small x). To convert the measured contact time to an adhesion force, one
would subtract the time to fall back to equilibrium from the contact
time, convert the remaining time into a displacement x, and then, using
the theory for deflection of a cylindrical beam, find the force necessary
to create that displacement. Although the calibration procedure is
straightforward, it was unnecessary in these preliminary experiments,
where we were concerned not with absolute force measurements but with the
grosser aspects of the data. Most of the initial data discussed here
were taken with uncalibrated fibers. For the data presented here we are
concerned only with the fact that F varies monotonically with t and that
larger t corresponds to larger contact times and forces.

RESULTS

    Our initial plan was to establish some basic curves for the
adhesion force as a function of humidity and fiber diameter which would
form the basis of more detailed investigations. However, our experiments
so far have turned out data that were unexpected and as yet not fully
explainable.

    Reproducibility is at the heart of these experiments. Even though
the adhesion force is a complicated function of many parameters, we
expect that for a fixed set of geometrical and environmental conditions
all contact times for the same two fibers should be equal except for
small statistical variations. Each set of data points should be
characterized by a statistical distribution about an average value with
the statistical variations dependent on the actual parameters used.

    Our experiments did not produce these expected results. In the
following figures we present some representative raw data which is a
small sample of the total accumulated. Later we will discuss their
implications and describe some investigations we did to help explain the
data. Then we will describe the additional work that has to be done.

    The quartz fibers we worked with varied in diameter between about
25 and 250 μm. The vertical fiber was approximately 6 cm long and the
crossed fiber was about 1 cm long. Data were taken at rates of 1 and 2
seconds per cycle. This was enough time to allow vibrations to dampen in
the vertical fiber. The fibers were held in a nitrogen atmosphere to
which water vapor could be added.

    Figure 2 (bottom) shows the kind of statistics expected from
repeated particle separation events. Essentially the contact time should
vary slightly about the "equilibrium" position giving some kind of a
standard statistical distribution. For directness, these data were taken
from the screen of an oscilloscope, but this does not permit measurement
of contact times as precisely as we can do using our timing circuit.

    The kind of statistical distribution produced by our new system is
shown in Fig. 3. It is a histogram showing the frequency with which a
particular contact time occurred. The inset figure is the actual record
of the contact times plotted directly onto a strip chart from the timer
memory. Contact times are indicated on the horizontal axis, the event
number goes vertically. Note that each individual event (contact and
separation) can be easily discerned. Both the histogram and data show two

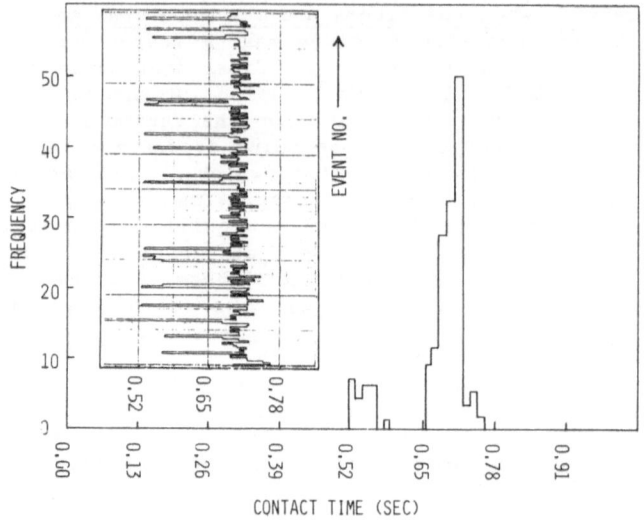

Figure 3. Histogram showing quantization of contact time. Inset: strip
chart record of contact times used to create histogram.

peaks in the contact times, one centered at 0.715 sec and the other at
0.552 sec. We refer to this phenomenon as "quantization." In using this
term we do not mean to imply that quantum physics is involved in the
adhesion process, but only that we record separate, discrete values of
the contact time, and thus the adhesion force.

     Data taken with the same fibers at a slightly later time and
different environmental conditions are shown in Fig. 4. Here two
distinct peaks occur at the same times measured for Fig. 3, but with more
scatter in the data. The histogram in Fig. 5 from another fiber pair in
a third set of conditions shows at least six distinct quantized contact
times. What is striking is that adjacent to very large peaks (high
frequency occurrences of contact times) the frequencies are zero. A
section of the associated strip chart contact times is shown in Fig. 6.

     Figure 7 shows another peculiarity in the data from crossed fibers.
Here contact time was measured as a function of humidity which increases
from right to left. The curve shows quantization, contact time
increasing with increasing humidity, and an oscillation in the contact
time. The curve A at the top progresses to the left and is continued in
curve B at the bottom.

     Figure 8 shows a general quantization feature of the adhesion force
as a function of humidity. Here too, contact time rises with increasing
humidity. Not all contact times are quantized, but it seems that for a
particular set of conditions certain quantized contact times are likely.
As the humidity changes, the probability of their occurrence changes. The
quantization appears as horizontal bands in the data. When the crossed
fiber geometry is changed, entirely different quantized contact times
appear. The quantization is highly dependent on the fiber geometry.

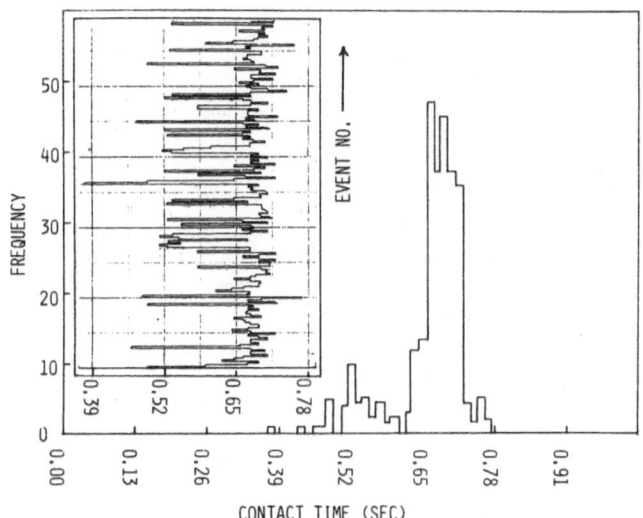

Figure 4. Histogram showing quantization of contact time, but at a later time than Fig. 3. Note broadening of both histogram peaks. Inset: strip chart record of contact times used to create histogram.

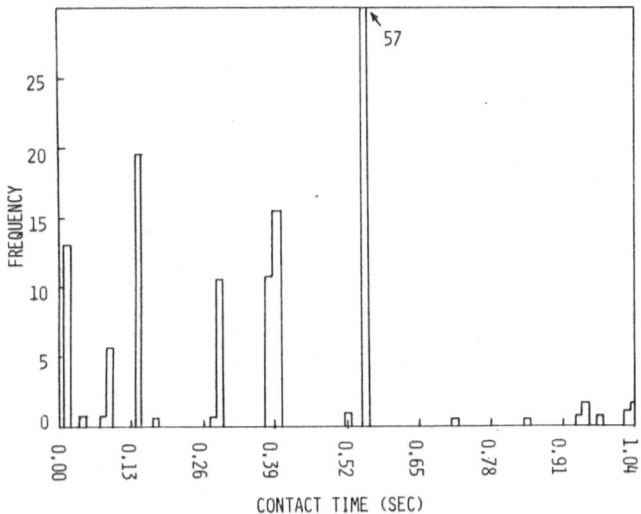

Figure 5. Histogram showing multiple quantization of contact time.

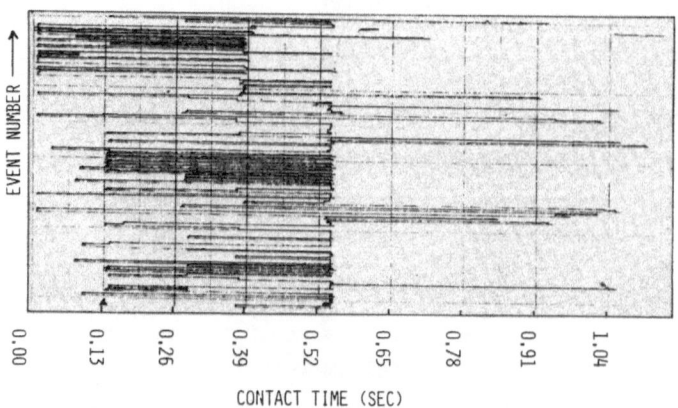

Figure 6. Strip chart record of contact times used to create histogram in Fig.5

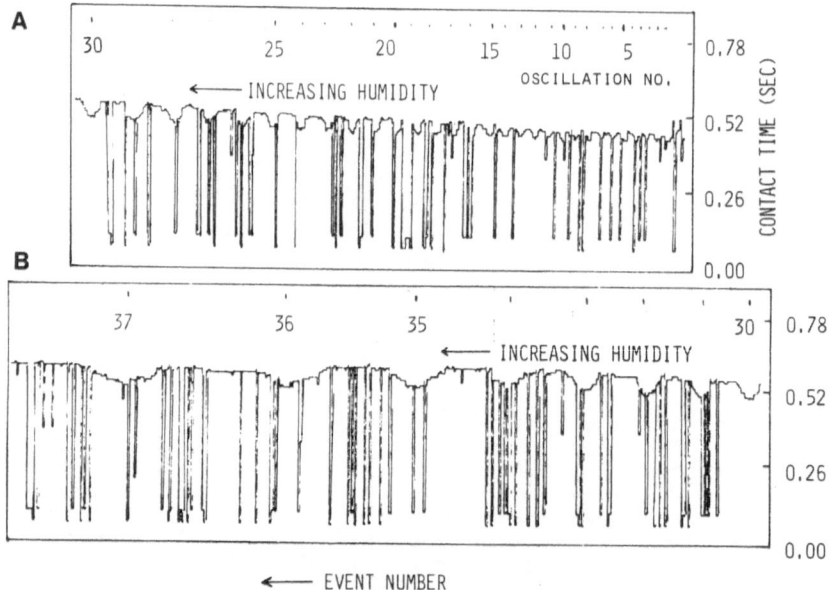

Figure 7. Strip chart record of contact times showing oscillations in contact times whose period increases with increasing humidity. Periods are numbered.

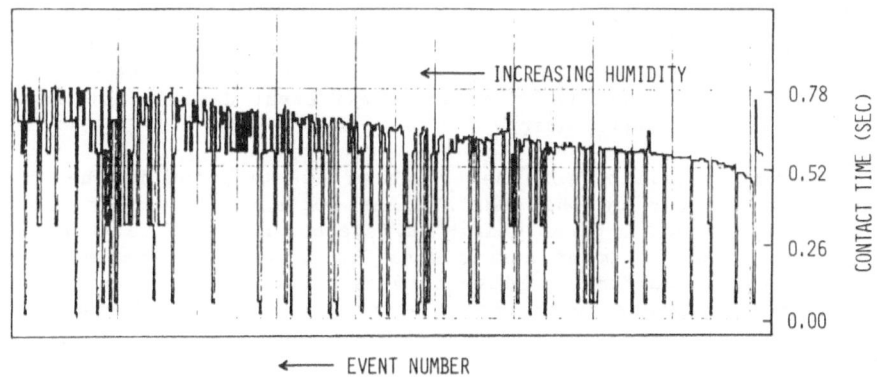

Figure 8. Strip chart record of contact times showing both quantization and increase in contact time with humidity.

DISCUSSION

In order to show that the quantization was not an artifact of the apparatus, optics or electronics, we took data with different fiber radii, lengths, and materials, and humidities and repetition rates. We isolated the experimental chamber from room vibrations and also decoupled the cam and motor from the chamber. We ruled out resonant vibrations excited within the fibers as a source of noise by monitoring the fiber's position in the laser beam. We cleaned the fibers with ethanol and acetone and inspected their surfaces for particulate contamination with an optical microscope. Further cleaning of the fibers had no effect on the results. Through all of this, quantization continued to appear in all of our data, and was easily seen by direct monitoring of the vertical fiber's motion. We have also observed quantization between crossed fibers in preliminary experiments with a new apparatus of entirely different design. At present we believe quantization to be an interesting and previously undetected aspect of particle adhesion.

The main thrust of our current research program is a continuing study of the quantization phenomenon. Zimon,[13] in his literature review of particle adhesion in a gas medium, refers to a number of researchers whose adhesion force measurements vary over a wide range, and cites lack of agreement among them. Many researchers measure the adhesion force for large numbers of particles at once, seeing what percentages are removed for various applied forces. If quantization is highly geometry dependent, the slight variations in contact geometry in a collection of "identical" particles adhering to a surface would be enough to blur the effect. Similarly, if only a few measurements are made on the same two particles, quantization may not be evident in such small numbers of measurements, or perhaps it might appear as "spurious" data points. Our apparatus, which can easily make thousands of repeated measurements with the same two particles, is ideally suited for observing the adhesion force quantization statistics.

At present, the origin of the quantization phenomenon is unknown, but we can speculate on its cause. One possibility we are investigating is intermittent sliding (slip-stick motion) between the two fibers. Studies of adhesion and frictional forces during intermittent sliding are reported by Roselman and Tabor,[14] Briscoe and Kremnitzer,[12] and Briscoe, Winkler, and Adams.[15] As our horizontal fiber pulls the vertical fiber sideways, the horizontal fiber will move in a circular arc, and the vertical fiber will preferentially move on a non-circular arc. The contact point between the two may then alternately slip and stick until contact breaks. However, we do not expect this to occur for small deflections of the vertical fiber. Also, slip-stick motion is not possible in our new apparatus of entirely different design, but in both cases we can observe quantization.

A second possibility for the source of quantization is the water adsorbed from the air by the quartz fiber surfaces. Pashley and Kitchener[16] report that a clean quartz surface can adsorb a water film of a thickness of up to 150 nm at high humidities. They also report that contaminated quartz surfaces achieve film thicknesses less than 10 nm and have large contact angles with water, which would support formation of individual droplets.

A continuous film of water on a quartz fiber would form a cylindrical sheath. Figure 9 shows some fiber sheath geometries that can occur during contact. Figure 9(A) shows two crossed fibers with sheaths prior to contact. Figure 9(B) shows them in contact but separated by two sheath thicknesses. Figure 9(C) shows separation of one sheath thickness, and 9(D) shows intimate contact between the fiber surfaces. These different geometries, which could occur in our experiments, would each give a different adhesion force. Figure 9(E) shows the sheath situation when condensation occurs to form water droplets. This case would give a larger statistical variation, but quantization could occur if one, two, or more drops randomly take part in the adhesion process.

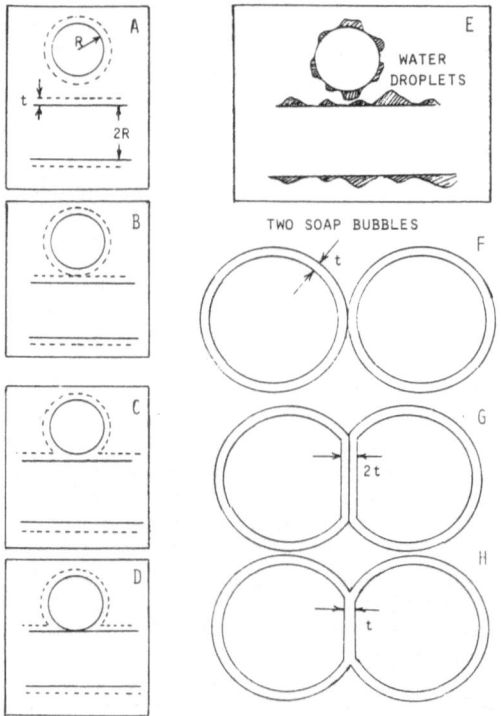

Figure 9. (A): Condensed water vapor forming sheath on quartz fibers in crossed cylinder geometry. (B, C, D): Fibers in contact, separated by two, one, and no sheath layers. (E): Water droplets on fibers in crossed cylinder geometry. (F, G, H): Two soap bubbles pushed together (F) showing their contact area to be two (G) or one (H) soap film layers thick.

We have observed that when two soap bubbles are brought into contact, the contact area can maintain its two-film thickness or, depending on the force or time in contact, can jump to a one-film thickness as shown in Figs. 9(F-H).

We can observe quantization between our crossed fibers even after they have been maintained in a dry nitrogen atmosphere for many hours. If condensed water vapor does play a role in quantization, the tiny amount remaining on the hydrophillic surface in such a dry environment must still be enough to have an effect. Both intermittent sliding and water vapor only partially explain the quantization we have observed. We suspect that quantization is due to a combination of these and perhaps other effects.

The role of humidity in adhesion is an old and complex problem, and a number of experimentalists have tried to quantify various aspects of it. McFarlane and Tabor[4] showed that in moist air, the observed adhesion between surfaces was due primarily to the presence of adsorbed water. The adhesion is dependent on relative humidity,[4,5] and on contact time.[5,18] Capillary condensation which occurs after contact can add to any water already present in surface layers to increase the adhesion force.[19,20,21] In general, the adhesion of solid particles is greatly influenced by humidity, and the effects are dependent on many factors. Much remains to be understood about this phenomenon.

CONCLUSIONS

We have built and tested an apparatus which can quickly and repetitively measure the adhesion force between a pair of crossed fibers, allowing us to study the statistics of the adhesion force. Cylindrical fibers crossed at right angles to each other provide a good model for sphere-sphere or sphere-plane contact. In our apparatus, one of the fibers acts as its own force gauge.

The crossed fiber apparatus has produced unexpected results. The force with which the fibers adhere to each other is often quantized, that is, it has certain discrete values. This "quantization" shows up as sharp peaks in frequency histograms of the data. We have performed a variety of tests which show that the quantization is not an artifact of the apparatus, and are researching possible sources for it. Our apparatus also has shown an expected increase in contact force with humidity.

ACKNOWLEDGEMENTS

We wish to thank the University of Arizona's Center for Microcontamination Control for their support of this research, and John Pattison for his expert help with the electronics.

REFERENCES

1. R. S. Bradley, Phil. Mag., 13, 853-862 (1932).
2. K. N. G. Fuller and D. Tabor, Proc. Royal Soc. A, 345, 327-342 (1975).
3. K. L. Johnson, K. Kendall, and A. D. Roberts, Proc. Royal Soc. A, 324, 301-313 (1971).
4. J. S. McFarlane and D. Tabor, Proc. Royal Soc. A, 202, 224-243 (1950).

5. W. J. Whitfield in "Surface Contamination: Genesis, Detection, and Control," K. L. Mittal, Editor, Volume 1, pp. 73-81, Plenum Press, New York, 1979.

6. H. Krupp, Adv. Colloid Interface Sci., $\underline{1}$, 111-239 (1967).

7. K. Kendall, J. Adhesion, $\underline{7}$, 55-72 (1975).

8. B. V. Deryagin, N. A. Krotova, and V. P. Smilga, (R. K. Johnston, Translator), "Adhesion of Solids," Consultants Bureau, New York, 1978.

9. G. A. Tomlinson, Phil. Mag., $\underline{6}$, 695-712 (1928).

10. N. Gane, P. F. Pfaelzer, and D. Tabor, Proc. Royal Soc. A, $\underline{340}$, 495-517 (1974).

11. J. N. Israelachvili, Contemp. Phys., $\underline{15}$, 159-177 (1974).

12. B. J. Briscoe and S. L. Kremnitzer, J. Phys. D: Appl. Phys., $\underline{12}$, 505-516 (1979).

13. A. D. Zimon, (R. K. Johnston, Translator), "Adhesion of Dust and Powder, 2nd Edition," Chapter 4, Consultants Bureau, New York, 1982.

14. I. C. Roselman and D. Tabor, J. Phys. D: Appl. Phys., $\underline{9}$, 2517-2532 (1976).

15. B. J. Briscoe, A. Winkler, and M. J. Adams, J. Phys. D: Appl. Phys., $\underline{18}$, 2143-2167 (1985).

16. R. M. Pashley and J. A. Kitchener, J. Colloid Interface Sci., $\underline{71}$, 491-500 (1979).

17. L. R. Fisher and J. N. Israelachvili, Colloids Surfaces, $\underline{3}$, 303-319 (1981).

18. M. Corn, J. Air Pollution Control Assoc., $\underline{11}$, 566-575, 584 (1961).

19. T. Gillespie and W. J. Settineri, J. Colloid Interface Sci, $\underline{24}$, 199-202 (1967).

20. F. M. Orr, L. E. Scriven, and A. P. Rivas, J. Fluid Mech., $\underline{67}$, 723-742 (1975).

21. J. N. Israelachvili, "Intermolecular and Surface Forces with Applications to Colloidal and Biological Systems," Chapter 14, Academic Press, London, 1985.

SENSITIVE NEW METHOD FOR THE DETERMINATION OF

ADHESION FORCE BETWEEN A PARTICLE AND A SUBSTRATE

G. L. Dybwad

AT&T Bell Laboratories
777 N. Blue Parkway
Lee's Summit, MO 64063

We have used an AT-cut quartz resonator with metal electrodes as the substrate. The material properties of the quartz are well characterized, and the resonate frequency can be determined accurately (1 part in 10^8 or better). When a particle was placed on the electrode, the resonant frequency increased, contrary to mass loading theory. If the particle-resonator system is modeled mechanically as a coupled oscillator system, indeed the resonant frequency should increase. The increase is determined by the bonding force constant between particle and substrate, which we calculate. Experimental data on the autohesion force constants of Au spheres on etched Au electrodes in air are presented. Also, a brief description of non-linear modeling of the system is given.

INTRODUCTION

Presently, there are very few experimental methods available for the study of macroscopic particle-substrate interactions which are nondestructive to the bond.[1] By most methods, the particles are removed from the substrate which complicates any attempt to understand the attachment bonds themselves. The most common method involves spinning the substrate at various rpm and analyzing the number and size of removed particles from photographs, i.e., the centrifuge method.[2]

This paper introduces an accurate dynamic method of studying weak bonding of a small particle lying on an electrode of a piezoelectric quartz resonator.[3,4]

METHOD

The substrate for this method was a piezoelectric resonator. In these experiments an AT-cut, etched single crystal quartz resonator with Au electrodes was used. It was chosen because its physical properties are well characterized, it is a very stable device with time and temperature, and its resonant frequency can be measured to 1 part in 10^8 or better. The resonators here operated in the MHz region being driven through a simple resistive transmission network from a variable frequency generator of fixed amplitude. The fundamental vibration mode is thickness shear as shown in Fig. 1.

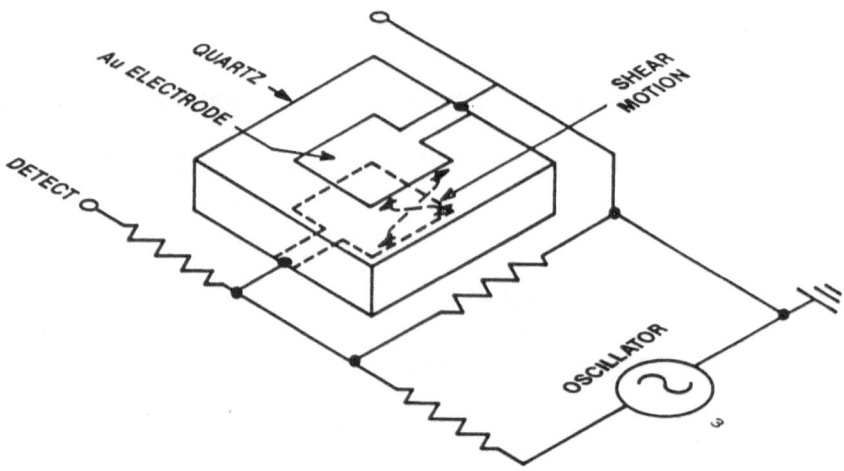

Figure 1.   Driven Quartz Test Resonantor

The particles were perfect Au spheres, 10-50μm in diameter.[5] The mass, m, of each particle was thus easily determined from its diameter and density. A few particles were simply dropped from a distance of about 2 cm onto the horizontal resonator electrode surface by first dipping a finely pointed glass rod in the Au powder, and then lightly tapping the end of the rod with the tip held over the electrode surface. All particles, but one, were then removed from the resonator surface using a single stiff fiber and a microscope to improve viewing.

We have modeled the particle-resonator system mathematically as a pair of coupled oscillators (Fig. 2).

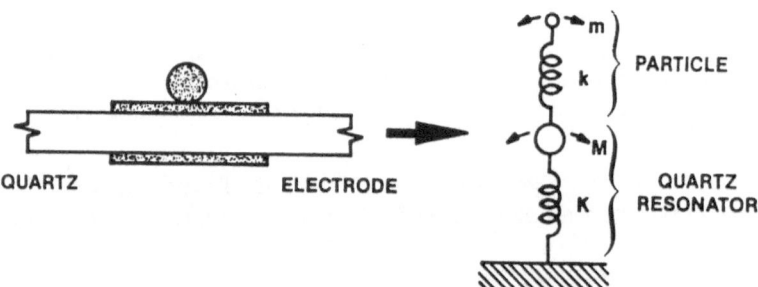

Figure 2.   Equivalent Mechanical Model of the Particle - Resonator System.

Although the resonator itself is usually modeled electrically, we have chosen a one-dimensional mechanical model in order to more easily gain physical insight into the attachment properties and behavior. In this model, the quartz resonator can be thought of as a mass M attached to a spring of force constant K. K is determined from the appropriate elastic constants of quartz, and since the initial resonator angular frequency ω is known, M is uniquely determined using

$$\omega_0^2 = K/M. \tag{1}$$

The particle of mass m is assumed to be attached by an unknown spring with force constant k to the resonator with mass M. This coupled oscillator system will now resonate at a new frequency ω determined using the same fixture as shown in Fig. 1, and ω (rad/sec) $= 2\pi f$ (Hz).

Using Newton's equations,

$$Mx_1 + Kx_1 - kx_2 + kx_1 = P_0 \sin \omega t \tag{2}$$
$$mx_2 + kx_2 - kx_1 = 0 \tag{3}$$

ω can be found in terms of k, m, K, and M:

$$2\omega^2 = \left[\frac{K}{M} + \frac{k}{M} + \frac{k}{m}\right] \pm \left[\left[\frac{K}{M} + \frac{k}{M} + \frac{k}{m}\right]^2 - 4\frac{K}{M}\frac{k}{m}\right]^{1/2} \tag{4}$$

Equation (4) is plotted with measurable system frequency ω versus the unknown particle attachment force constant k, in Fig. 3. For weak particle binding, the system frequency should increase. This contradicts normal mass loading theory which states that resonant frequency should decrease as mass is deposited on a resonator. That is, from Eq. (1), as the resonator total mass M increases, the resonant frequency ω will decrease. The maximum coupled oscillator frequency is reached when the natural frequency of the particle equals the natural frequency of the quartz resonator ω. For stiffer values of k, the frequency asymptotically approaches a value lower than ω. This asymptotic value equals the frequency decrease the resonator would have experienced if the mass of the particle was an integral, total, part of the quartz resonator and not loosely bound to the resonator. That is, for k=K, the quartz resonator spring constant, Eq. (4) becomes approximately.

$$\omega^2 \simeq \frac{K}{M + m},$$

which is the mass loading of Eq. (1) with the particle, m, rigidly attached.

RESULTS

Experiments have confirmed the coupled oscillator model for varying values of the attachment parameter k. Table I shows this for some of the experiments conducted here; the relative location of each calculated k with respect to the vertical discontinuity shown in Table I,

$$(k = m\omega_0^2), \tag{5}$$

can be determined using the values listed in the lower table.

Other experiments which have used a few angstroms of evaporated Au to "cement" particles to the substrate produced negative frequency shifts, implying firm attachments ($k \simeq K$), as shown on the right in Fig. 3.

Table I. Experimental Determination of the Bonding Force Constant, k

EXPERIMENT	$\Delta\omega$(ppm)	PARTICLE DIAMETER (μm)	CALCULATED k(d/cm)
1	+0.1	10	6×10^6
2	+0.5	25	1×10^7
3	+1.5	50	4×10^7

MAXIMUM POSSIBLE k FOR POSITIVE $\Delta\omega$ $k = m\omega_0^2$			
		10	3×10^7
		25	4.1×10^8
		50	3.2×10^9

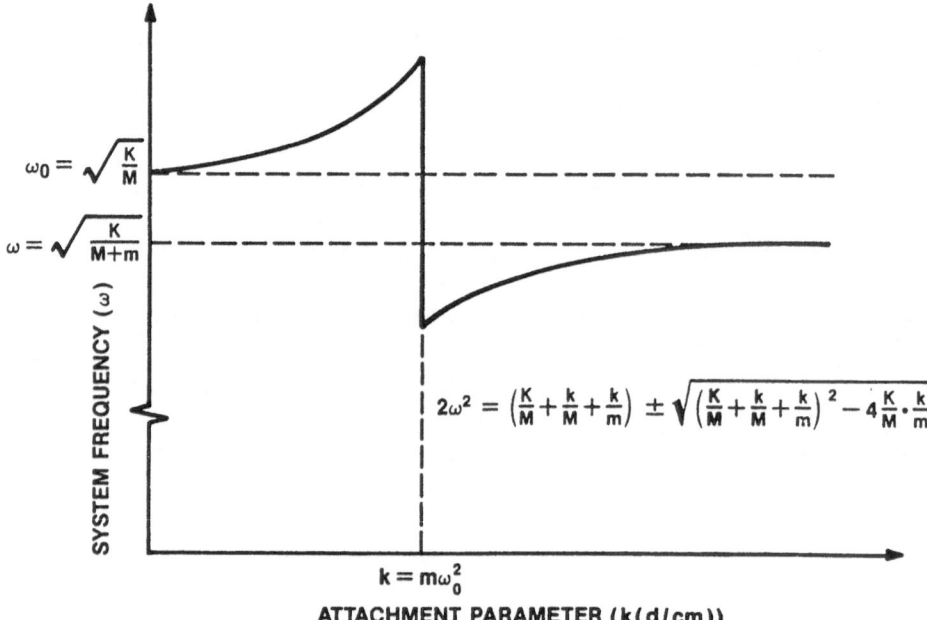

$$2\omega^2 = \left(\frac{K}{M}+\frac{k}{M}+\frac{k}{m}\right) \pm \sqrt{\left(\frac{K}{M}+\frac{k}{M}+\frac{k}{m}\right)^2 - 4\frac{K}{M}\cdot\frac{k}{m}}$$

Figure 3:   Ideal Particle - Resonator Frequency Behavior

The placement of Au spheres on Au electrodes quartz resonators in air produced positive frequency shifts of 0.1-3 ppm depending on surface cleanliness, surface roughness, etc. Analysis of the coupled oscillator frequency Eq. (4) shows that this implies an attachment force constant k of 10^6 to 10^7 dyne/cm in shear. Since shear constants obtained from various molecules using IR spectroscopy[6] range from 10^2 to 10^5 dyne/cm, the bonds found here represent about a thousand atomic bonds or ten thousand van der Waal bonds. This correlates with the known fact that the area of contact of two contiguous solids is very much less than their geometrical area.

EXPANDED MODEL WITH RESULTS

It was clear from our experimental data of single particles on quartz resonators with gold electrodes, that other effects besides an increase in resonant frequency occurred. For instance, the Q or "quality factor" for the quartz resonator, an inverse measure of the damping, was changed when a particle was placed on the resonator. Generally, the Q decreased, sometimes by more than an order of magnitude, implying an increase in system damping (Fig. 4).

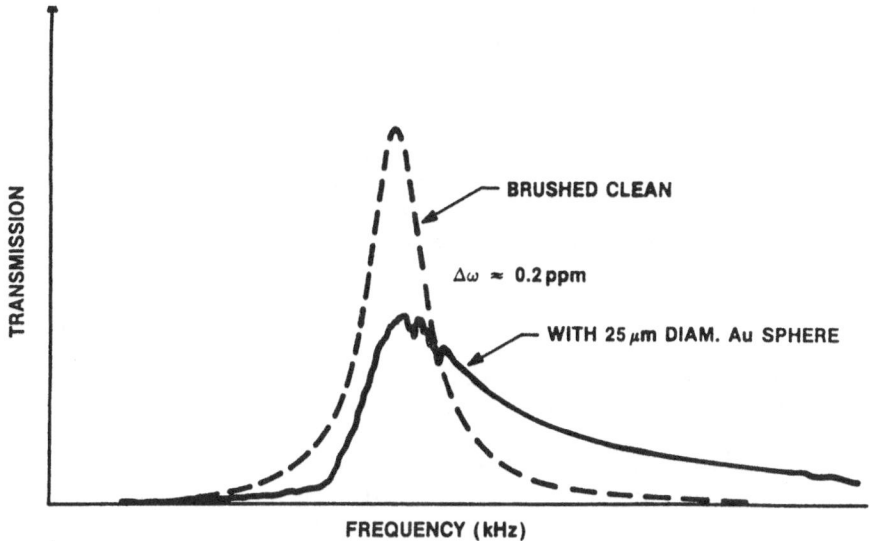

Figure 4.   Response of Resonator to a Au Dust Particle.

Also, it was found that the resonant frequency of the resonator-particle system changed with the amplitude of the resonator sinusoidal driving voltage, i.e., the coupled system was non-linear. These observations implied that Newton's Equation for a coupled system (equation 2 and 3) should be expanded to better describe the observations. A variety of physical models and their mathematical representations in Newton's equations were investigated. We have added four terms which best describe our laboratory observations: $\gamma_0 x_1$, dashpot damping of the quartz resonator; $\gamma_2 (x_1 - x_2)$, dashpot damping of the particle; $\pm$ F, Coulomb dry friction damping of the particle; and k' $(x_2^3 - x_1^3)$, a non-linear reduction or "softening" of the particle attachment spring constant, k, at high vibration amplitudes.

We have solved this system using the method of Bogaliubov and Mitropolsky for non-linear systems.[7] The results agree with our experimental observations:

1. The system resonant frequency depends on the drive level amplitude, P_o.

2. The particle will break away from the resonator at large vibration amplitudes, viz., when $X^2 \geq k/k'$.

3. The system resonant frequency can still increase for particle force constants, k, which are small, as long as the particle damping, γ_2 and F, are not too large.

4. The total system damping, Γ, is also drive level amplitude dependent:

$$\Gamma = \gamma_0 + \gamma_2 + \frac{4F}{\pi X \omega}$$

5. The Coulomb damping, F, has a profound effect at low amplitudes, x, preventing some resonators with particles from vibrating altogether.

FUTURE CONSIDERATIONS

To study a wider range of values of the coupling force constants found here, a variety of quartz resonant frequencies should be used rather than the single frequency used here. This would require the use of other AT-cut resonators, which operate best in the MHz range, as well as other quartz cuts which offer resonant frequencies outside the MHz range. Some additional ideas for experimentation include use of different resonator electrode materials, different resonator surface finishes and cleaning methods, resonator operation in vacuum rather than air to eliminate loading effects, and evaluation of coupling force constants with time (aging)[8], static charge[8] and temperature.

SUMMARY

We have introduced a sensitive method of studying particle attachments by employing an AT cut shear mode quartz resonator with well characterized physical properties. A simple coupled oscillator model of the particle-resonator system predicts positive and negative frequency shifts, which have been confirmed by experiment. The frequency shift with particle in place has been related to an equivalent particle attachment shear force constant, k. This shear force constant was found to be about 10^6 to 10^7 dyne/cm for small gold spheres on gold electrodes in air. Additional experimental results with particles on resonators show that the resonant frequency and damping depended on the drive level amplitude, and that large drive level amplitudes can remove particles. An expanded version of the original coupled oscillator model which includes a non-linear particle restoring force constant and two types of damping confirmed these experimental results.

ACKNOWLEDGEMENT

The author would like to thank T. R. Meeker for helpful discussions concerning particle-resonator interactions.

REFERENCES

[1]   W. G. Bohme, et. al., "Haftung Kleiner Teilchen an Feststoffen", Zeit. Angew. Phys., 16, 186 (1962).
[2]   T. Allen, "Particle Size Measurement", Chapman and Hall, 1981, Chapter 12, p.350.
[3]   G. L. Dybwad, Bull. Am Phys. Soc., 19, 1082 (1974).
[4]   G. L. Dybwad, J. Appl. Phys., 58, No. 7, 2789 (1985).
[5]   Metz Metallurgical Corp., 3900 S. Clinton Avenue, S. Plainfield, New Jersey.

[6] G. Herzberg, Infrared and Raman Spectra (Van Nostrand, New York, 1956), pp.172-174.

[7] N. Bogaliubov and Y. Mitropolsky, "Asymptotic Methods in the Theory of Non-linear Oscillations", Gordon and Breach, New York, 1961.

[8] W. Kothler, H. Krupp, and H. Rabenhorst, "Adhesion and Electrically Charged Particles", Zeit. Angew. Phys., 24, No. 4, 219 (1968).

PART III. PARTICLE DETECTION, ANALYSIS AND
CHARACTERIZATION

DETECTION OF PARTICLES ON CLEAN SURFACES

Josef Berger

VLSI Standards, Inc.
2660 Marine Way
Mountain View, CA   94043

Optical particle sizing by surface particle detectors using PSL
spheres and monochromatic laser light on silicon wafers is non-
monotonic and multi-valued. The substrates with different optical
properties have a significant effect on optical particle sizing.
It is important that surface particle detectors be calibrated using
PSL spheres on all the layers which are important in a given
process. The effect of a substrate with different reflectivity or
different optical properties cannot be represented, even in the
first order theory, by just changing the scattering cross section
by the reflectance factor.

INTRODUCTION

The importance of particles in processing of Very Large Scale Integrated
(VLSI) circuits has been recognized for many years; however, quantitative
work only started after the first surface particle detectors became
available, which was about five years ago. Since that time, process
engineers have been able to see the number and the size of particles
deposited on silicon wafers[1]. In this paper, we would like to discuss
and experimentally illustrate some of the pitfalls and difficulties in
detection and sizing of particle by optical means.

PARTICLES IN FREE SPACE

The referenece particles used for instrument calibration are polystyrene
latex (PSL) spheres. These have been recognized by the industry as the
reference particles for many years. PSL spheres were orginally used to
challenge filters and to calibrate airborne particle counters [2,3]. Today
their use is being extended to calibration of surface particle detectors.
The theory of light scattering by dielectric spheres was developed by
Professor Gustav Mie in 1908, and this theory represents a good starting
point for our discussion.

Let us take a look at the nature of light scattered by small particles.
First, let us look at very small particles, much smaller than the
wavelength of icident light. Such particles are called Rayleigh
scatterers and light scattered by these particles is characterized by the
fact that the intensity of forward scattered light and back scattered

light are the same. In other words, these small particles scatter light
in a very symmetrical fashion. Larger particles, in the range of half a
micrometer to several micrometers, are characterized by the fact that the
light is scattered predominatly in the forward direction. This is an
important fact because surface particle detectors collect the scattered
light above the wafer. It is clear that in order for all the light to be
collected, it first has to interact with the substrate. The ability of
particles to scatter light is characterized by a coefficient which is
called the "scattering cross section". Its dimension is area and the
scattering cross section is simply the optical size of the particle. The
scattering cross section is determined by integrating the light intensity
over all angles other than the angle of the reflected light. At this
point, it is useful to define a so-called pratical scattering cross
section, which will be determined by a pratical instrument which collects
light only over a certain size solid angle. The following experimental
results are based on measurements taken using Tencor Instruments'
Surfscan 4000. Figure 1 shows a schematic of collection optics of such
an instrument. We see that this instrument collects light in a solid
angle starting from about 3.5 degrees of normal to 45 degrees of normal,
except for a slot in the light collection sphere also shown in the Figure
1. All of the experimental results are based measurements of light
collected in this solid angle.

PARTICLES ON BARE SILICON

PSL spheres deposited on bare silicon represent a reference case. In
our experiments, we have deposited various sizes of PSL particles on bare
silicon wafers. The particles are deposited in random fashion over the
wafer in a monodisperse population. This means the particles are not
clustered and care was taken to minimize the density of additional
particles which could be introduced in the deposition process.

PARTICLES ON THIN LAYERS

Table 1 shows in the first two columns latex spheres of different
diameters deposited on a bare silicon wafer. This is a reference case.
The scattering cross sections for given latex size spheres are listed in
the second column. The scattering cross section is a non-monotonic,
mukti-valued function of the latex diameter. There is a significant
minimum in the neighborhood of the latex sphere of 0.76 micrometer
diameter. This effect is common for all surface particle detectors
which are using monochromatic, circularly polarized light. This
phenomenon was discussed in several papers in the literature [5,6]. As was
mentioned, the PSL spheres on bare silicon are the reference case. The
remaining columns of Table 1 show the measured value of the scattering
cross section of PSL spheres of a given diameter deposited on oxidized
silicon wafers or wafers containing other surface layers, normalized to
the scattering cross section of the same size PSL shperes on bare silicon.
Let us examine the behavior of silicon dioxide layers on siliconof
different thicknesses. The third column shows results of the measurement
of particles on 380 Å of silicon dioxide. The behavior of latex spheres
on this layer is very similar to the behavior of latex spheres on bare
silicon. The data show variation from 70% to about 104% of the amount
of light scattered by PSL spheres of various diameters on bare silicon.
The next layer thickness 1080 Å, represents a quarter wavelength of
incident light (6328 Å); the refractive index for silicon dioxide is
1.46. This case shows much more dramatically that the scattering cross
section of latex spheres on this layer is significantly affected by the
silicon dioxide layer. The smaller size latex spheres scatter only about
50% of the amount of light they scatter when deposited on a bare silicon
wafer and the minimum ratio occurs for a 0.9 micrometer latex sphere.

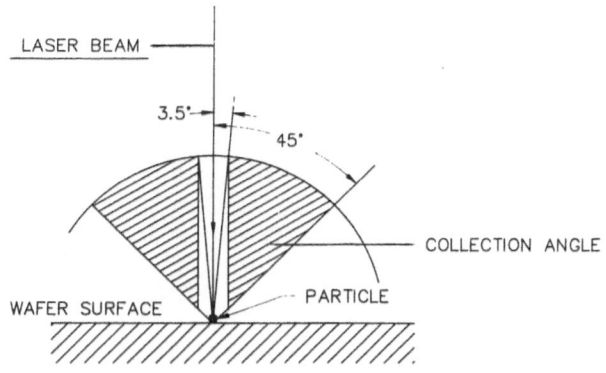

Figure 1. Schematic of collection optics.

The column shows the ratios for a silicon dioxide layer of 2160 Å where the variations become even much more wide spread. The next layer, which is a one wavelength layer of 4329 Å, is very interesting. There is a significant variation in the amount of light scattered by particles on such silicon dioxide layer compared to the reference case. The amount of light scattered varies from about 54% to 270%.

Silicon dioxide on silicon is a very common layer and in Figure 2 we see the relative scattering cross section of 1.1 micrometer PSL spheres on silicon dioxide layers of varying thicknesses. We have examined many silicon dioxide thicknesses on the range of up to 1 micrometer. The plot shows that the behavior of the relative scattering cross section is quite complex, and in order to understand it, many more measurements in much finer steps of silicon dioxide thicknesses would be needed. Columns 7 and 8 of Table 1 describe the behavior of latex spheres on silicon nitride

TABLE 1

LATEX SPHERES ON SILICON AND THIN LAYERS

COLUMNS:

1	2	3	4	5	6	7	8	9
PSL µm	SCS µm²	OXIDES: 380 Å	1080 Å	2160 Å	4320 Å	NITRIDES 825 Å	1400 Å	CHROMIUM 1200 Å
0.364	0.24	0.96	0.46	0.71	0.54	ND	0.80	2.02
0.5	0.61	0.93	0.55	0.97	0.84	ND	0.84	1.82
0.624	0.54	0.69	0.52	1.28	1.24	ND	1.07	1.85
0.76	0.20	0.80	0.49	1.80	2.70	ND	1.06	1.40
0.9	0.74	0.93	0.24	0.35	-	ND	0.49	-
1.1	0.93	0.84	0.69	1.25	0.99	ND	1.03	1.90
2.02	2.1	1.01	0.61	0.90	0.73	0.10	0.77	2.24
5.0	6.74	1.04	0.68	0.82	0.59	0.22	0.48	2.14

ND...not detectable

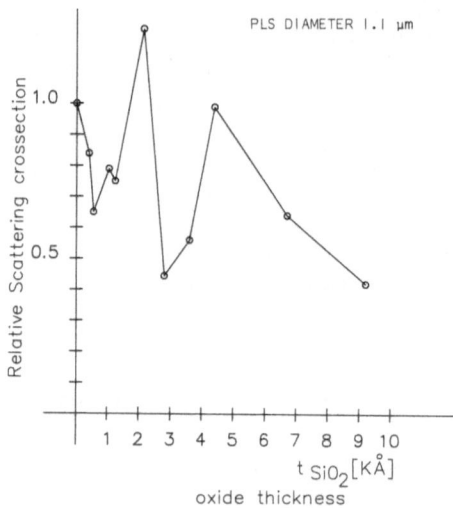

Figure 2. Relative scattering cross section as a function of oxide thickness.

layers. It is very interesting to note as shown in column 7 that a silicon nitride layer of 825 Å thickness causes a dramatic change in the relative scattering cross section. The result is that 2 micrometer diameter latex spheres are just barely detectable. All PSL spheres smaller than 2 micrometers fall at or below the detection limit of our instrument. The importance of this fact is very significant for VLSI technologies, since layers similar to 825 Å are routinely used for masking of field oxidation. The following column shows silicon nitride of 1400 Å. Latex spheres all the way down to 0.364 micrometer are again detectable.

Column 9 of Table 1 shows scattering ratios of latex spheres on chromium coated wafers. As can be expected, the PSL spheres on a shiny chromium surface scatter about twice as much as light as the PSL spheres on bare silicon. These data have significant implications for VLSI processes because we have demostrated that in order to understand the behavior of a surface particle detector on silicon wafers with different layers, it is important to calibrate the instrument for the specific layer thickness and its optical properties.

CONCLUSION

In this paper, we have shown that optical particle sizing by surface particle detectors using PSL spheres and monochromatic laser light on silicon wafers is non-monotonic and multi-valued. This effect was reported previously in the literature [5,6] and it is common for all surface particle detectors which are using monochromatic laser light. We have also shown that substrates with different optical properties have a significant effect on optical particle sizing. For example, an 825 Å silicon nitride layer on silicon changes the particle optical size by almost an order of magnitude. It is important that surface particle detectors be calibrated using PSL spheres on all layers important to a given process. Specular metal layers increase the apparent particle size. The effect of a substrate with different reflectivity or different optical properties cannot be represented, even in the first order theory, by just changing the scattering cross section by the reflectance factor.

250

REFERENCES

1. J. Berger and R. Ervin, Contamination in Semiconductor Processing, 1986 Proceedings of the Institute of Environmental Sciences, p. 424.
2. D.M. Garvey and R.G. Pinnick, Response Characteristics of the Particle Measuring Systems Active Scattering Aerosol Spectrometer Probe (ASASP-X), Aerosol Sci. Technol., $\underline{2}$, 477 (1983).
3. J. Gebhart, et al, Counting Efficiency and Sizing Characteristics of Optical Particle Counters, 1985 Proceedings of Institute of Environmental Sciences, 31st Annual Technical Meeting, p.102
4. P. Gise, Application of Laser Scanning for Wafer and Photoplate Inspection, Microcontamination, 41, (October/November 1983).
5. R.G. Knollenberg, The Importance of Media Refractive Index in Evaluating Liquid and Surface Microcontamination Measurements, 1986 Proceedings of Institute of Environmental Sciences, p.501.
6. B.R. Locke, R.P. Donovan, and D.S. Ensor, The Detection of Polystyrene Latex Particles on Polished and Oxidized Silicon Wafers: An Instrument Characterization, 1986 Proceedings of Institute of Environmental Sciences, p.487.

DETECTION OF PARTICLES DOWN TO A "FEW" MICROMETERS ON NON-SPECULAR MICRO-
ELECTRONIC SUBSTRATES AND OTHER SURFACES

Clarence Smith and Ted Ross

IBM Corporation
East Fishkill Facility
Hopewell Junction, NY   12533

Detection of contamination "where it falls" whether it be on
substrates, work surfaces or in tools, can challenge ones in-
genuity. The emphasis in this paper is on practical techniques
which have been useful for contamination measurement in a substrate production environment. Grazing angle light (dark
field) techniques are used in the detection of particles on
non-specular surfaces. These techniques are useful on
particles down to a few micrometers, depending on the surface
roughness and topology.

A range of illumination sources are described which are applicable to various substrate geometries. Methods of observing,
sizing and recording these particles range from the unaided
eye to step and repeat image analyzers. Tools and fixtures
utilizing the above principles are shown and described. Techniques for calibration are discussed as well as are tool
enhancements for rapid particle location in follow-on analytical activity.

INTRODUCTION

Control of contamination has become increasingly important as packag-
ing density has increased and feature sizes have become smaller. As processes mature, contamination becomes a major loss mechanism.

Since the critical feature dimensions of substrates are in the tens of
micrometers, the emphasis of contamination control must be on particles of at
least this size and larger. Usually, the minimum particle size to be monitored and controlled is considered to be one-third to one-tenth of the critical feature dimension.

Particles of these sizes tend to migrate within clean rooms on surfaces.
Therefore, detection and control of debris on surfaces become of prime importance.

Typical contamination migration sources whose surface contamination must
be controlled are: product carriers, work stations, carts, tools, materials
and walls. Airborne particle detection methods using particle counters or

air sample filtration are not effective since particles of these sizes are not normally airborne.

Often measurement of surface contamination is accomplished by rinsing the surface and measuring the quantity of particles contained in the rinse solution. This is accomplished either by a liquid particle counter, if size is appropriate, or by filtration and then counting and sizing the particles on the filter. The main uncertainty with this method is the percentage of surface contamination which transfers to the solution.

Another method sometimes used is the indirect approach of placing test plates in the area of interest and determining the area particle deposition rates. Test plates are selected such that the contamination level can be determined by the most suitable method, depending on requirements and technology at hand. Techniques which can be appropriate are densitometers, automatic defect measurement equipment, image analyzers and other automatic particle counting equipment.

Instead of the above indirect approaches, grazing-angle light techniques will be discussed in this paper which allow direct detection of this surface contamination. To understand detection capabilities and limitations, we must first discuss the principles of this technique.

GRAZING ANGLE LIGHT TECHNIQUE

Light is directed across the surface at a grazing angle such that particles on the surface will scatter light but the surface will be barely illuminated, if at all (see Figure 1). An observer will see this scattered light as points of light against a dark surface, as shown in Figure 2. Particle sizes down to a few micrometers can be seen using this technique. Sensitivity depends on the light energy and ambient light as well as the particles shape, texture and index of refraction.[1] Flatness and finish of the surface on which the particle is resting will also affect the sensitivity,[2] as these parameters affect both the light energy reaching the particle and the light scattered from the surface, thereby affecting the signal-to-noise ratio. Due to the wide ranges of possible values of these parameters, the sensitivity in specific situations is difficult to predict.

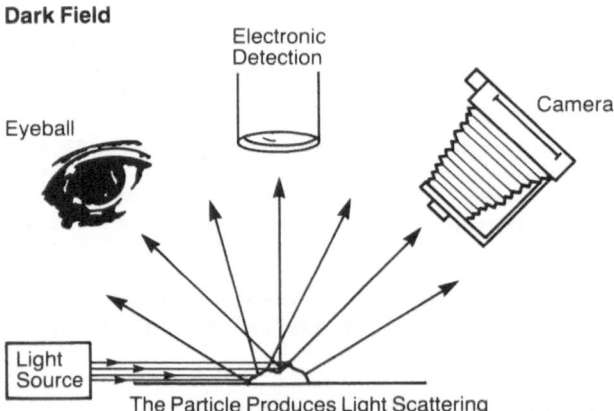

Figure 1. Principles of Grazing Angle Light particle detection.

Figure 2. Dark Field Display of particles illuminated by grazing angle light.

Fig. 3. Particle detection at output of a cleaner using low angle lamp illuminator.

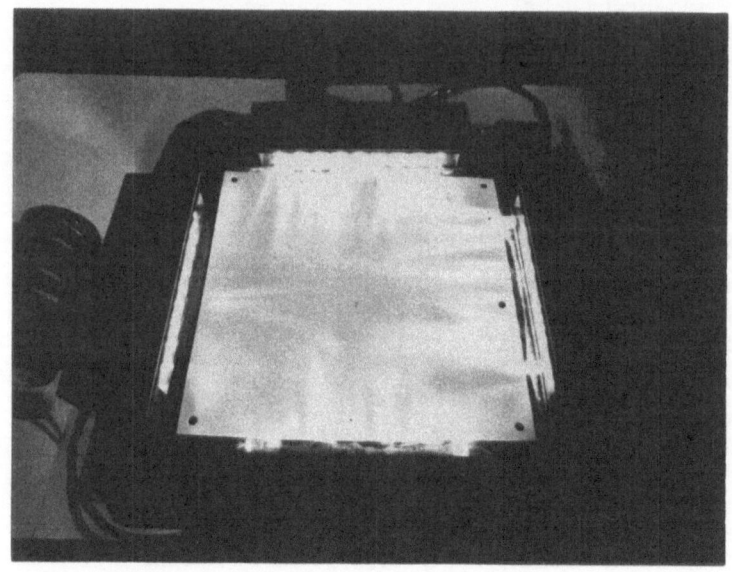

Figure 4.   Particle inspection station using 4-way multiple lamp
illumination.

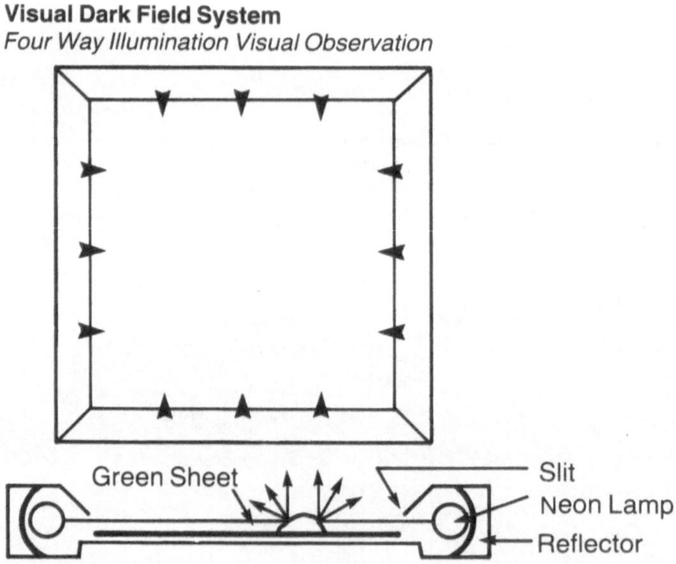

Visual Dark Field System
Four Way Illumination Visual Observation

Green Sheet

Slit
Neon Lamp
Reflector

Figure 5.   Diagram of inspection fixture using fluorescent or neon tube
illumination.

The "size" of the particle, as determined by the intensity of the scattered light, appears to be much larger than actual size due to the high energy of the illumination source.

Techniques for achieving grazing angle illumination can combine a variety of lenses and mirrors, with light source types varying from adjustable neck lamps to lasers. Detection techniques range from eyeballs to image analyzers.

LIGHT SOURCES

Simple microscope lights or flexible neck lamps are often adequate for small area use. Figure 3 shows such a system used to determine the effectiveness of a ceramic sheet-cleaning tool. For larger areas, multiple lens-end lamps can be used (Figure 4). However, multiple lamps usually result in an uneven illumination field. Slide projector systems offer a relatively low cost light source with optics. A slit aperture slide can provide broad grazing angle illumination.

For larger areas, confined space or limited access, fluorescent lamps can be used to advantage. When combined with a slit opening and cylindrical lens or mirror, a compact unit can be achieved, as in Figure 5. A fluorescent system will often require a darkened inspection station due to the relatively low light intensity provided.

Neon tube light sources offer an improvement over the fluorescent lamp in both intensity and flexibility. A neon tube can be bent such that a grazing angle light can be provided completely around the object to be inspected, Figure 6.

For specific applications fiber optics are useful. Figure 7 shows two fiber optic illumination heads. Fiber optics offer the advantage of allowing the light source convection currents and cooling fans, which create and distribute contamination, to be remote from particle detection surfaces. For higher intensity requirements, mercury vapor lamps and scanning lasers can be used.

PARTICLE DETECTION METHODS

The simplest method of particle detection is visual. Highest sensitivity is obtained when the field is darkest and ambient light is low, i.e., high signal-to-noise ratio. Particle detection capability with the unaided eye can range from 50μm down to 2μm. To achieve 2μm detectability, the viewing surface must be highly polished (0.05μm surface finish) and the viewing position must be optimized. The use of magnifying lenses and microscopes will improve sensitivity.

Photographing the dark field display with an instant camera can again improve the sensitivity while at the same time provide a photographic record, (see Figure 8). Notice the blooming of the particle "images" and the resultant apparent large particle size. Photographing the image can remove some subjectivity from the particle count. Size distribution can also be obtained from the "spot" size on the photograph.

Image Analyzers can be employed to directly read the scattered light surface. Figure 9 is a diagram of a standard system. A TV type system is easily adaptable and has the advantage of providing hard copies of the screens for later analysis. Particle sizes histograms are also useful when

Figure 6.   Inspection tool, using neon tube, as shown in Figure 5.

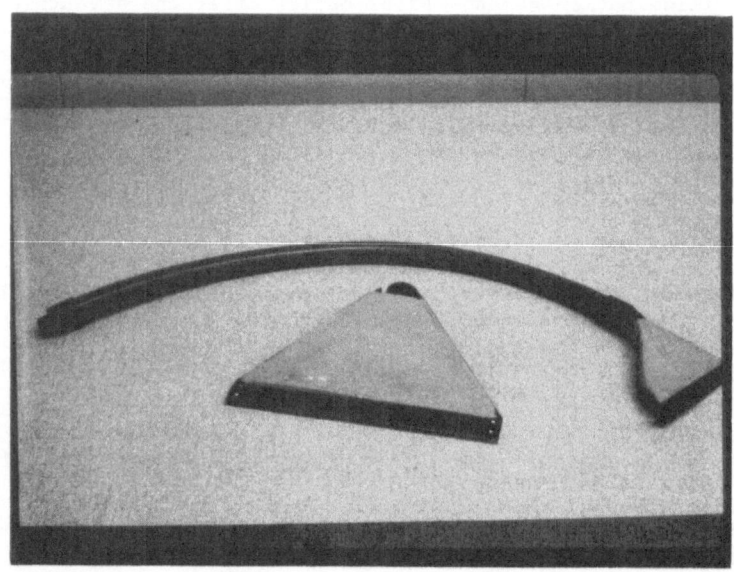

Figure 7.   Fiber Optic, slit illuminator heads.

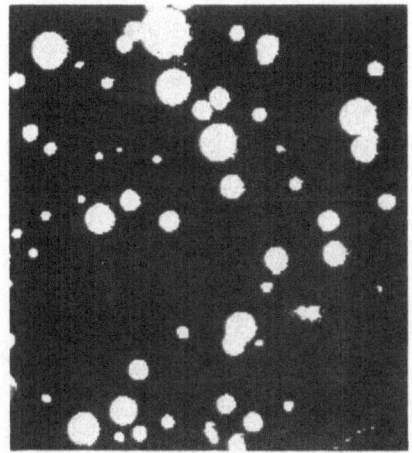

Figure 8. Photographic record of dark field, scattered light display.

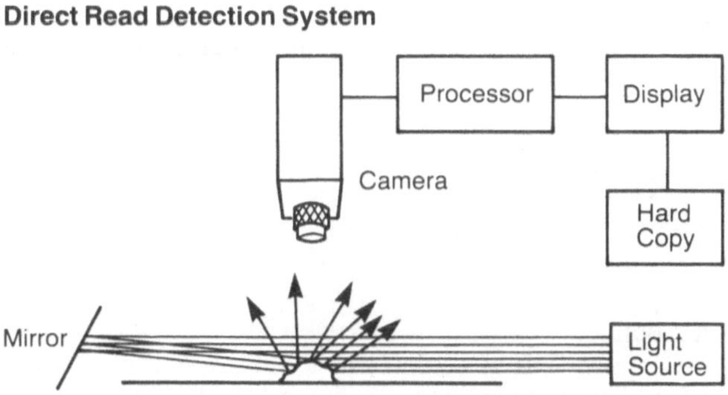

Direct Read Detection System

Figure 9. Diagram of TV type image analyzer detection system.

different products are sensitive to different particle sizes. This tool can provide rapid total count and size distribution information. An image analyzer can also be used to read photographs in those instances where photography is more practical than direct read with an analyzer. The purpose of the analyzer in this case is the removal of subjectivity from total count and size distribution data.

There are, of course, many other variations which can be used, such as microscopes with "TV" attachments for observation or further image processing and analysis.

Figure 10. Grazing angle tool with 4-way neon illumination and adjustable height capability.

PARTICLE DETECTION TOOL

A 4-way illumination system set up for visual inspection is shown in Figure 10. This system has a variable base height for adjusting the inspection surface for best dark field observation. A similar system with a dark box is shown in Figure 11. The dark box provides lower particle size sensitivity by providing a better signal-to-noise ratio.

Figure 12 shows the setup for a two way illumination system using a slide projector system for the light sources. This system is mobile and fitted with camera. The actual system is shown in Figure 13.

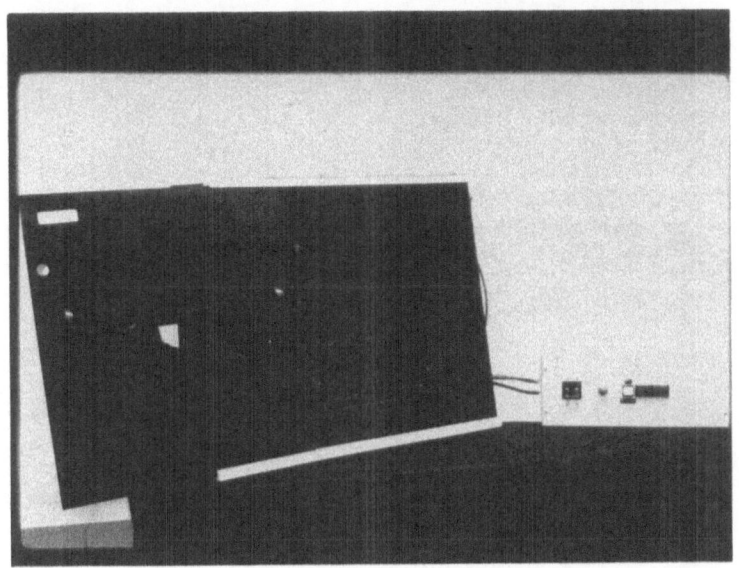

Figure 11.   Prototype 4-way neon illuminated system in a "dark" box.

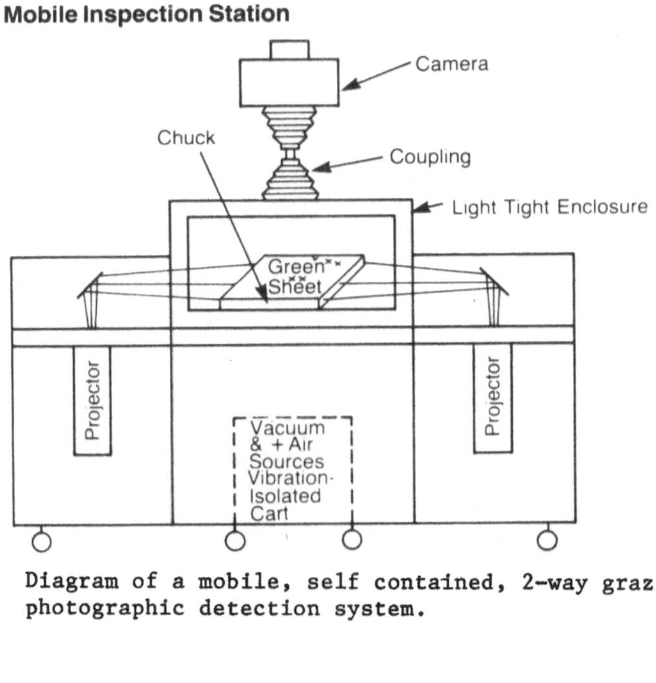

Mobile Inspection Station

Camera

Chuck

Coupling

Light Tight Enclosure

Green
Sheet

Projector

Projector

Vacuum
& + Air
Sources
Vibration-
Isolated
Cart

Figure 12.   Diagram of a mobile, self contained, 2-way grazing angle,
photographic detection system.

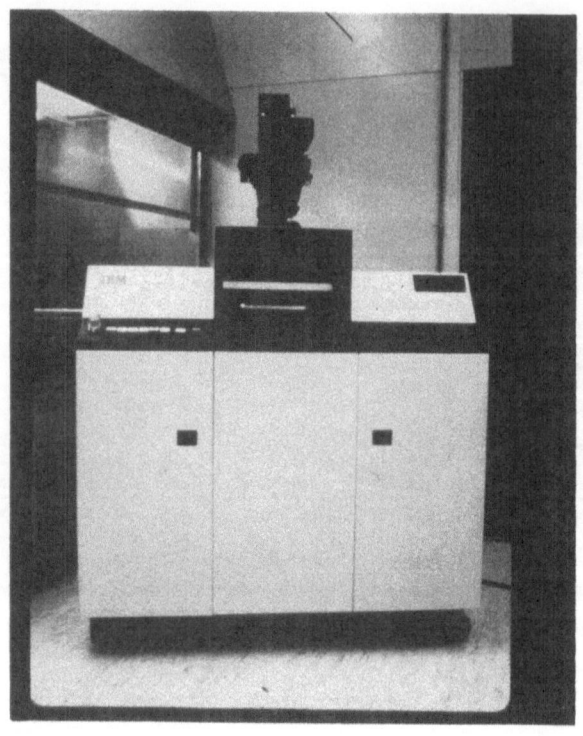

Figure 13. Photographic particle detection system as shown in Figure 12.

Portable Camera Unit

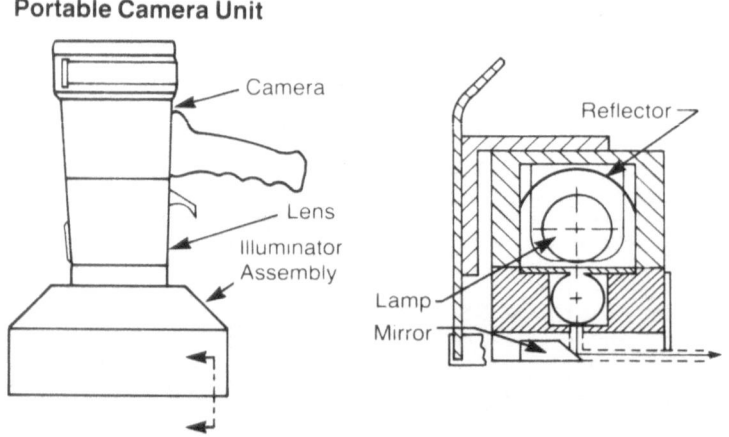

Fig. 14. Drawing of portable particle detection camera unit.

Fig. 15. Portable photographic particle detection tool as described in
Fig. 14.

   A drawing and photograph of a portable photographic system is shown in
Figures 14 and 15. This system is very handy for miscellaneous jobs such as
inspection of workstation cleanliness, equipment surface contamination,
walls, product containers, settling plates, etc.

   A four-way inspection and cleaning station is shown in Figure 16. This
system is equipped with peripheral exhaust such that the product can be
cleaned, brushed or whatever, while on the inspection table, until it is
visually free of particles.

TOOL CALIBRATION

   Practical calibration is accomplished for visual systems by placing
known particle sizes on surfaces to determine what can be "seen". For a
photographic system the spot size relationship can be determined for various
known particle sizes such that an idea of the particle size distribution can
be obtained.

   There are several sources of particles which can be used for calibration. Latex spheres and pollen have been most useful. For sizes up to a few
micrometers, latex spheres are sprayed onto a smooth surface (see Figure 17).
For larger sizes 11μm mulberry pollen, Figure 18, and 32μm lycopodium pollen,
Figure 19, are used.

   It should be remembered that what the observer/detector is actually
"seeing" is the light scattered from the particle. With a given illumination
source and viewing surface, the amount of light scattered is not only due to
the particle size but also due to its shape, orientation, surface texture and
index of refraction. Therefore, "calibration" is useful mainly for establishing a basic relationship between actual and observed sizes and maintaining the repeatability and stability of the grazing angle instrument. After

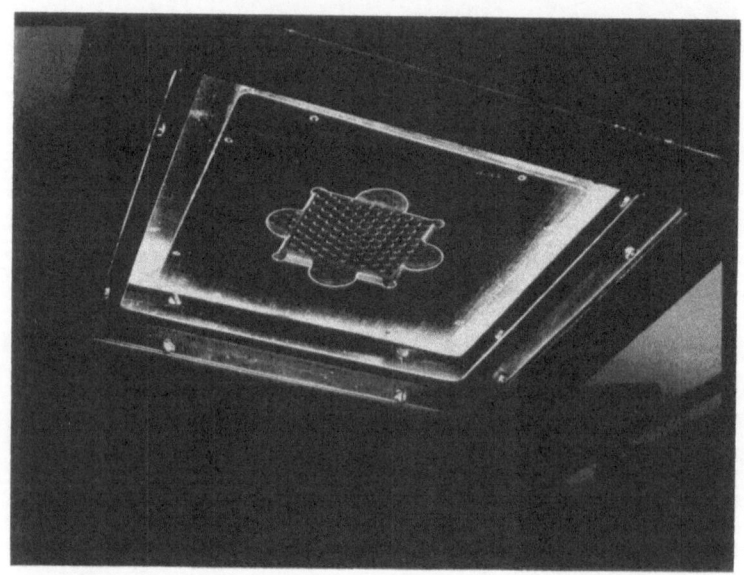

Figure 16. Manual 4-way illuminated cleaning and inspection station using peripheral exhaust.

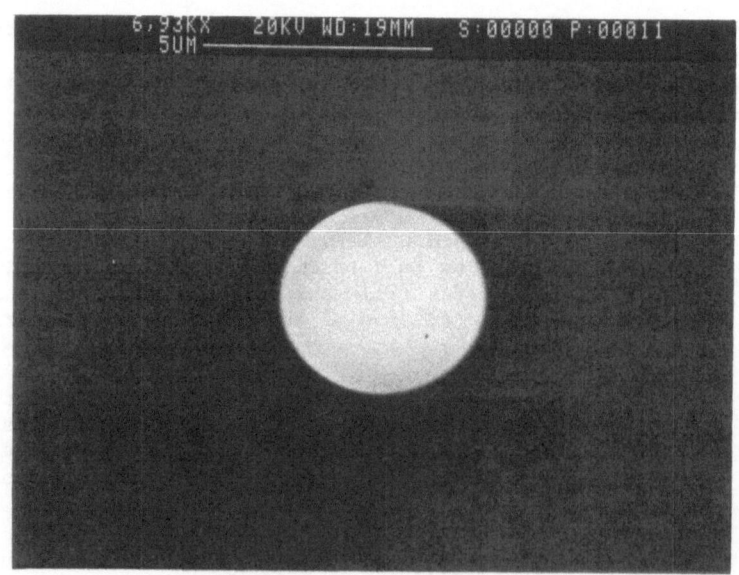

Figure 17. Latex Sphere.

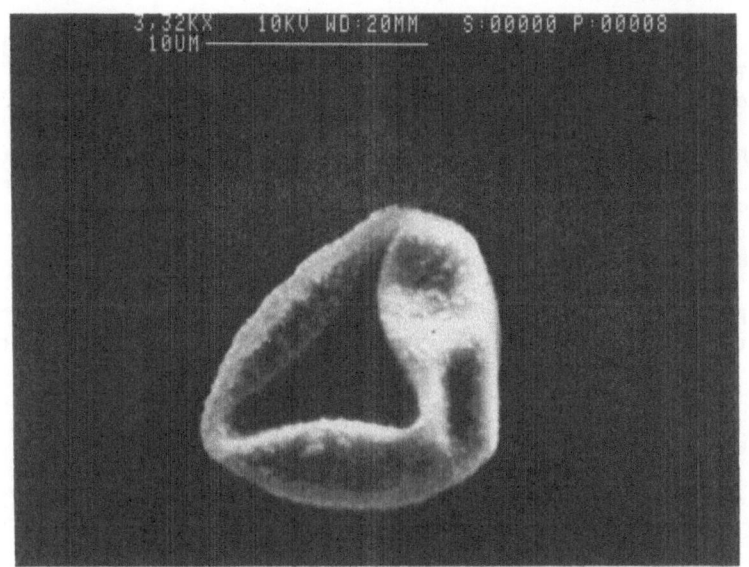

Figure 18.　Mulberry Pollen.

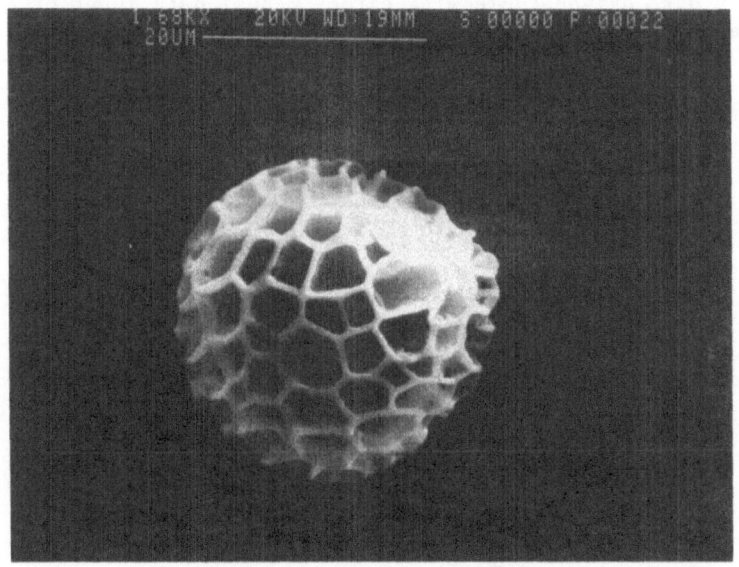

Figure 19.　Lycopodium Pollen.

all, the debris one measures is also not uniform in shape, orientation, texture and index of refraction.

LARGE AREA MEASUREMENTS

Increasing the area of inspection usually results in a reduction of sensitivity. This is not only due to the fall off in light intensity as the beam is widened but also due to increased distance from particle to source.

If a photographic system or image analyzer system is used to read the larger dark field display, a further loss in sensitivity occurs due to the resolution of the system. Sensitivity can be improved by going to a step and repeat operation whereby the field of view is again reduced and the resolution of the system regained.

A stepping operation using area sampling can be used to advantage to reduce inspection time, by sampling, and to avoid topological light scattering.

RAPID PARTICLE LOCATION FOR ANALYSIS

A basic method for rapid particle location is to simply establish the coordinates of the particles on the sample and then drive the sample to those coordinates under the analyzer.

Particle coordinates can be established simply by a grid on the inspection station, Figure 20, a grid over a photograph or by using the data coordinates from an analyzer.

The sample can be located under a microscope or SEM by a manual indicator table; a data driven table with either manual data input or a direct data link to the analyzer; or a simple sliding fixture, Figure 21.

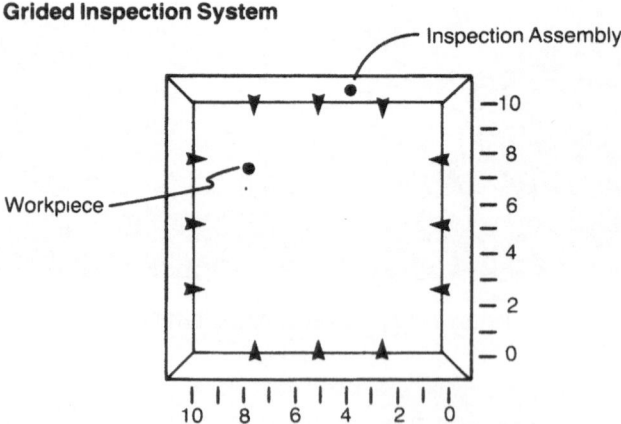

Grided Inspection System

Figure 20. Inspection station using grid system for location.

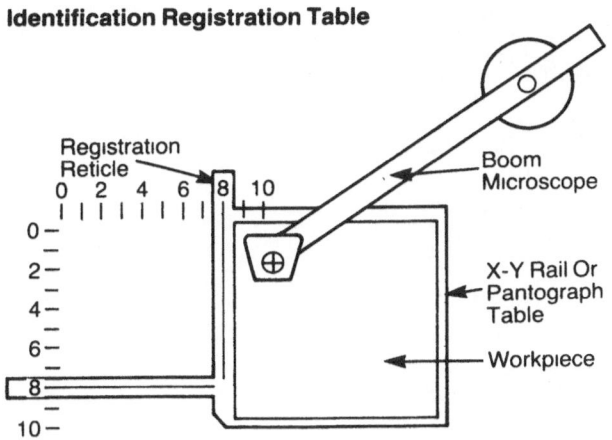

Identification Registration Table

Figure 21.   Particle locating table using grid registration system.

SUMMARY

The wide variety of techniques and configurations available for the application of grazing angle illumination inspection can be a very versatile method for the detection of particles down to a few micrometers in the day to day requirements of contamination detection.

REFERENCES

There are many references in the literature on light scattering of particles of different sizes and shapes and of arbitrary shapes, as well as references on light scattering from surfaces of differing roughness. However, there is little treatment of light scattering from particles on rough surfaces. Two general references are provided below:

1.   M. Kerker, "Scattering of Light and Other Electromagnetic Radiation", Academic Press, New York, 1969.

2.   P. Beckmann and S. Spizzichino, "The Scattering of Electromagnetic Waves from Rough Surfaces", MacMillan, New York, 1963.

ACCURATE PARTICLE COUNTING FOR BARE SUBSTRATE INSPECTION

L. Galbraith, G. Kren, A. Neukermans and G. Pecen

Tencor Instruments
2400 Charleston Road
Mountain View, California 94043

The factors affecting the particle counting accuracy of laser scanners for semiconductor wafer inspection are reviewed. The errors which may be expected with the present algorithms are discussed. Generally they give rise to substantial overcounting as satellite counts are generated. A new technique is described, which is demonstrated to give rise to very accurate counting as verified with surface contamination standards.

INTRODUCTION

Inspection for particulates on unpatterned wafers is rapidly becoming a necessary part of the semiconductor manufacturing process. With the present emphasis on manufacturability and yield enhancement in the semiconductor industry, measurement and control of defects, especially particulates, has become a major concern. As the critical dimensions (CD's) of the devices are decreasing, in some instances becoming smaller than one micron, it is clear that the maximum sizes of the contaminating particles which can be tolerated are getting correspondingly lower.

There is at present no clear understanding, at least in the published literature, what the maximum size of the contaminants may be with respect to the critical dimensions of the I.C. Estimates of the critical particle size run from one half to one tenth[1] of the dimension of the line width. Moreover, in general, one can expect the size to be dependent on the nature of the particle and the process step under investigation. For example, in optical lithography, a particle with an index of refraction n near 1.5 (typical of a wide class of contaminants) will affect the printing process significantly less than a silicon chip (n = 3.83 at λ = 0.6328 μm) of the same size. In contrast, in reactive ion etching or in resist spinning, the index of refraction is of little importance and the physical size and shape may be the most relevant factors. In metalization processes, the electrical conductivity of the particle may be the crucial parameter. Clearly the relationship between particle contamination and subsequent defect generation is quite intricate, and the best one can hope for is that a set of empirical rules will gradually emerge. In many semiconductor facilities, "particle" budgets allowed for various critical steps are still undefined and appear to be either unrealistically demanding or too lax.

Besides the obvious dictum that fewer and smaller is better, there seem to be little numerical data. Many semiconductor houses are now putting out specifications for six inch wafers for fewer than 0.025 particles per cm^2 of diameter of 0.5 µm and above.

Hence it is becoming increasingly important in wafer processing to get an exact count of contaminating particles of a given size. Scanning a focused laser beam across the wafer surface allows the use of detectors with large solid angles for collection of light scattered from defects. The solid angle advantage over imaging detectors such as TV cameras results in less uncertainty in particle size measurement when scattering is angle dependent as is often the case. In addition, the limited dynamic range of such instruments severely limits the spatial resolution that can be obtained when attempting to count a wide range of particle sizes on the same substrate. However, the nature of the laser scanning process with a Gaussian beam profile introduces some subtleties into the defect counting process and can lead to incorrect counting if no provisions are taken. Either over-counting or under-counting can occur, depending on the particle density, the detection threshold setting, the pixel size of the display and the nature of the particle distribution. This article describes a signal-processing technique which eliminates virtually all of these inherent problems, and results in an easily verified correct count.

A. THE LASER SCANNING PROCESS

In scanned-laser defect mapping systems, detection and measurement of contaminants is based on the amount of light scattered into a detector from a surface anomaly encountered by a scanning laser beam on a highly polished substrate, such as a semiconductor wafer or blank photomask. This scheme is analogous to dark-field microscopy, in which specularly-reflected light is ignored and only light scattered out of the specular beam is collected. The scatterer need not be approximately spherical, as usually associated with a particle. It may even be in the form of a perfectly flat layer; all that is required is that its reflectivity be slightly different from the surrounding substrate. For example, pinholes in chrome photomasks can be detected as easily as particulates, but a single detection scheme cannot differentiate between the two.

Although this technique is conceptually very simple, it is surprisingly powerful and can visualize defects which are otherwise very difficult to detect. For example, a waterspot a few microns in diameter may have a thickness on the order of 300 Å and a refractive index of 1.5. Such a defect on silicon may have a scattering power, as measured by its scattering cross-section*, of 0.01 µm^2 and can be reliably detected by state-of-the-art instruments such as the Surfscan 4500. Yet such a spot may be very difficult to detect in a scanning electron microscope because of the low surface relief of the defect and the resulting poor contrast.

---

*The scattering cross-section is the term used to express the strength of a scatter. In general, it is defined as the ratio of the power removed by the scatterer from the beam through all processes (absorption, reflection, diffraction, refraction), to the power incident per cm^2. It is the product of the scattering efficiency, and the geometric cross-section of the particle, and is expressed as an area. For our purposes, we adopt a slightly modified version, and define it as the ratio of the power scattered in the detector, to the incident power density. Hence, the scattering cross-section depends on the detector solid angle and orientation, scatterer structure, index of refraction and the nature of the supporting substrate. Since only the scattering cross-section can be measured by a dark-field instrument, size measurements must be inferred from a known relationship between size and scattering cross-section for the particles under scrutiny, on a given substrate.

This particular scan technique discussed here is rectilinear raster scanning, in which the laser beam is scanned linearly across the wafer in one direction (Y axis) and the wafer is simultaneously translated mechanically in the perpendicular (X) direction.

For scanning purposes, it would be ideal to have a beam with a rectangular cross-section with constant intensity per unit area. In this case, successive scans could be simply linked together. At each particular location on the wafer, the beam would dwell only once, and the illumination intensity would be uniform over the entire wafer (see Figure 1a). In practice, such a system is very difficult to implement and a Gaussian beam is used.

The natural Gaussian profile of a CW laser beam remains Gaussian as the beam propagates through aberration-free optics, despite magnification changes, focus errors and diffraction[2]. This makes it desirable to accept the Gaussian profile and to find ways to circumvent the deleterious effects of beam nonuniformity.

Let us assume that the dimensions of a scatterer are much smaller than the characteristic size of a Gaussian-profile illuminating beam (typically the diameter measured at the e^{-2} intensity points). Since the instantaneous scattered-light power is proportional to the incident light intensity at the scatterer, the detector signal is a Gaussian function of time when the beam is uniformly scanned. Raster-scanning the beam results in a succession of Gaussian detector pulses, generally of unequal amplitude, from the same scattering object (see Figure 2). If we plot intensity against the X position of the beam (Figure 1b), we find that an error ΔI can occur in the scattering cross section, depending on the particle location with respect to the peak of the Gaussian. The closer the spacing of the overlapping scans, the lower the measurement error will be, but the scan time for the wafer will increase. Therefore a sensible tradeoff between the scantime and sizing accuracy should be made. The smaller the beam diameter, the higher will be the peak intensity of the scattered light, but maintaining the error ΔI at reasonable levels requires proportionally smaller X-direction translation steps.

In the Tencor "Surfscan" series of particle detectors on bare wafers, the wafer is scanned with a Gaussian beam 100 μm in diameter (e^{-2} points) with a scan to scan increment of 25 μm. As seen from Figure 1.b, the maximum illumination intensity variation over the wafer is less than 11%, but the average particle is exposed to 96.7% of the peak laser intensity. To arrive at this figure, it is assumed that the particles are randomly distributed between the illumination peaks.

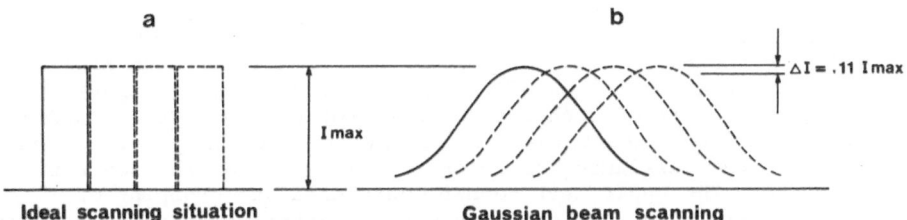

Figures 1a,b. In an idealized scanning arrangement with a constant intensity beam, no overlap is required. (Fig. 1a). In contrast, with a Gaussian beam scan it is necessary to overlap the beams closely, in order to illuminate each part of the surface at nearly the same peak intensity. In the Tencor Instruments designs, the beam diameter is 100 μm, and the scan separation 25 μm, giving $\Delta I/Imax = 0.11$.

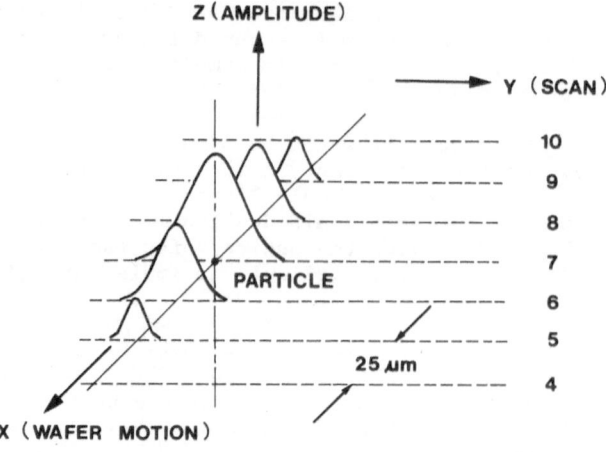

Figure 2. As the Gaussian beam scans across the point scatterer, the overlapping scans in time give rise to series of Gaussian waferforms (effectively cross-sections of the intensity profile of the beam), a scan width apart.

As a result of this scanning configuration, it is obvious that the particle will be observed by the detection system several times. At each successive scan, the light pulse produced will be Gaussian in time, but its amplitude will vary, as the center of the beam moves over the particle. These light pulses can be considered as cross-sections of the three dimensional beam profile, taken one scan (25 μm) apart.

Because of inevitable noise in the detector signal arising from various sources, particle scanners employ a signal threshold above which a detector pulse is registered and below which it is ignored. In a raster-scanned Gaussian beam system, a defect may be detected and will commonly be detected several times depending on the ratio of the signal threshold, T, to the maximum possible scattering signal P (peak). Basically, the detection probability approaches zero as T/P approaches 1.0. At lower T/P ratios (lower thresholds), the detection probability rapidly approaches unity for single detection events. At still lower T/P ratios, multiple detections of the same particle are possible. The situation is summarized in Figure 3 for the case of a 100 μm diameter scanning beam and a 25 μm scan pitch. From one to seven above-threshold peaks (hits) per particle are commonly generated at typical instrument threshold settings.

Multiple threshold crossings can be readily displayed using the "Zoom" feature available on the Tencor Surfscan 3000 and 4000 series. This feature allows the result of every above-threshold scan to be displayed in effect magnifying a local area of the normal display map. Figure 4 illustrates the result of such a scan on a contamination standard from VLSI Standards[3]. The latter contains four groups of scattering centers arrayed in square patterns, designed for integer ratios between the scattering cross sections. The upper left corner contains a 5x5 grouping of scattering centers, each one of which has a scattering cross-section of 0.54 μm^2. With a threshold setting of 0.2 μm^2 in this measurement, the threshold/peak ratio is 0.37, and from Figure 3, we would expect on the average three hits on successive scans, as indeed appears to be the case. The scattering centers on the bottom left have scattering cross-sections of 7.5 μm^2; the threshold/peak ratio is 0.0267. From Figure 3, we derive an average of six detections as observed.

272

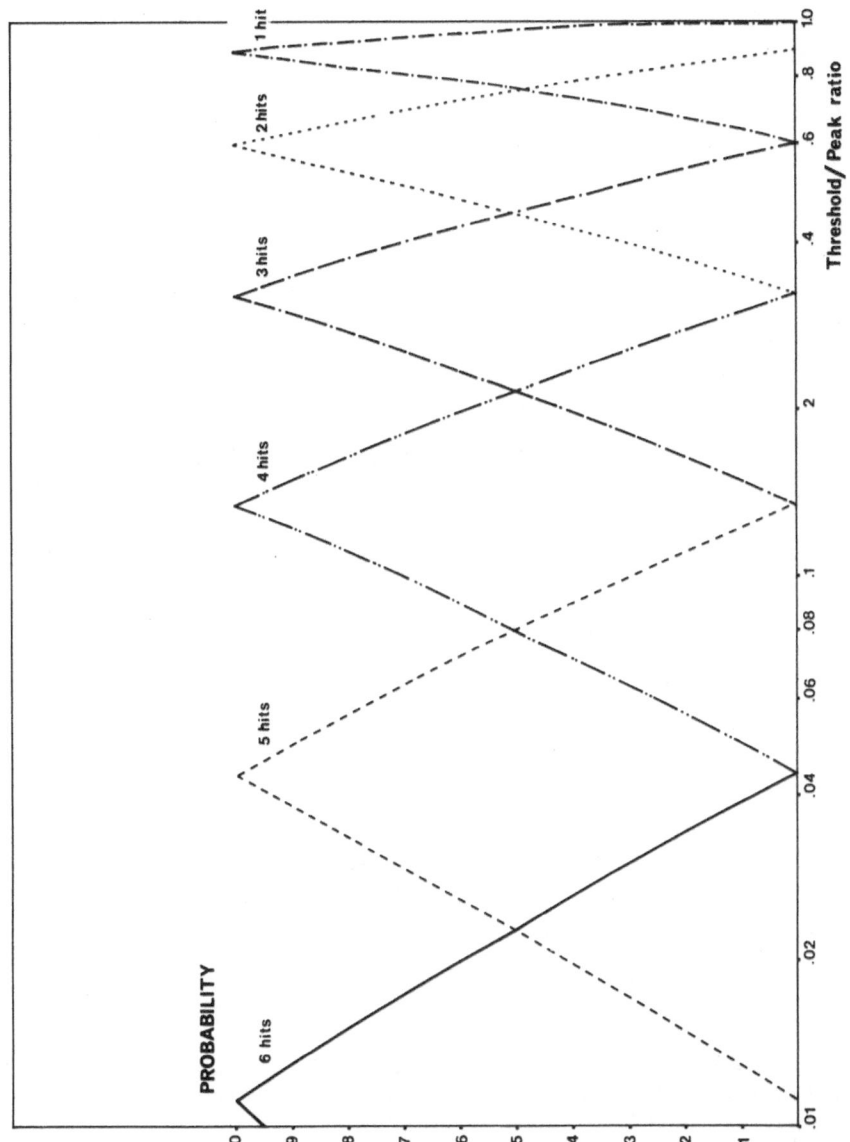

Figure 3. This graph gives the probabilities of detecting the particle n times (the number of threshold crossings which will occur in successive scans), for a given threshold to peak ratio, the larger the number of detections. The particle will be detected no more than once for $0.89P < T < P$; from one to two times for $0.606P < T < 0.89P$, etc.

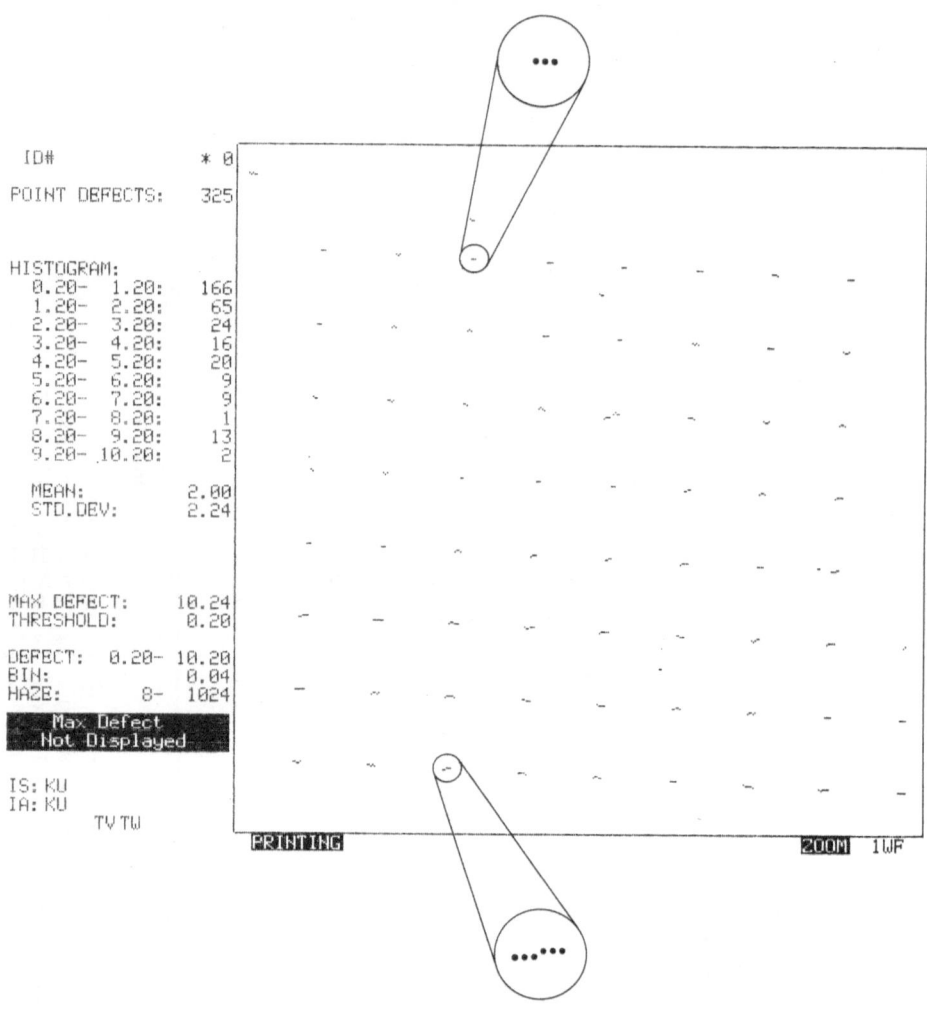

Figure 4. This illustrates the multiple detections of defects which can be demonstrated with a special zoom feature on the Surfscan 4500. In the upper left-hand corner, an array of 25 small scatterers produces on the average three hits in successive scans, whereas underneath 16 times larger scatterers produces an average 6 hits. (This display is not normally accessible to the user).

Clearly, a simple count of above-threshold detector pulses is not an accurate representation of the actual particle count. Fortunately, signal-processing schemes are possible which greatly reduce this count discrepancy.

B. PIXEL COUNTING

All laser scanning particle detectors process the detector signals to produce a video or printer display representing the positions of the particle on a map of the substrate surface. The substrate is divided into area elements which are represented as picture elements, or pixels, on the particle map. Pixels are also represented by memory locations in the

274

instrument and are given values (which can be simple binary or more complex magnitude data) representing particles occurring in the corresponding substrate locations. By pixel counting, we mean counting those pixels in which particles are found. Two pixel-counting schemes will be discussed here.

In the simplest and most straightforward implementation, the display pixel will be lit to designate a particle whenever the detection threshold is exceeded while the area corresponding to the pixel on the wafer is being scanned. The number of lit pixels is then counted and the count taken to represent the number of particles. This turns out to be a powerful technique for reducing the count ambiguity which occurs with multiple detection. The pixel is lit the first time the threshold is exceeded, and succeeding above-threshold pulses, or "hits", in the same pixel area do not change the situation. However, if two or more particles are present in the same pixel, no indication will be given that more than one is present, since the pixel can only be lit once. Hence this counting algorithm can give rise to undercounting if the number of particles is significantly larger than the number of pixels. Since this number is in the thousands to tens of thousands in typical instruments, undercounting will be a problem in cases of grossly contaminated local sites on the wafer.

A more serious problem with this simple pixel-generation scheme occurs when multiple hits cross over pixel boundaries, which happens with increasing probability as the threshold/peak ratio is lowered.

Figure 5 illustrates the situation with a particle which is physically located in pixel B. If the particle has a scattering cross section much higher than the detection threshold and is sufficiently close to the boundary of pixel A, the tail of Gaussian beam illuminates the particle and produces a threshold crossing while the beam is still scanning pixel A. This hit will be associated with pixel A. Subsequently, pixel B will produce a hit as the center of the beam moves over the particle. Hence in this case two pixels will be lit, although only one particle is present. In fact, one particle can be associated with not just two, but also three and four pixels if the particle is located close enough to the corner of a pixel. This is graphically illustrated in Figure 6, which outlines the loci of scatterer positions which produce hits in one, two, three or four pixels. A certain percentage of randomly distributed particles will be overcounted even at low particle densities.

The overcount probability can be calculated from the ratios of the areas of the multiple-pixel loci in Figure 6 to the total pixel area. These loci are defined in terms of effective beam radius, R_{eff}, which is the distance from the scanning beam center to the points where the beam intensity/center intensity equals the threshold/peak ratio for the particle in question. R_{eff} is proportional to the beam $1/e$ diameter and decreases monotonically as T/P increases, becoming zero at T/P=1.0. Clearly, the probability of multiple hits is a function of the threshold setting with respect to the peak signal, and of the beam size to the pixel size.

Figures 7a, 7b, and 7c illustrate the probabilities for generation of multiple pixel counts per particle for three different pixel to beam diameter ratios (2.5, 4.0 and 6.0 respectively). The larger the pixel is with respect to the beam, the smaller will be the probability of overcounting. These curves can be used to derive a count correction factor if the scatters are the same (monodisperse) and the scattering peak-to-threshold ratio is known.

This correction factor can be quite significant. Assume for example that the pixel size is 400 μm and the beam diameter 100 μm (as in Figure 7b). A measurement of monodisperse particles at a threshold of 0.025 times the peak cross section yields a correction factor: 0.436x1 + 0.449x2 + 0.0248x3 + 0.091x4 = 1.772. The various factors are the probabilities for one, two, three and four pixels designated per particle. In this case the defects are overcounted by 77%. If the particle distribution is not monodisperse, then the overcount correction factor becomes a variable whose effect on the particle count depends on the nature of the scatterer size distribution.

Even when the histogram of the particle number versus size is known, an accurate correction factor is not obtained through straight-forward application of the above reasoning.

Better accuracy can be obtained with pixel schemes if peak detection is applied to the scattered-light detector signal. An above-threshold event is not recorded in a pixel unless the peak of the detector signal occurs in the pixel area. The scanning beam acts as a horizontal line segment of width 2 R_{eff}, so that a particle can now designate only one or two pixels (Figure 8.) For a particle located in pixel A, the peak signal will always occur at the same vertical location and can, therefore, occur only in pixel A or in pixels horizontally adjacent to A. Hence only pixel B will register satellite hits in this case; pixels C and D are outside the signal peaks.

As a result, only single or double counts are possible. Figure 9 illustrates the probability of double counting for the same pixel size to beam diameter ratios as in Figures 7a-c (2.5, 4.0, 6.0). Note that the overcounting is substantially reduced. For the same case as discussed above, (400 μm size, 100 μm beam diameter, monodisperse particles and a threshold/peak ratio of 0.025) the number of designated pixels will be 1.25 the number of particles. The overcount of 25% is a substantial improvement over the previous case (77%), but the error is still significant.

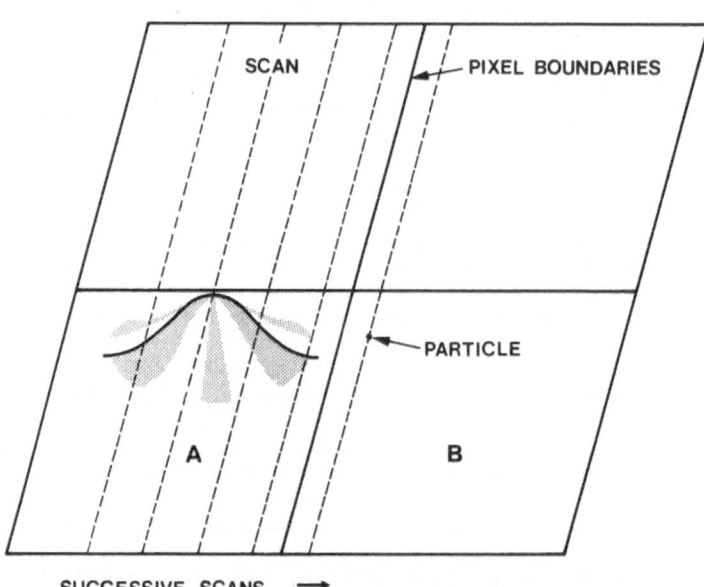

Figure 5. Schematic illustration of a particle situated in pixel B, but large enough to be detected while the beam scans still in pixel A. If the signal is larger than the threshold, pixel A will be lit, even though it contains no defect.

Histograms, generated by these counting methods, are also affected by the same deficiencies which affect the counting accuracy. Each bin of the histogram also contains counts from signal peaks which have spilled over into adjacent "satellite" pixels from larger defects. Figure 10 shows how the satellite pixels add a "tail" to the ideal single bin histogram expected for a monodisperse particle population. If all the pixels in the tail down to the threshold level are counted along with the main bin, the result will be the overcount depicted in Figure 9.

C. COUNTING WITH PULSE POSITION CORRELATION (PPC)

We have seen how peak detection in the vertical (Y) direction reduces the spillover errors inherent in simple pixel-designation schemes. This reduces the scanning beam from a circle of radius R_{eff} to a line segment. If the peak-detection process could be extended in the X direction, pixel spillover errors would be totally eliminated, enabling a correct count to be generated. Particles could also be localized within the precision allowed by the horizontal scan pitch and the vertical sampling interval of the instrument's A-D converter. Particle localization would no longer be dependent on the display pixel size.

This process is accomplished in the Surfscan 4500 by a Pulse Position Correlator (PPC) circuit. Essentially the circuit processes signal peak magnitude and location data from successive scans of a single particle and selects the highest peak. The magnitude and position is stored in memory as one byte for scattering cross section magnitude and two bytes for the location. After this filtering process, the particle map pixel grid can be thought of as an overlay on the finer grid defined by the scan pitch and the vertical A-D converter sampling interval. Since the defect information

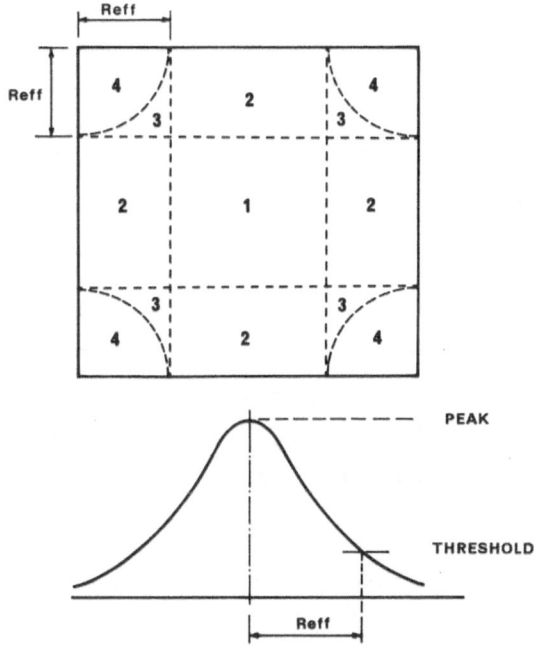

Figure 6. This figure illustrates the various regions of pixel in which a point scatterer will produce above-threshold hits in 1,2,3 and 4 pixels for a given characteristic dimension R_{eff} (R_{eff} is defined as the radius where the intensity of the Gaussian beam relative to the center equals the threshold-to-peak ratio of the point scatterer).

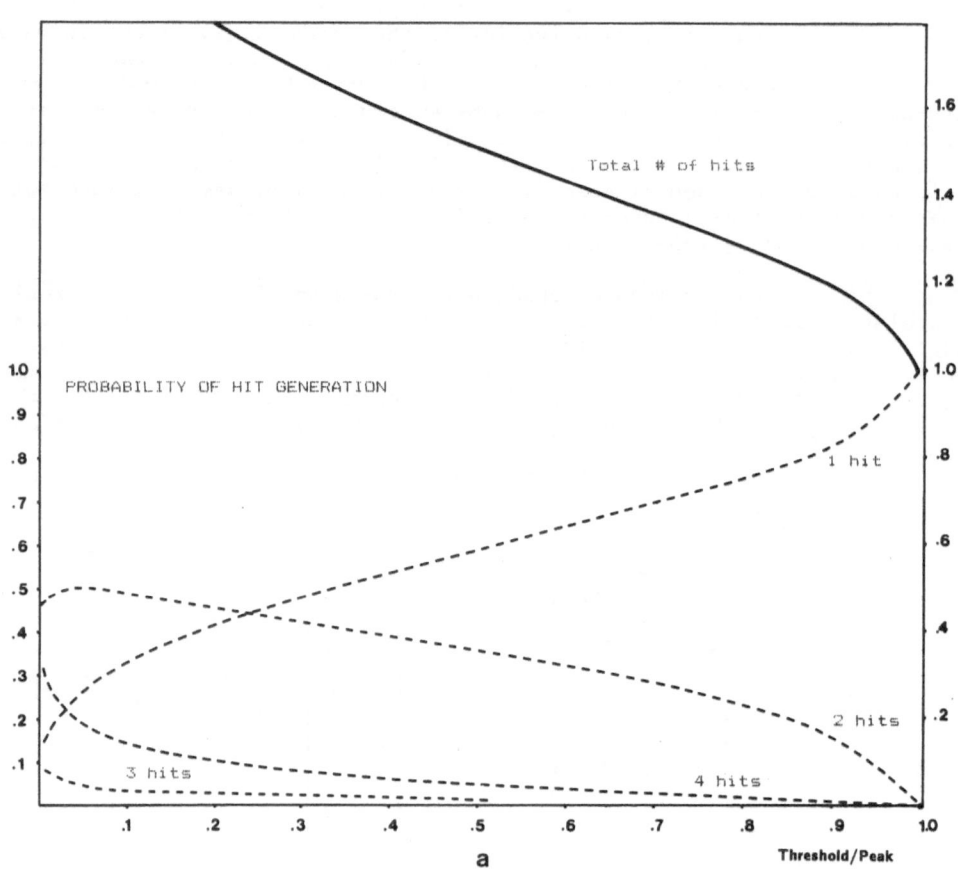

Figure 7a,b,c. These curves list the probability of assigning one, two, three and four pixels from a single defect, as a function of the threshold to peak amplitude of the defect. The curves are drawn for three different pixel-size-to-beam ratios (2.5, 4 and 6). The larger the pixel, the less likely the "satellite" pixel production. The left axis represents the probability of various hits, the right axis the "cumulative" hits from one count.

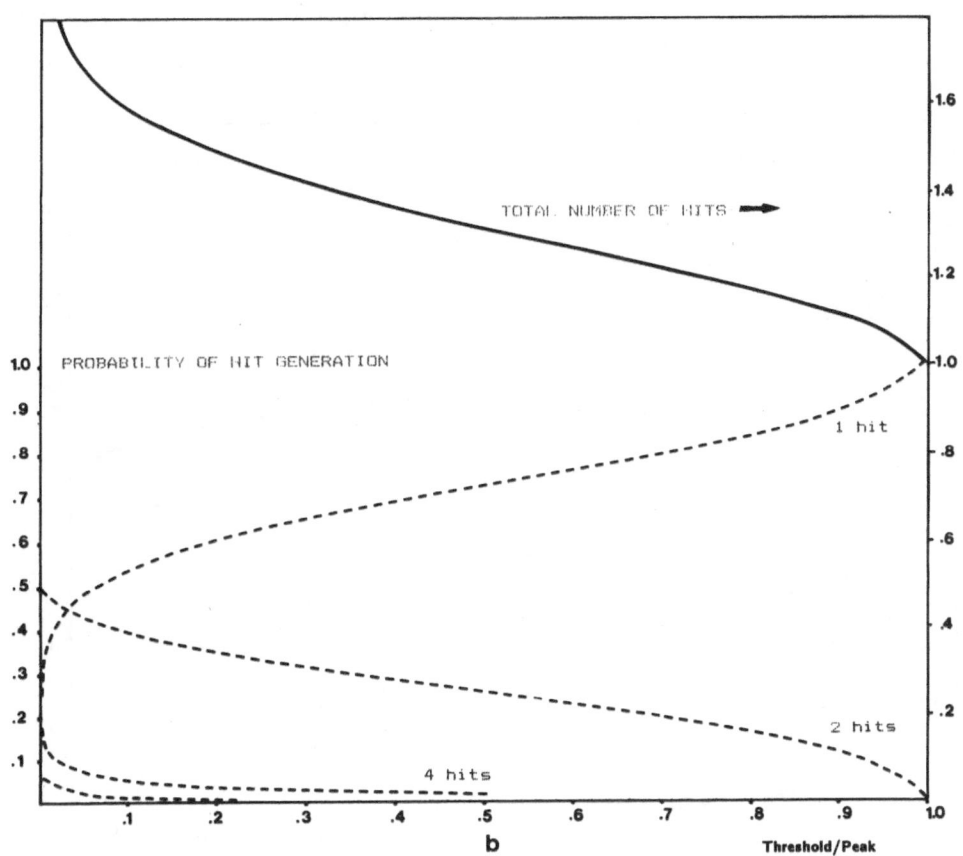

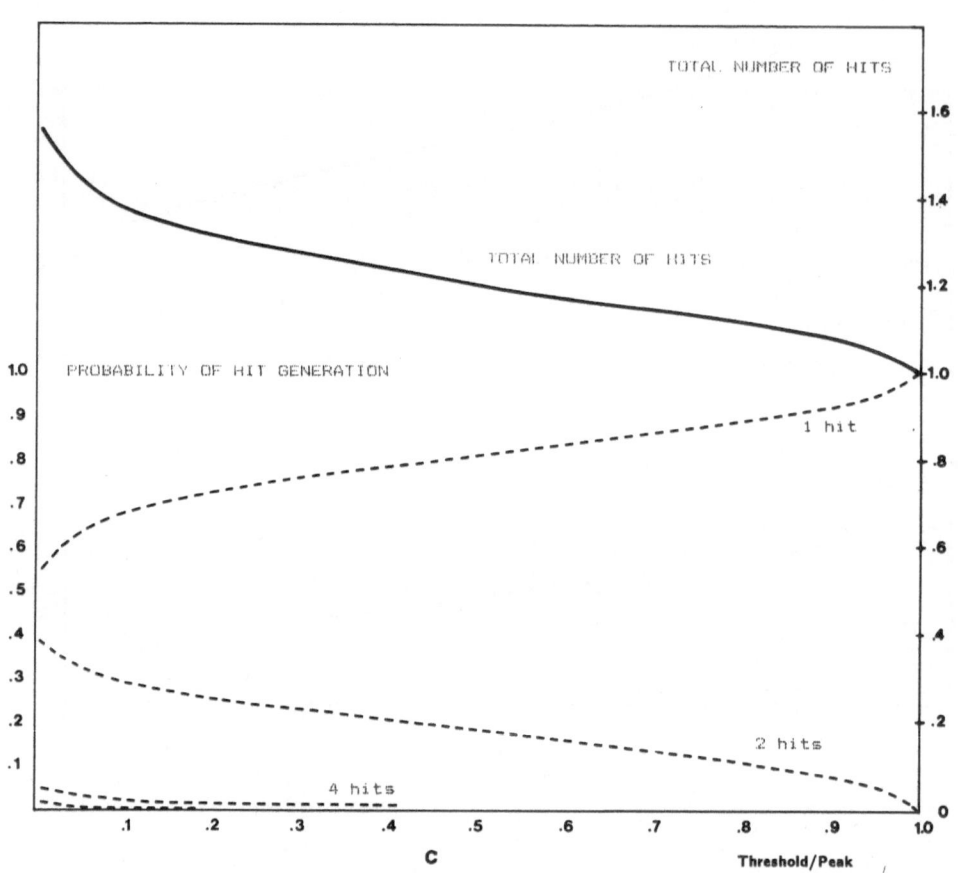

TOTAL NUMBER OF HITS

is reduced to single points on this grid, there is no ambiguity as to which pixel any one hit belongs. If there are two particles in a pixel, the pixel can only be lit once, but the particle count will nevertheless still be correct.

One can verify correct defect counting of the PPC with the VLSI Standards wafer described earlier. This wafer is fabricated with 2100 standardized defects in four sets of 525 of equal scattering magnitude. A measurement with the PPC on a slightly contaminated wafer yields a count of 2118 and four clean, distinct histogram peaks, as can be seen in Figure 11. The 2100 defects of the original pattern can be readily identified since they form a distinctive geometric pattern, and the remaining 18 can be easily identified as contamination. A "zoom" closeup of one pattern is illustrated in Figure 12. Note that each defect, regardless of its size, now registers only one "hit". In contrast, the same wafer measured at the same threshold but without the PPC shows 2541 defects (Figure 13a), an overcount of 20%. The defect map (Figure 13b) shows the effect of satellite hits, which add tails to the histogram peaks as described earlier in connection with Figure 10. Without the PPC, the particle count is very threshold sensitive. For example, raising the threshold from 0.2 to 0.4 μm^2 decreases the count from 2541 to 2427. This difference does not represent real particles situated between the two thresholds, but rather a reduction in satellite hits from the larger scattering centers. With the PPC, a similar threshold excursion produces a count difference of 4, and this can be accounted for entirely by contaminating particles with scattering cross-sections between the two thresholds.

The presence of satellite pixels without the PPC is apparent in Figure 13a (note the groups of two horizontally adjacent pixels). Because the defects are not randomly placed on this wafer but have a spatial period close to an integer multiple of the instrument pixel size, the overcounting theory developed above may not apply strictly to this wafer; nevertheless the agreement with experiment is good in this case.

The VLSI Standards contamination standard is a very powerful tool for evaluation defect counters on substrates as will be elaborated further.

The theory of particle overcounting without the PPC can be easily verified with randomly-placed monodisperse spheres. Figures 14a and 14b

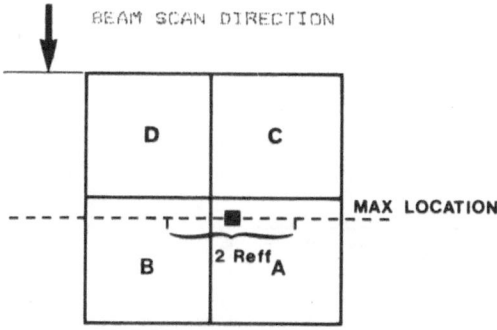

Figure 8. When a peak detection scheme is used, only horizontally adjacent pixels can record satellite hits, because all successive signals from the particle will peak at the same location vertically in the scan. Hence even with a scatterer located as illustrated, at most A and B will record hits, not C & D.

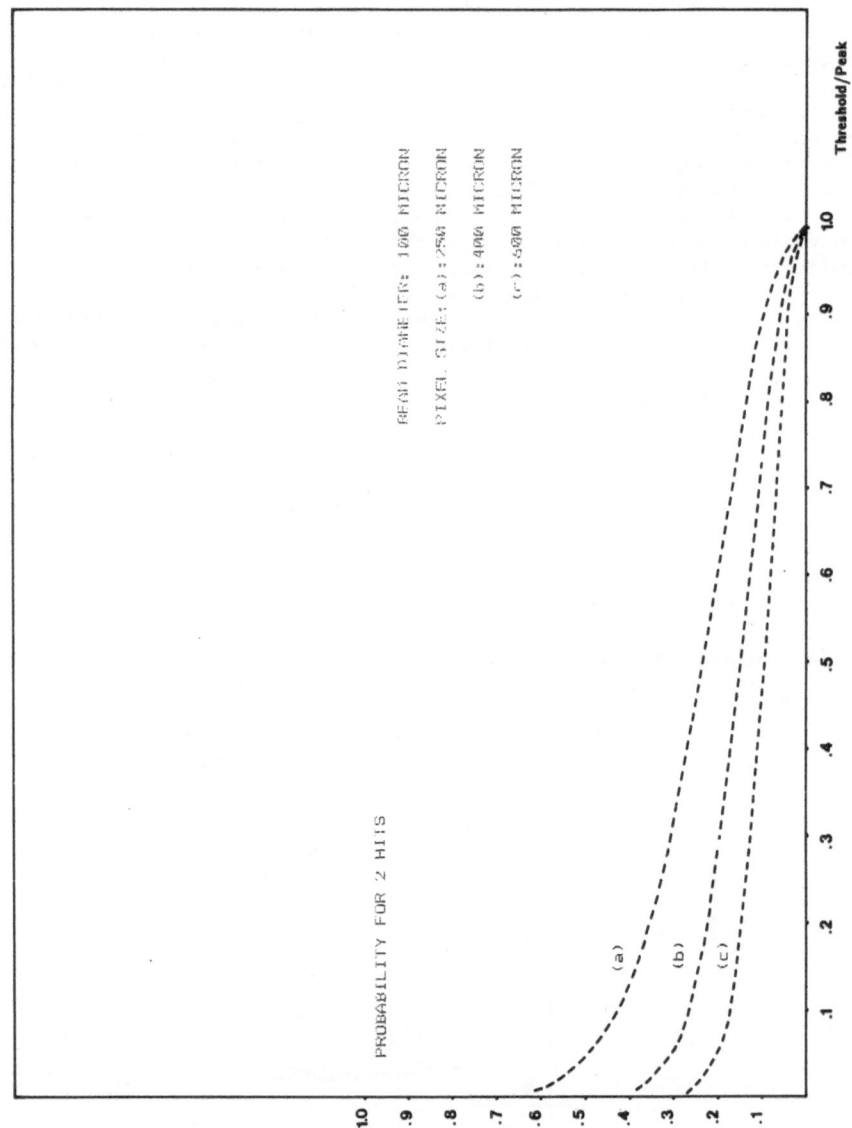

Figure 9. The graph illustrates the probability of generating double hits as a function of the threshold-to-peak ratio for pixel to beam ratios of 2.5 (lower), 4 (middle) and 6 (upper). Since only one or two pixels can be designated by a defect, the probability of two hits is also the total overcount factor.

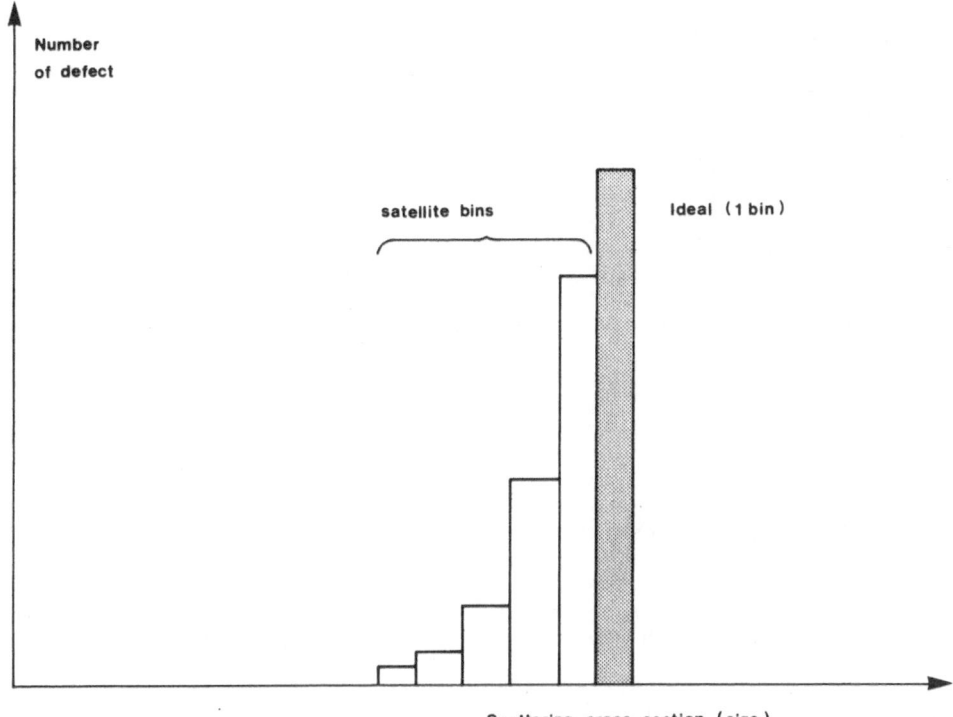

Figure 10. Representation of a scattering cross-section histogram that
would be generated by a monodisperse defect distribution. It shows the
"tail" on the low side produced by spillover of defect "hits" into adjacent
pixels in the peak-detection pixel counting scheme. The area under the
"tail" corresponds to the overcount factor depicted in the previous figure.

show the total count above threshold and histograms from a wafer containing
monodisperse 2.0 μm diameter polystyrene latex (PSL) spheres, both without
and with the PPC respectively. Note the clean peak obtained with the
latter; the smaller defects shown represent true contamination, not
satellite hits. (We note here that the scattering cross-section for
1.0 μm diameter PSL spheres on silicon is nearly 1 μm^2 and that for 2.0 μm
spheres is about 2 μm^2; this does not imply a linear relationship between
particle diameter and scattering cross section!)

    Table 1 lists the particle count measured as a function of threshold
setting both with and without the PPC. Assuming the former to be correct,
we can calculate the overcounting as a function of the threshold to peak
setting. This is plotted in Figure 15 which shows the agreement with the
theory used to generate the overcounting curves in Figure 9.

    The ability of a PPC-equipped scanned-laser instrument to count point
particles accurately, together with standardized features in a recognizable
geometric pattern such as are present on the VLSI Standards wafer, provides
very accurate determination of the sensitivity of the instrument. Only if
the counting can be guaranteed correct and the number of particles exactly
known a priori can one determine the detection probability for a particle
of a given size. Most manufacturers of defect detectors specify a lower
particle size limit without specifying the level of confidence for
detecting these particles. A lower size specification without confidence
limits is useless. Because of instrumental noise there are fluctuations in

Table 1. Particle count versus threshold with and without PPC for 2 μm polystyrene latex spheres on silicon.

Threshold (μm²)	No PPC	With PPC	Ratio No PPC	Threshold/Peak Ratio*
0.2	2808	2199	1.27	.11
0.24	2718	2190	1.24	.13
0.28	2651	2167	1.22	.15
0.32	2611	2143	1.22	.17
0.36	2573	2125	1.21	.20
0.4	2243	2119	1.20	.22
0.52	2465	2091	1.18	.28
0.6	2421	2074	1.17	.33
0.72	2355	2052	1.15	.39
0.8	2325	2043	1.14	.43
1.0	2269	2038	1.11	.54
1.2	2202	2027	1.09	.65
1.4	2128	2015	1.06	.76

*Peak scattering cross section = 1.84 μm²

the detector signal from scan to scan which result in a finite probability that a particle signal, however small, will exceed the set threshold.

One can use a VLSI Standards wafer with a light attenuation filter in front of the scattered-light detector to simulate a calibration wafer with arbitrarily small features. Since these scattering centers are arrayed in easily recognized pattern, their presence or absence is easily discerned. Contaminants on the wafers can be recognized and eliminated from the count. This is a more meaningful test than could be performed with randomly-distributed latex particles, since the exact positions of such particles are rarely known a priori and hence an accurate baseline particle count cannot be established for comparison with the instrument data. The test with the patterned calibration wafer can be easily performed in most laboratories without the requirement of latex sphere deposition equipment and the skilled personnel needed to operate it properly. Care must be taken to bring the instrument noise back up to the level obtained without the light attenuation filter in order to avoid measuring unrealistically good performance. This can be done by injecting electronic noise into the detector preamplifier or introducing an appropriate low level of ambient light. A patterned calibration wafer used in this way can produce accurate performance specifications based on percentage detection of particles and noise counts as a function of threshold level.

D. NOISE REDUCTION WITH THE PPC

The Pulse Position Correlator possesses a feature which allows a significant rejection of instrumental noise at the cost of a slightly larger minimum detectable defect size. Through selection from the operating software menu, the user can require that each valid particle detection be represented by at least two successive hits rather than the normal one or more. Hence, at the same Y position in the scan, a threshold crossing has to occur on two consecutive scans before a particle is registered. If the threshold is chosen judiciously, the likelihood that two noise pulses will give rise to threshold crossings on adjacent scans at exactly the same Y coordinate is very low (coincidence within a clock cycle, representing a spatial distance of 20 μm on the substrate, is required.)

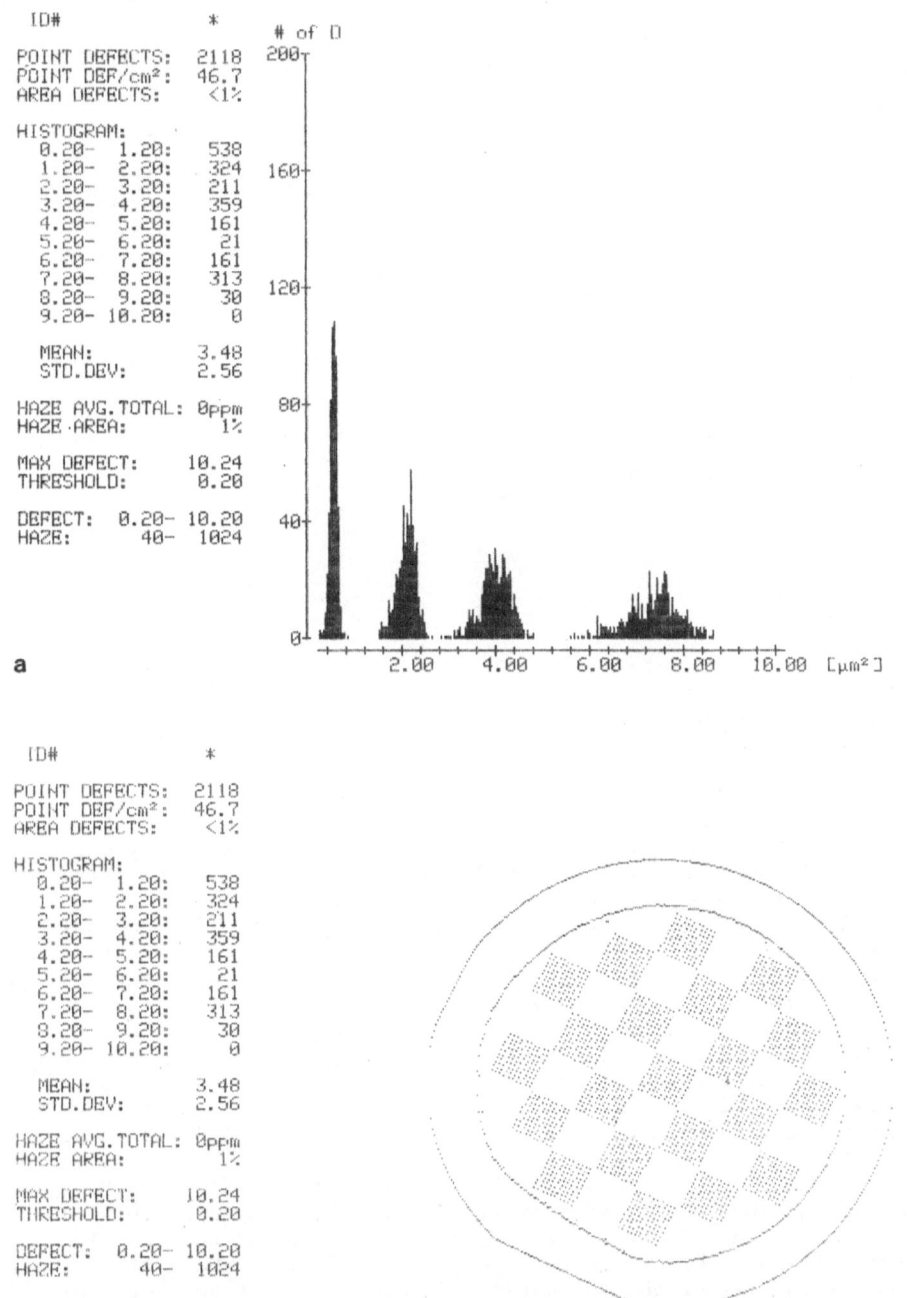

```
     ID#              *        # of D
                               200
  POINT DEFECTS:    2118
  POINT DEF/cm²:    46.7
  AREA DEFECTS:      <1%

  HISTOGRAM:
    0.20-  1.20:     538
    1.20-  2.20:     324    160
    2.20-  3.20:     211
    3.20-  4.20:     359
    4.20-  5.20:     161
    5.20-  6.20:      21
    6.20-  7.20:     161
    7.20-  8.20:     313
    8.20-  9.20:      30    120
    9.20- 10.20:       0

    MEAN:            3.48
    STD.DEV:         2.56

  HAZE AVG.TOTAL: 0ppm      80
  HAZE AREA:         1%

  MAX DEFECT:      10.24
  THRESHOLD:        0.20

  DEFECT:   0.20- 10.20     40
  HAZE:       40- 1024
```

a

```
                 2.00    4.00    6.00    8.00   10.00  [μm²]
```

```
     ID#              *
  POINT DEFECTS:    2118
  POINT DEF/cm²:    46.7
  AREA DEFECTS:      <1%

  HISTOGRAM:
    0.20-  1.20:     538
    1.20-  2.20:     324
    2.20-  3.20:     211
    3.20-  4.20:     359
    4.20-  5.20:     161
    5.20-  6.20:      21
    6.20-  7.20:     161
    7.20-  8.20:     313
    8.20-  9.20:      30
    9.20- 10.20:       0

    MEAN:            3.48
    STD.DEV:         2.56

  HAZE AVG.TOTAL: 0ppm
  HAZE AREA:         1%

  MAX DEFECT:      10.24
  THRESHOLD:        0.20

  DEFECT:   0.20- 10.20
  HAZE:       40- 1024
```

b

Figure 11a,b.   An example of the defect count accuracy and well-resolved
histogram peaks which are produced on a calibration wafer in a PPC-equipped
instrument.   The other contaminants can be readily identified because they
are not part of the geometric pattern of standard defects.

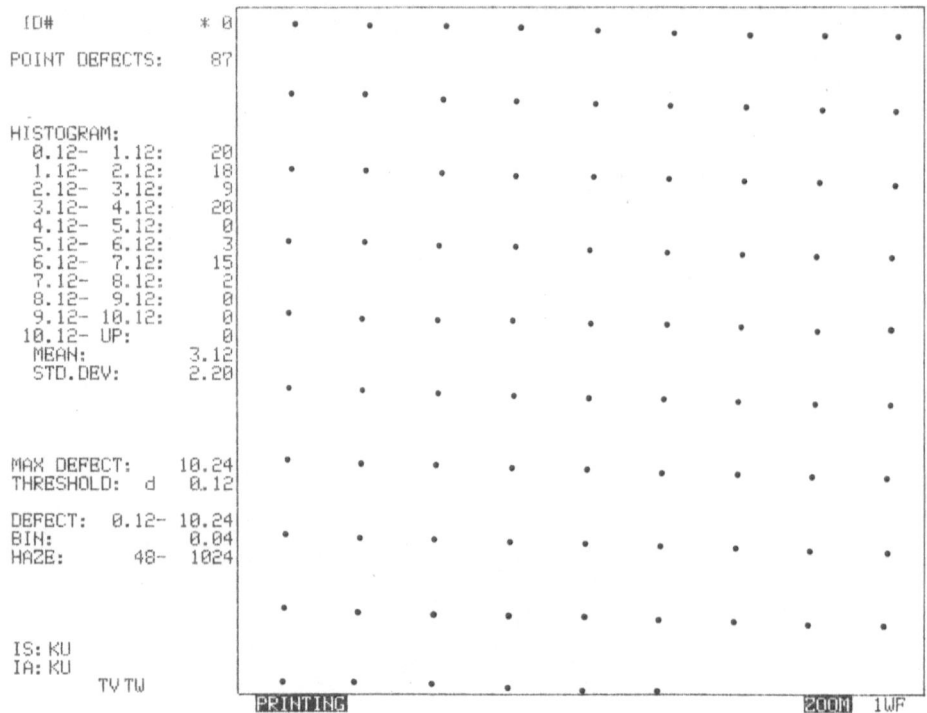

ID#   * 0
POINT DEFECTS: 87

HISTOGRAM:
  0.12- 1.12:   20
  1.12- 2.12:   18
  2.12- 3.12:   9
  3.12- 4.12:   20
  4.12- 5.12:   0
  5.12- 6.12:   3
  6.12- 7.12:   15
  7.12- 8.12:   2
  8.12- 9.12:   0
  9.12- 10.12:   0
 10.12- UP: 0
  MEAN: 3.12
  STD.DEV: 2.20

MAX DEFECT: 10.24
THRESHOLD: d 0.12

DEFECT: 0.12- 10.24
BIN: 0.04
HAZE: 48- 1024

IS: KU
IA: KU
        TV TW

PRINTING   ZOOM 1WF

Figure 12. A "zoom" closeup of Figure 11 shows the accurate delineation of the geometric standard defect pattern. Notice that only single hits are present, as a result of the removal of redundant defect data by the PPC.

If we require a minimum of two successive hits for particle detection and maintain the original threshold setting, those particles large enough to have a threshold/peak value of 0.606 or lower will be detected (see Figure 3) and all single hits will be rejected. The price one pays for double-hit detection is that the particle scattering cross section must be at least 1/0.606 or 1.65 times the threshold setting to be detected with 100% probability. This compares with 1/0.88 or 1.13 times the threshold in the single-hit case. Roughly speaking, the particle cross section must be about 1.5 times higher relative to the threshold level in the double-hit mode versus the single-hit case for the same level of detection confidence. However, the reduction of noise hits in the double-hit mode allows the user to lower the threshold by 1.5x and regain the original particle size lower limit.

If the instrument noise amplitude distribution function were Gaussian, the instrument's performance would be about the same in the single-hit and double-hit modes, taking into account the different minimum threshold levels for the two cases. However, noise from photo-multiplier detectors tends to contain a significant number of larger-than-average pulses; the noise is "spiky" compared with Gaussian noise. As a result, the frequency of noise hits in the single-hit case is higher than one would calculate from Gaussian noise statistics, but the spikes are of short duration and thus are particularly well rejected by the double-hit PPC.

Figures 16a and 16b show an experimental verification of this point. A calibration wafer was made so that the smallest features would be close to the instrumental noise limit. Measurements were made at a threshold of 0.006 μm^2 in the single-hit mode (Figure 16a) and at 0.004 μm^2 in the double-hit mode (Figures 16b). Note that the probability of detection of the smallest features is virtually identical in both cases, but there is a net reduction in noise counts of about 280 in the double-hit case. It is apparent that the double-hit PPC is a powerful tool for noise reduction in scanned-laser surface inspection instruments.

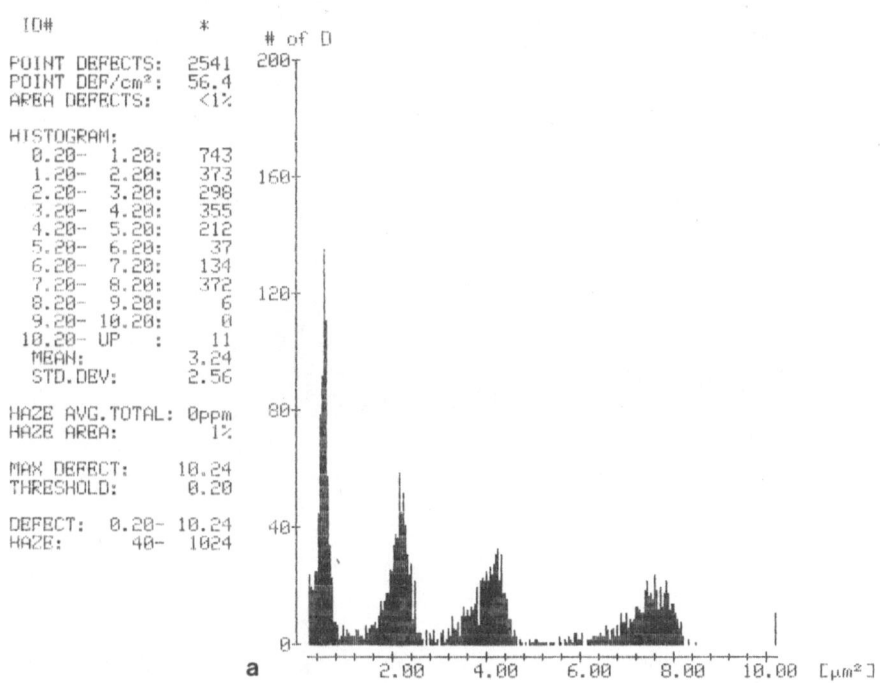

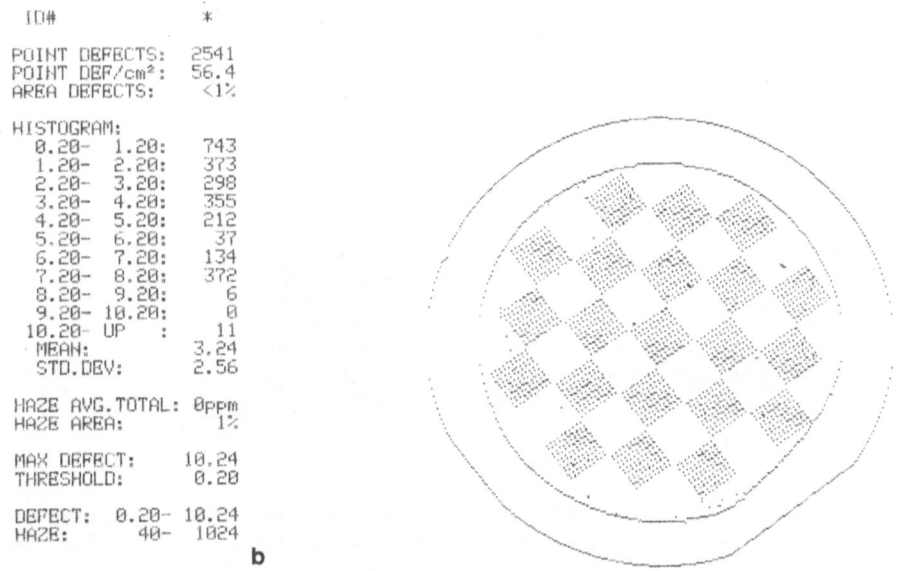

Figure 13a,b. The wafer of Figure 11, as measured without a PPC. Notice the double-pixel groups in the defect pattern where satellite hits are occurring, and the "tails" on the low side of each histogram peak.

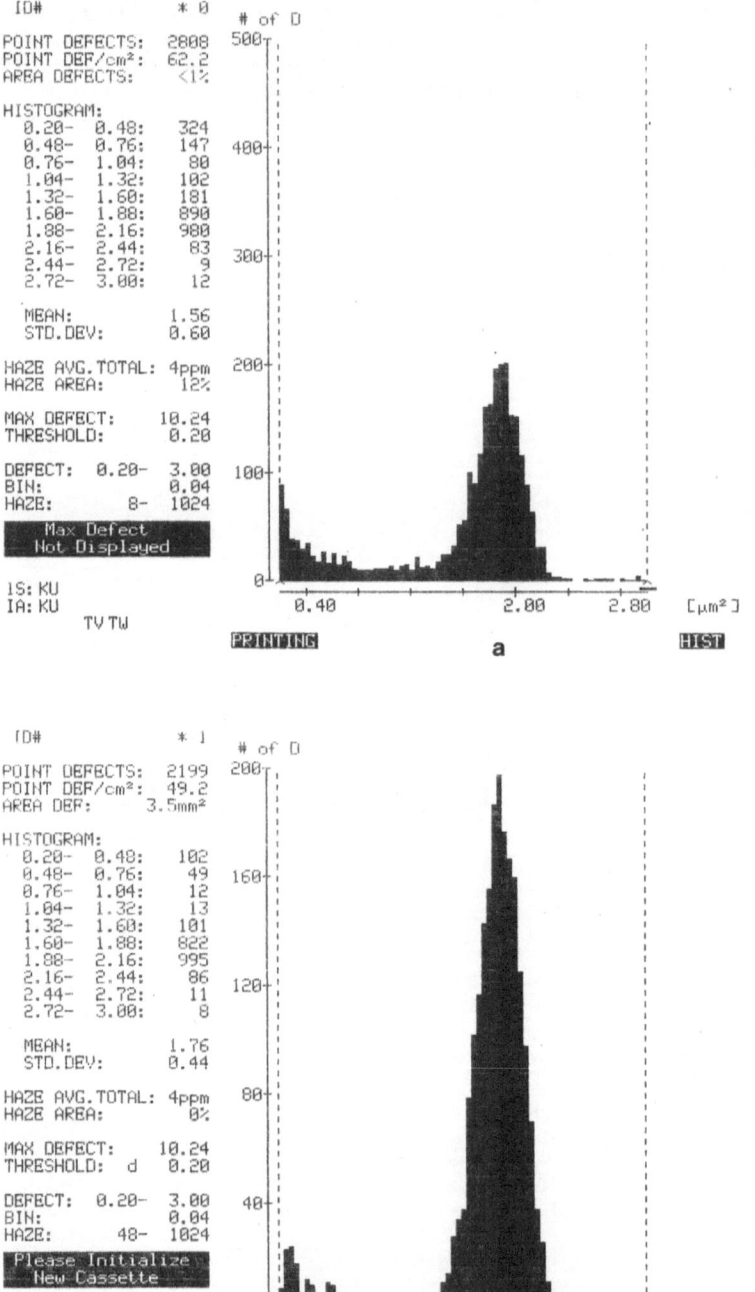

Figure 14a,b. Histograms for 2 μm polystyrene latex (PSL) spheres, with
and without the PPC. The particle count without PPC is 2808; with the PPC
it decreases to 2199. (The scale change was performed by the instrument
because of the lower defect count in the PPC run.) Assuming the PPC count
is correct, then with a threshold to peak ratio of 0.02/1.84, we expect
from Figure 9, curve (b), a count of 2781 without the PPC, close to the
observed count.

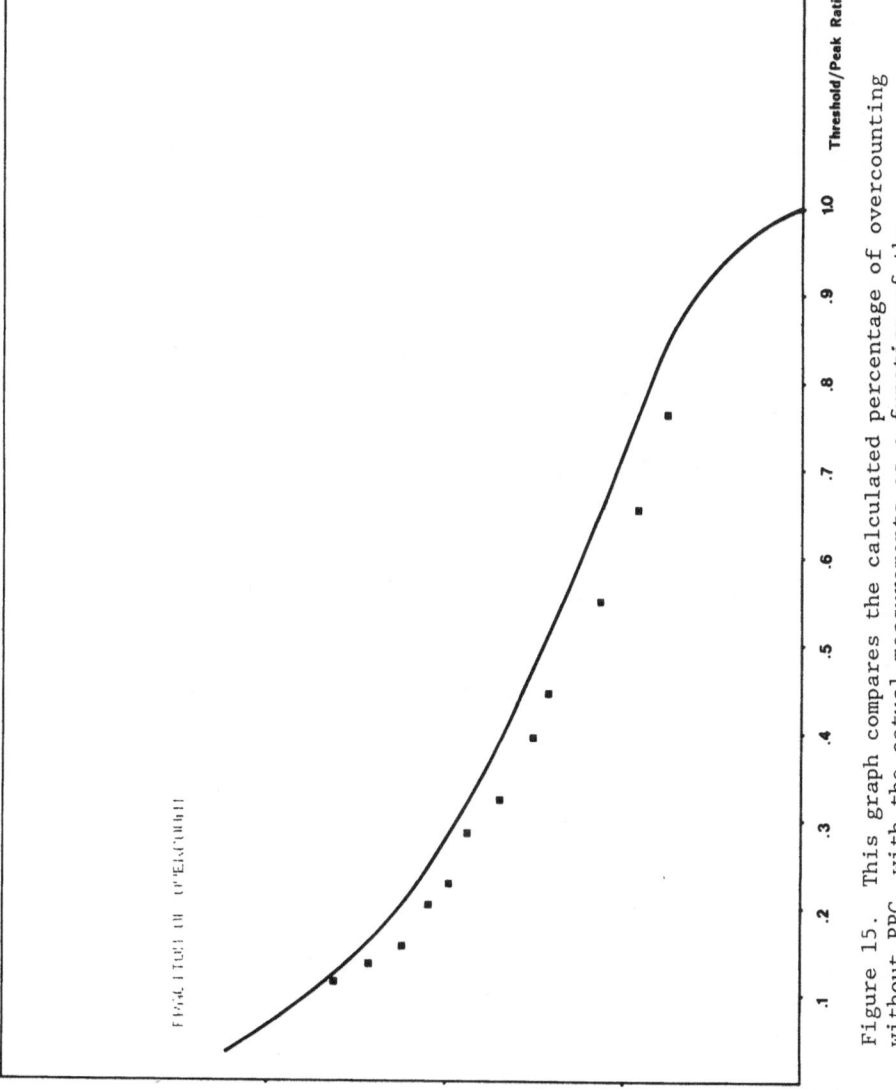

Figure 15. This graph compares the calculated percentage of overcounting without PPC, with the actual measurements as a function of the threshold/peak ratio on the 2.0 μm PSL wafer. The agreement is quite good over the whole threshold range.

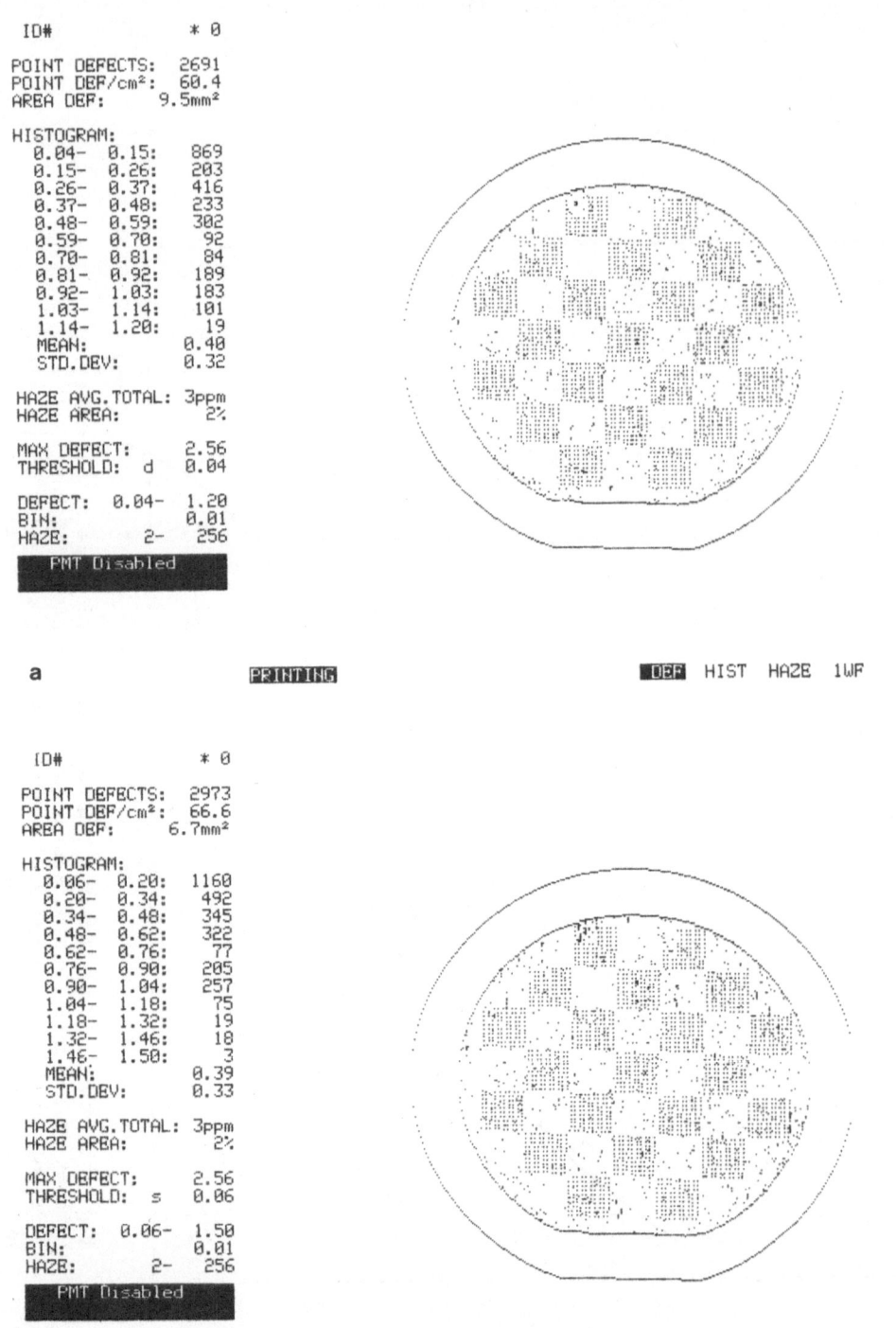

```
ID#             * 0

POINT DEFECTS:  2691
POINT DEF/cm²:  60.4
AREA DEF:      9.5mm²

HISTOGRAM:
  0.04-  0.15:   869
  0.15-  0.26:   203
  0.26-  0.37:   416
  0.37-  0.48:   233
  0.48-  0.59:   302
  0.59-  0.70:    92
  0.70-  0.81:    84
  0.81-  0.92:   189
  0.92-  1.03:   183
  1.03-  1.14:   101
  1.14-  1.20:    19
  MEAN:          0.40
  STD.DEV:       0.32

HAZE AVG.TOTAL: 3ppm
HAZE AREA:        2%

MAX DEFECT:     2.56
THRESHOLD:   d  0.04

DEFECT:  0.04-  1.20
BIN:           0.01
HAZE:       2-  256
```

PMT Disabled

a   PRINTING   DEF HIST HAZE 1WF

```
ID#             * 0

POINT DEFECTS:  2973
POINT DEF/cm²:  66.6
AREA DEF:      6.7mm²

HISTOGRAM:
  0.06-  0.20:  1160
  0.20-  0.34:   492
  0.34-  0.48:   345
  0.48-  0.62:   322
  0.62-  0.76:    77
  0.76-  0.90:   205
  0.90-  1.04:   257
  1.04-  1.18:    75
  1.18-  1.32:    19
  1.32-  1.46:    18
  1.46-  1.50:     3
  MEAN:          0.39
  STD.DEV:       0.33

HAZE AVG.TOTAL: 3ppm
HAZE AREA:        2%

MAX DEFECT:     2.56
THRESHOLD:   s  0.06

DEFECT:  0.06-  1.50
BIN:           0.01
HAZE:       2-  256
```

PMT Disabled

b   PRINTING   DEF HIST HAZE 1WF

Figure 16a,b. Demonstration of the noise reduction possible in the two-hit
mode (b) of the PPC versus the single-hit mode (a), using a calibration
wafer. Note the lower threshold setting in (b) and the nearly equal detection probability of the smallest defects in the two cases.

290

E. SPATIAL RESOLUTION WITH THE PPC

If two defects are separated horizontally by two scan pitch lengths (50 μm), and the threshold is set so that a hit occurs when the beam center encounters each particle, but not during the scan in between, then the PPC will resolve the hits as two distinct defects (in the single-hit mode). In this case the spatial resolution of the instrument is actually much better (50 μm) than the display pixel size (400 μm) or even the beam diameter (100 μm). This is especially evident when the display is in the Zoom mode with the PPC installed; separate defects which appear in the same pixel at normal display magnification are readily picked out. This demonstrates the high level of defect information that can be generated by scanning-laser instruments, when coupled with powerful signal processing techniques.

CONCLUSIONS

For those particles which have an amplitude substantially greater than the collection threshold, the counting accuracy is exact and repeatability perfect. However, as the size of the particle nears the collection threshold, statistical effects become important, and the variability of the count will increase; a discussion of these effects will be given elsewhere.

It should be realized that a real increase in particle count of 20% on critical levels (the kind of error which is readily encountered) can have rather dramatic effects on the overall I.C. yield. Correct defect counting is, therefore, of the utmost importance when trying to establish correlations between yield and particle densities.

REFERENCES

1. B. Tullis, Microcontamination, 3, No. 11, 72, (Nov 1985).
2. A. Siegman, "Introduction to Lasers and Masers," McGraw Hill, New York, 1972.
3. V.L.S.I. Standards, Inc., 2660 Marine Way, Mountain View, CA 94043

AUTOMATED SEM/EDS IMAGE ANALYSIS OF PARTICLES ON FILTER BLANKS

Garvin J. Stone

IBM Corporation
San Jose, CA 95193

The automated Scanning Electron Microscope (SEM) Energy
Dispersive Spectrometer (EDS) Image Analysis (IA) system is
used to measure the compositions and numbers of particle contaminants found on virgin 0.4 μm polyester membranes. This
type of filter is commonly used in the laboratory for trapping
particles borne in aerosols and liquids (1). This paper
describes a baseline for the level of particle contamination
on filter blanks. Establishing a baseline on the blanks is
critical for providing needed accuracy when sampling environments containing low particle concentrations. The results
from the blanks are compared to a set of filters used in
measuring the particle levels in IBM disk files.

INTRODUCTION

A common technique for the characterization of particles employs
trapping the particles onto filter membranes (polyester or polycarbonate)
using a pump and tubing. The membranes are then examined using SEM/EDS.
A standard procedure is to characterize the first 20 particles, usually
the most visible, and then to project a conclusion for the entire particle
population present on the filter (2,3,4). Early experience in our laboratory using these techniques demonstrated similar conclusions for samples
on filters as well as filter blanks, thus indicating the need for a study
such as the one described in this paper.

The automated SEM/EDS Image Analysis system eliminates human bias
and presents an accurate interpretation of particle populations on filter
membranes because the system does not seek particles. The system scans
a reasonably large amount of membrane area using many different frames
and only those particles encountered within each frame are analyzed.

The source of the particle contamination found on the filter mem-
ranes is mostly due to the manufacturing process. The membranes are made
using an irradiation/mask process followed by etching within heated
chemical baths. It is most likely that particles from the materials and
the chemicals used in the baths are transferring to the membranes during
the etching process.

Table I. Lotus 123 Spread Sheet showing Numbers of Each Particle Type per 200 Frames for Different Unused Filter Membranes. Sample IDs 1-18 Represent Different Different Blanks, Except for Numbers 5-8 Which Are 1-4 Analyzed a Second Time.

SAMPLE ID #	#1	#2	#3	#4	#5	#6	#7	#8	#9	#10	#11	#12	#13	#14	#15	#16	#17	#18
Elements	Blk 1-1	Blk 2-1	Blk 3-1	Blk 4-1	Blk 1-2	Blk 2-2	Blk 3-2	Blk 4-2	Blk A-1	Blk B-1	Blk B-2	LotB1DOA2B	SAME	LOTB1DOA41	SAME	Blk #1	Blk #2	Blk #3
No sp./Org	50	12	11	23	33	33	177	172					16		16	71	99	68
Na	0	0	0	0	0	0	0	0	0	0	0	0	0	0	0	0	0	0
NaCl	5	0	0	0	0	0	0	1	0	1	0	0	0	0	0	0	0	0
NaClK	0	0	0	0	9	0	0	0	0	0	0	0	0	0	0	0	0	0
NaS	0	0	0	1	0	0	0	1	0	0	0	0	1	0	0	0	0	0
NaSCa	0	0	0	0	0	0	0	0	1	0	0	0	0	0	0	0	0	0
Mg	2	0	0	0	1	0	0	0	0	0	0	0	1	0	1	0	0	0
MgSi	0	0	0	0	0	0	0	0	0	0	0	0	0	0	0	0	0	0
MgSiCr	0	0	0	0	0	0	0	0	0	0	0	0	0	0	0	0	0	0
MgCa	1	0	0	0	0	0	0	0	0	0	0	0	0	0	0	0	0	0
MgFe	0	0	0	0	1	1	0	0	0	1	0	0	0	0	1	1	0	37
Al	0	11	0	16	0	0	4	8	0	0	3	2	2	2	1	1	1	0
AlSi	0	0	0	1	1	1	0	1	0	1	0	0	0	0	0	0	0	0
AlSiSi	0	0	0	0	0	0	0	0	0	0	0	0	0	0	0	0	0	0
AlSiFe	0	1	0	0	0	0	0	0	0	0	0	0	0	0	0	0	0	1
AlSiCu	0	0	0	0	0	0	1	0	0	1	0	0	0	0	0	0	0	0
AlPCa	0	1	0	0	0	0	0	0	1	1	0	0	0	0	1	0	0	0
AlSCa	0	0	0	0	0	0	0	0	0	0	0	0	0	0	0	0	0	2
AlCl	0	1	1	0	1	1	1	1	0	0	0	0	1	0	1	0	0	1
AlCr	0	0	0	0	0	0	0	0	0	0	0	0	0	0	0	0	0	0
AlCrFE	0	1	0	0	0	0	0	0	0	0	0	0	0	0	0	0	0	1
AlSb	0	0	0	0	0	0	0	0	0	0	0	0	0	0	0	0	0	0
Si	6	2	0	3	4	5	3	5	1	1	2	2	2	0	14	1	2	0
SiScr	0	0	0	0	0	0	0	0	0	0	0	0	7	0	0	0	0	0
SiCl	0	0	0	0	0	0	0	0	0	0	0	0	0	0	0	0	0	0
SiClCr	0	0	0	0	0	0	0	0	0	0	0	0	0	0	0	0	0	0
SiCr	0	0	0	2	0	0	1	1	1	0	1	1	0	0	1	1	0	1
SiFe	0	0	0	0	0	0	0	0	0	0	0	0	0	0	0	0	0	0
SiCu	0	0	0	0	0	0	0	0	0	0	0	0	1	0	1	0	0	0
SiSn	0	0	0	0	0	0	0	0	0	0	0	0	0	0	0	0	0	0
P	0	0	0	0	0	1	1	2	2	1	4	1	1	0	3	0	0	0
PCa	12	5	10	5	19	10	39	51	8	1	0	1	9	0	3	25	16	18
PSb	0	0	0	0	0	0	0	0	0	0	0	0	1	0	0	0	0	0
S	0	1	0	1	0	0	0	0	0	0	0	0	0	0	0	1	2	0
SCa	2	2	2	0	5	2	0	0	3	0	0	2	2	0	0	0	0	0
SCr	0	0	0	1	0	0	0	0	0	0	0	0	0	0	0	0	0	0
ClK	1	0	0	0	12	2	0	0	0	0	0	2	1	0	0	0	0	0
ClCa	0	0	0	1	0	0	0	0	0	0	0	0	0	0	0	0	0	0
Ca	0	0	0	0	0	0	0	0	0	0	0	0	0	0	8	0	0	0
CaFe	7	2	4	1	5	11	3	4	1	0	0	4	2	0	0	3	2	0
Cr	0	0	0	3	0	0	0	1	0	0	0	0	0	0	0	0	0	1
CrFe	0	0	0	0	0	1	0	0	1	6	0	1	1	0	0	1	0	0
Fe	0	2	7	0	1	1	0	0	0	0	0	1	0	0	0	0	0	0
Sb	8	2	7	0	4	6	31	44	0	1	1	2	5	2	1	6	6	5
Total particles	74	40	35	58	95	73	260	292	15	11	11	22	49	12	47	108	127	135
Length	1.57	7.65	1.55	4.81	1.89	2.35	0.97	1.37	1.14	4.06	3.74	2.42	1.59	1.38	7.88	1.04	0.94	4.67
Width	0.45	2.68	0.43	0.73	0.54	0.4	0.34	0.37	0.57	0.71	0.48	0.57	0.59	0.48	0.88	0.42	0.4	0.49

INSTRUMENTATION

The filters were analyzed on an SEM/EDS Image Analysis system that includes the following components:
1. LeMont DA-10 Image Analysis data system
2. Hitachi S-520 with a Rath automated stage
3. Kevex 8000 EDS and data system

The SEM/EDS Image Analysis system is fully automated and capable of unattended operation for the analysis of six samples (5).

PROCEDURE

1. The blank filter membrane is attached to the carbon planchet with carbon low-resistance contact cement (also known as dag).
2. The planchet, with attached blank, is carbon-coated using a Fullum vacuum evaporator.
3. The analyses are performed using the SEM/EDS/IA system in the backscatter mode. Accelerating voltage 20 kV, emission current 52 μA, WD 15 mm, and 0 degree tilt.
   A. Each filter is scanned and divided into 200 equally spaced frames at a magnification of 1000X or 100X.
   B. Each particle in the frame is sized and an elemental spectrum acquired.
4. A summary is computed and provided for each filter.

RESULTS AND DISCUSSION

Comparisons between the common elements or groups of elements considered contaminants on the filters are presented in Table I, which is a spread sheet from Lotus 123 and was used to make all the figures presented in this paper. Filters 1 through 4 were re-analyzed at or near the same location and are represented as filters 5 through 8. Filters 12 and 13 were from a different lot, as were filters 14 and 15.

The total particles counted are presented in Figure 1 and show an increased count for filters 5, 6, 7, and 8 (replication of filters 1 through 4). This increase may be the result of the electron beam interacting with the surface of the filter during the first run. By altering the smooth surface of the filter during the first run, the SEM/EDS/IA system detects more organic or non-spectrum particles during the second run. Figure 2 shows the increase in organic or non-spectrum particles in the second scan. For this reason, the high-magnification analyses should be run first. The total organic or non-spectrum particles are in one of the following categories: (a) the sum of the carbon aggregates from the coating process, or (b) actual organic particles. Particles containing elements with atomic numbers lower than sodium can be detected and sized but will fall into this category because no elements are identified within the EDS spectrum. The threshold limits on the LeMont Image Analyzer determine the detectability of these particles during analysis in the backscatter mode.

As seen in Figure 3, chromium (Cr) is a major contaminant on the polyester filters. From SEM observations, the Cr appears as a crust integrated into the filter. In Figure 4, antimony (Sb) is seen as another major contaminant. Antimony is an extremely small particle (average L x W = 0.3 x 0.2 μm) and can only be detected by using backscatter imaging.

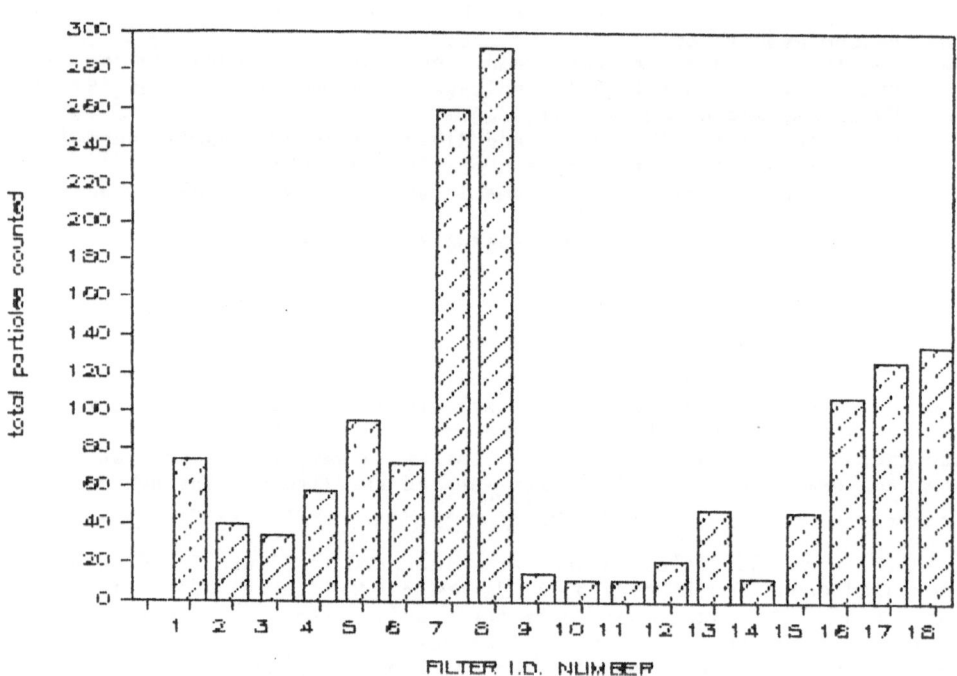

Figure 1. Total particles counted for each filter blank. Blanks 12 and 13 are from lot 81DOA28; blanks 14 and 15 are from lot 81DOA41. All other blank filters are from the same lot.

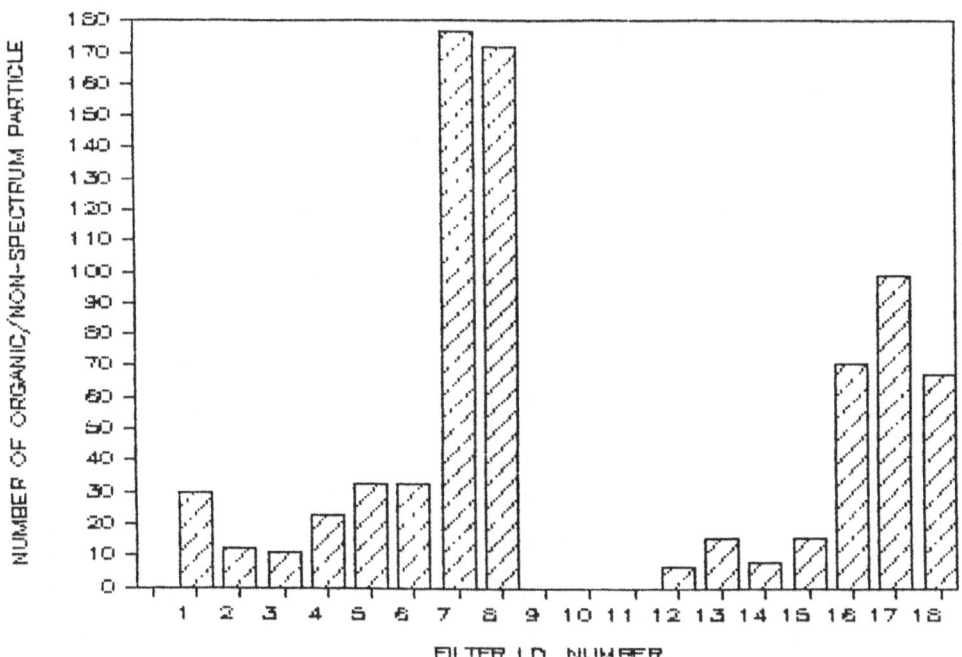

Figure 2.   Number of organic or non-spectrum particles. Non-spectrum
particles contain elements with atomic numbers lower than
sodium (EDS limitation).

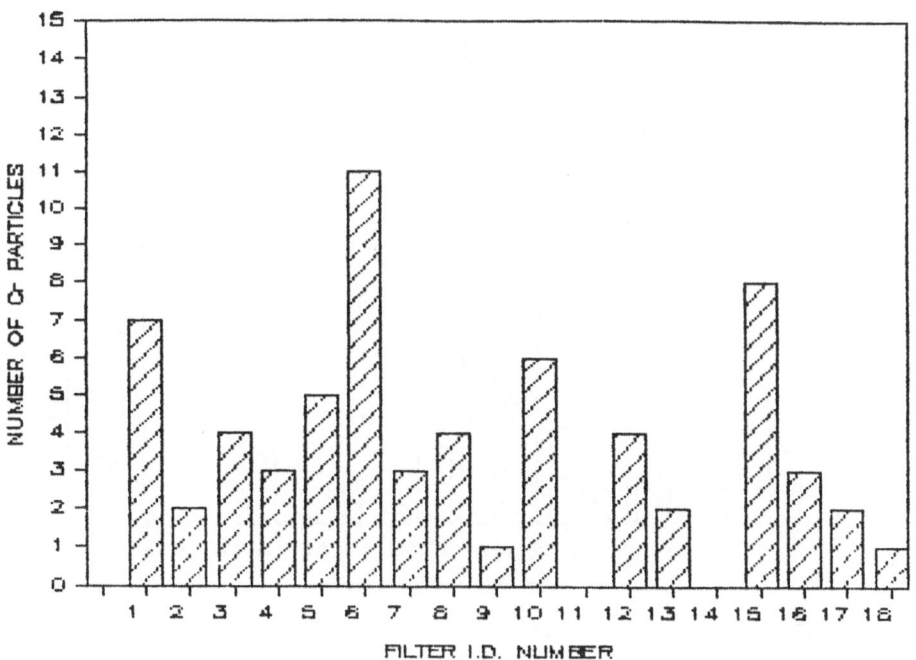

Figure 3.   Number of Cr particles.

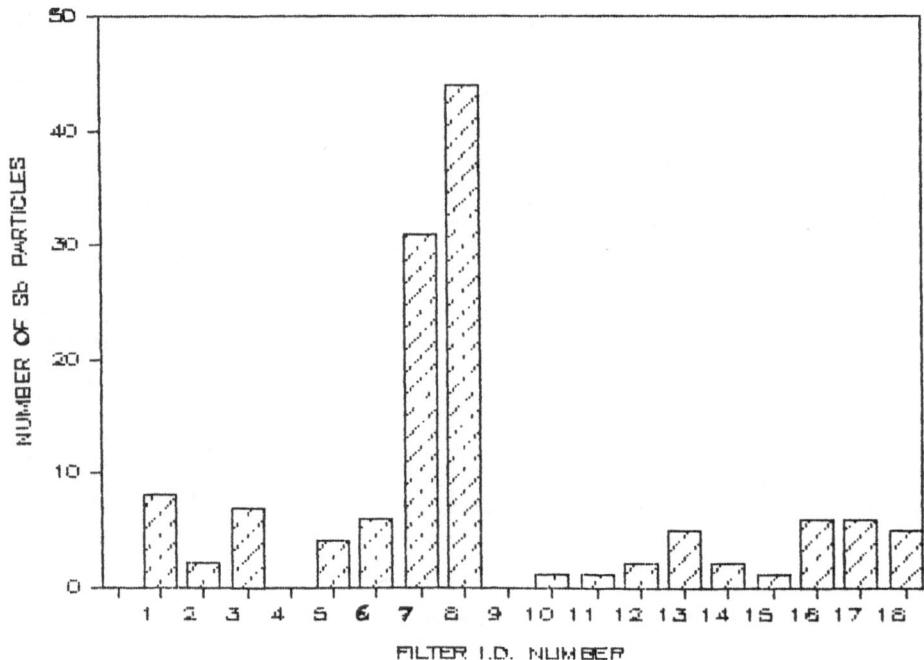

Figure 4.   Number of Sb particles.

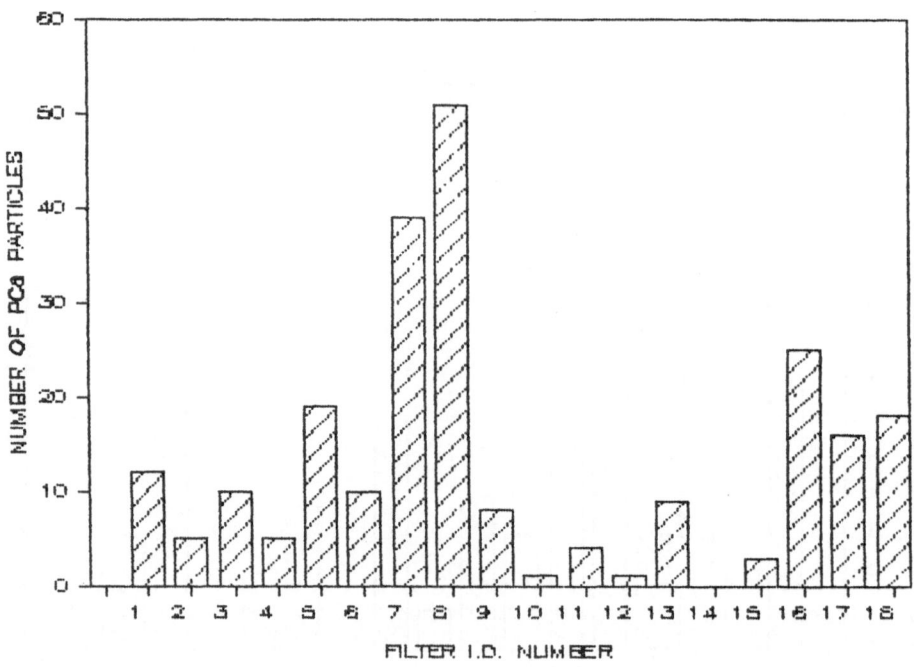

Figure 5.   Number of P/Ca particles.

Figure 5 presents the number of phosphorus/calcium (P/Ca) particles, which are also considered as a major contaminant. These particles are also small, and thus difficult to locate. Figure 6 shows silicon; its significance as a contaminant is not fully understood at this time.

Figure 7 represents the number of sulfur/calcium (S/Ca) particles. They appear in low numbers and irregularly, and thus may not be considered a contaminant. Aluminum appears to be a contaminant (see Figure 8). The rest of the chemical categories or particle types -- chlorine/potassium (Cl/K), sodium/chlorine/potassium (Na/Cl/K), aluminum/chromium (Al/Cr) -- are sporadic and low in numbers (see Figures 9, 10, and 11).

Figure 12 gives a measure of the average length by width (in μm) for all the particles found on each filter.

Figure 13 documents the small size of the P/Ca particles using some data taken in the field from two different disk files. The blank data are in columns 1 and 2; sample 1 data are in columns 3 and 4, and sample 2 data are in columns 5 and 6. Numbers 1, 3, and 5 are the analyses performed at 1000X. Numbers 2, 4, and 6 are the same analyses performed at 100X. The P/Ca particles have a much lower concentration in the lower magnification analysis, thus supporting their smaller size.

Figure 14 demonstrates a reverse situation for the Cr particles. The analysis was the same as that described for Figure 13. Here the Cr particles have the higher concentration in the lower magnification; thus the average particle is larger.

Figures 15 and 16 demonstrate an interesting phenomenon that is occurring during long-term sampling. Filters 1 through 4 are blanks. Filters 5 through 11 are of aerosol samples taken from different 3380 disk files in the field. Air taken from inside the operating disk files (air supplied by HEPA filters) was pulled through the filters continuously for three months. It appears that the fine P/Ca and Sb particles are forced into or even pulled through the filters, as seen by their non-appearance in columns 5 through 11. This phenomenon has not been reported previously.

Figure 17 shows the data for the Cr particles during this same sampling. The Cr, being a larger particle, is not affected by the sampling, and in the case of samples 5 and 8 could be a contaminant in the HEPA environment of the disk file.

When a contaminant is present in the environment being sampled, it shows up readily. Figure 18 shows the presence of molybdenum/chromium/cobalt (Mo/Cr/Co) particles on samples 6, 7, and 11. The sample collected onto filter 11 may very well represent a disk file headed toward failure.

CONCLUSION

The compositions and numbers of contaminant particles found on virgin filter blanks have been discussed in this paper. A baseline has been established that can lend accuracy to the data collected during sampling. Elemental categories considered major contaminants on unused polyester filters are Cr, Sb, P/Ca, and Al. Particles also present, but at less significant levels, included Si, S/Ca, Cl/K, Na/Cl/K, and Al/Cr. It has been shown that the various particles fall within two size categories, thus indicating the need to perform analyses at both high and low magnifications, with the high magnification first. Importantly, it has been found that fine particles (< 0.3 μm) can be forced into or through the filter during a long (> three-month) sampling period.

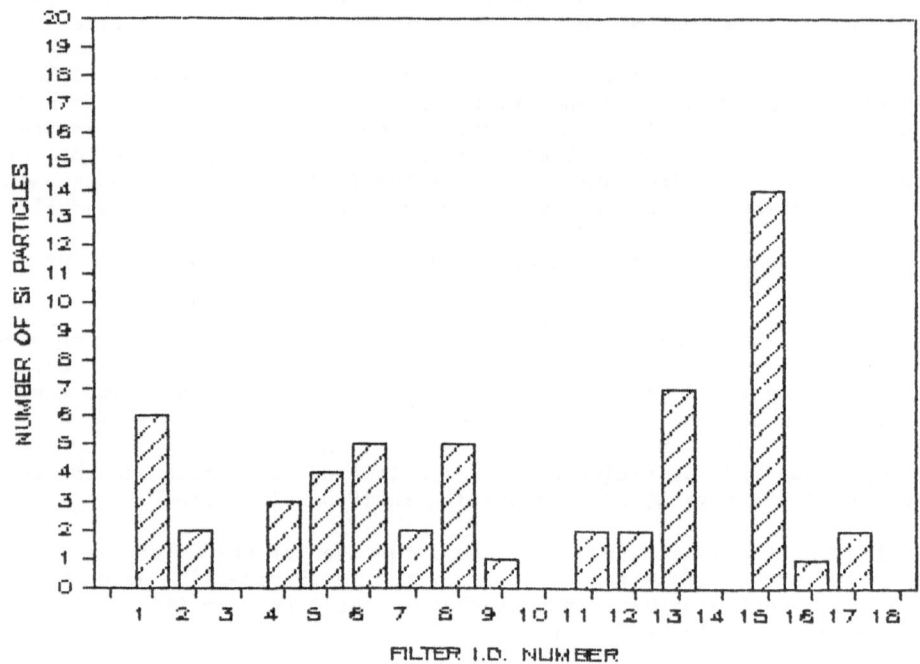

Figure 6.   Number of Si particles.

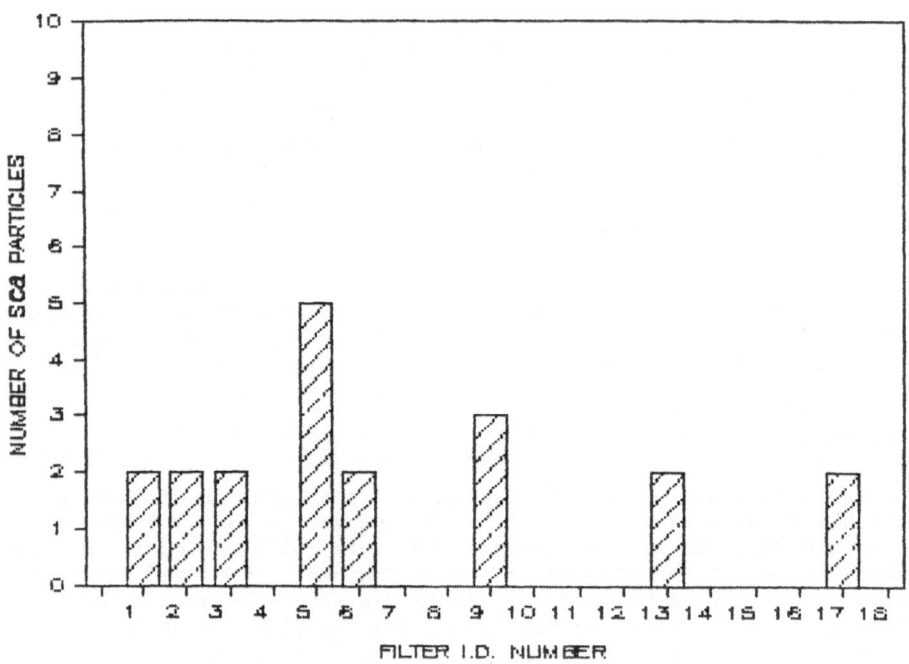

Figure 7.   Number of S/Ca particles.

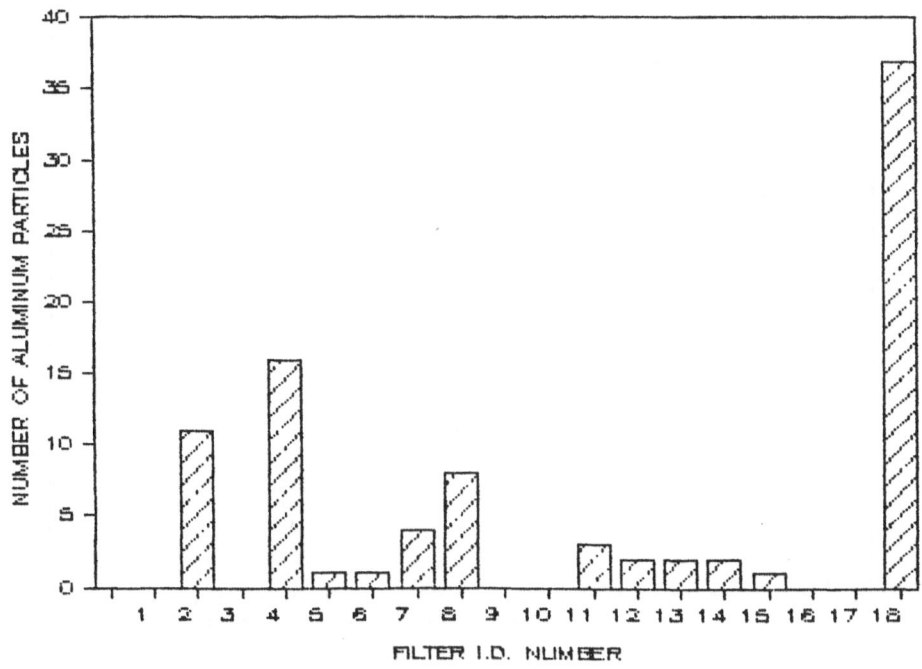

Figure 8. Number of Al particles.

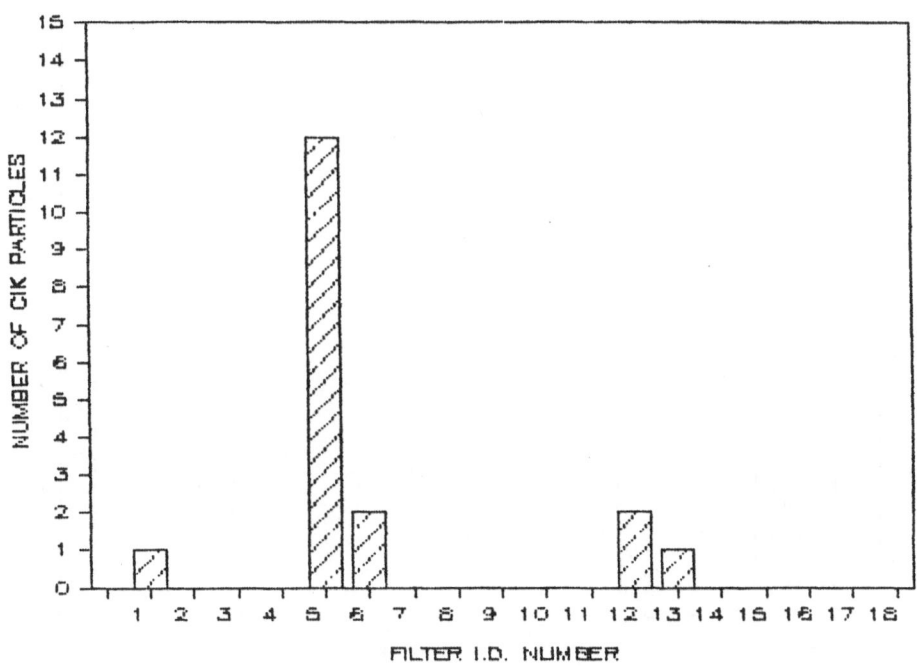

Figure 9. Number of Cl/K particles.

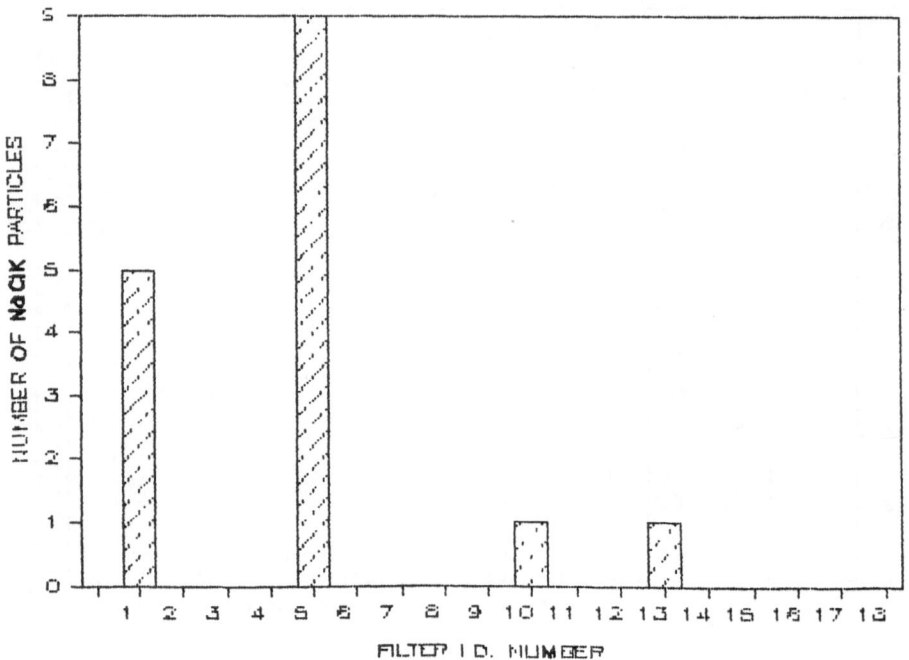

Figure 10. Number of Na/Cl/K particles.

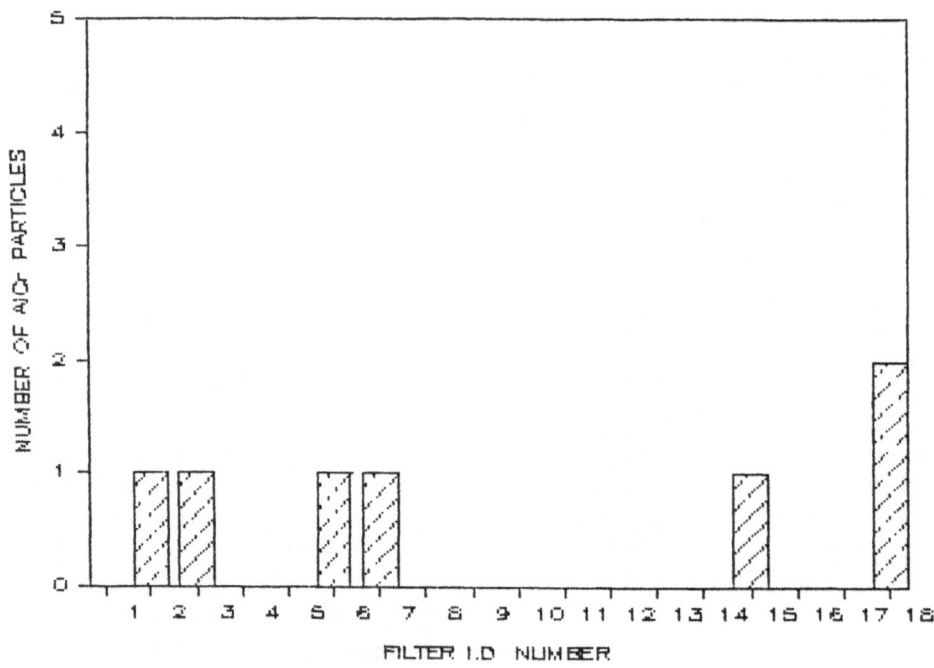

Figure 11. Number of Al/Cr particles.

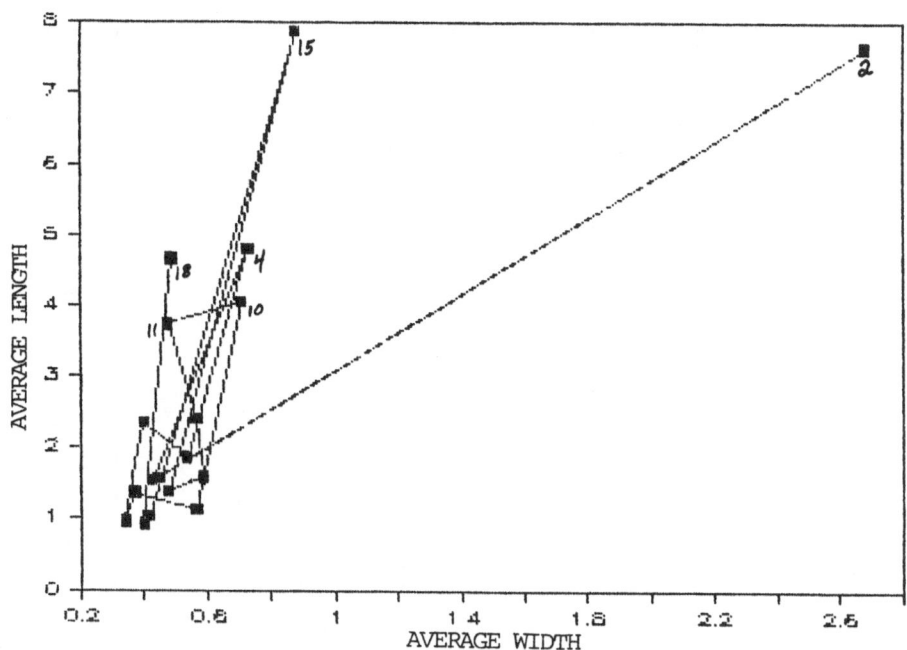

Figure 12. Average length and width for all particles.

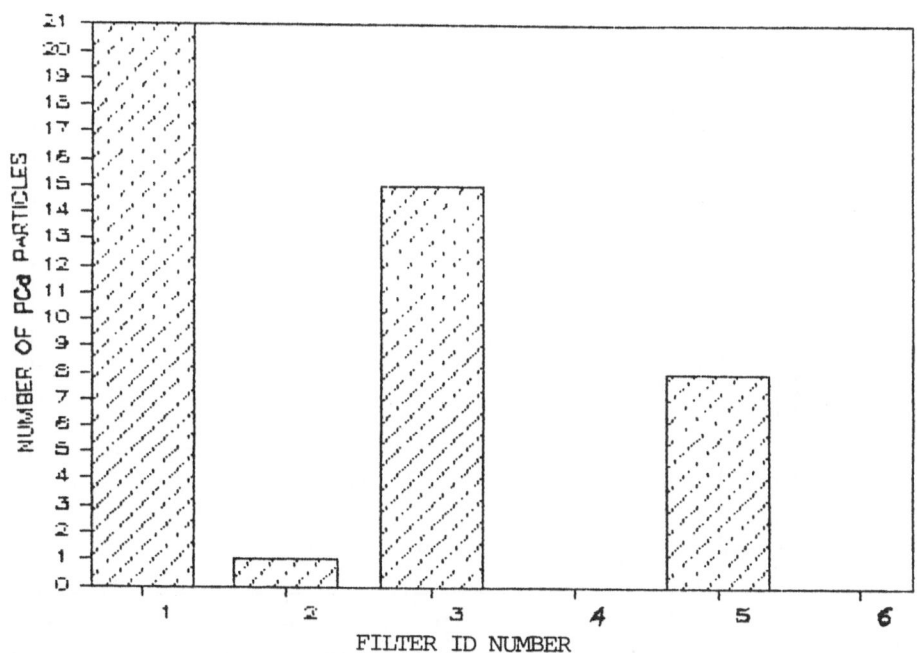

Figure 13. Number of P/Ca particles. Blanks (columns 1 and 2) compared
against two samples taken from disk files. Columns 3 and 4
are sample 1, and 5 and 6 are sample 2. Columns 1, 3, and 5
are the results from analyses at 1000X; 2, 4, and 6, at 100X.

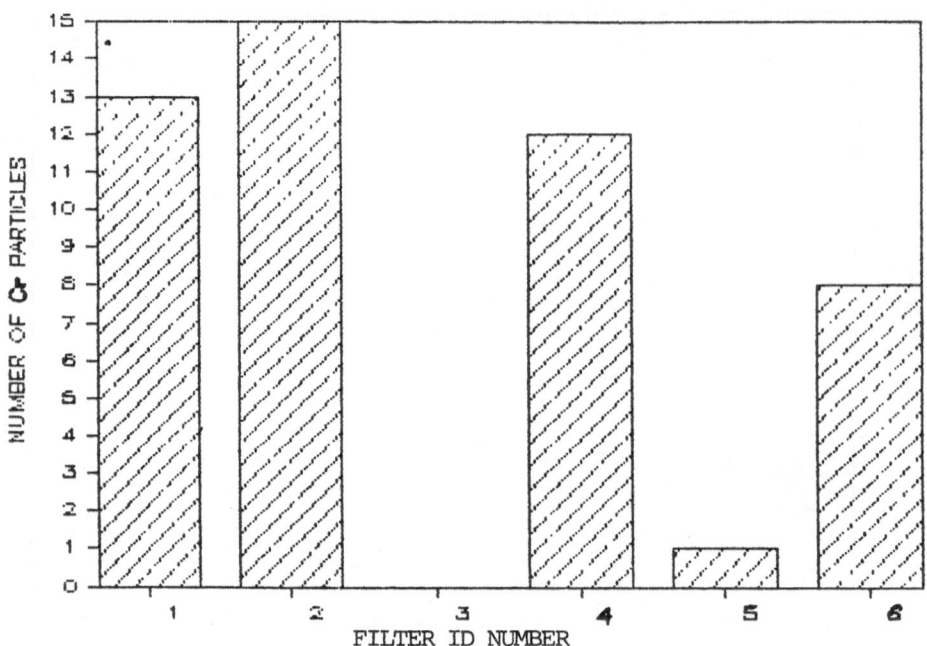

Figure 14. Number of Cr particles. Columns 3 and 4 are sample 1,
and 5 and 6 are sample 2. Columns 1, 3, and 5 are the
results from analyses at 1000X; 2, 4, and 6, at 100X.

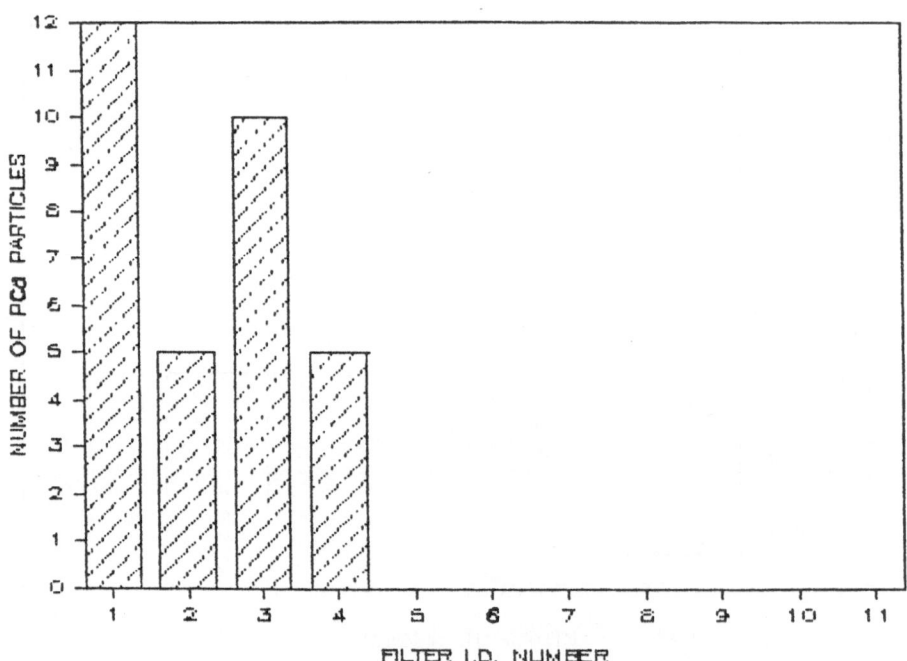

Figure 15. Number of P/Ca particles. Blanks (columns 1 through 4)
compared against continuous disk file samples (columns 5
through 11, which appear blank because no P/Ca particles
were detected during the analysis). Analysis at 1000X.

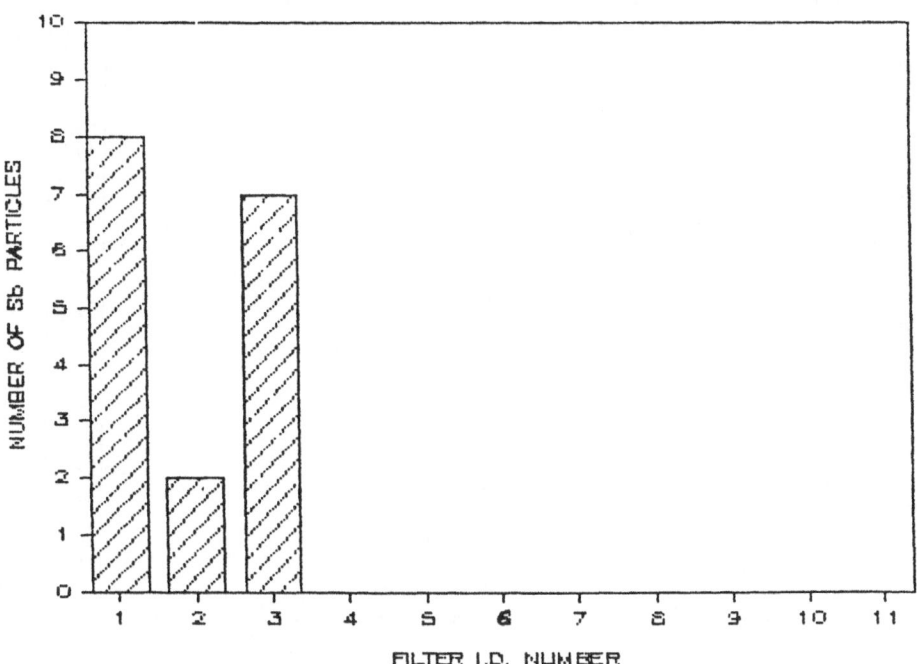

Figure 16. Number of Sb particles. Blanks (columns 1 through 4; no
Sb particles were encountered on blank 4) compared against
continuous disk file samples. Analysis at 1000X.

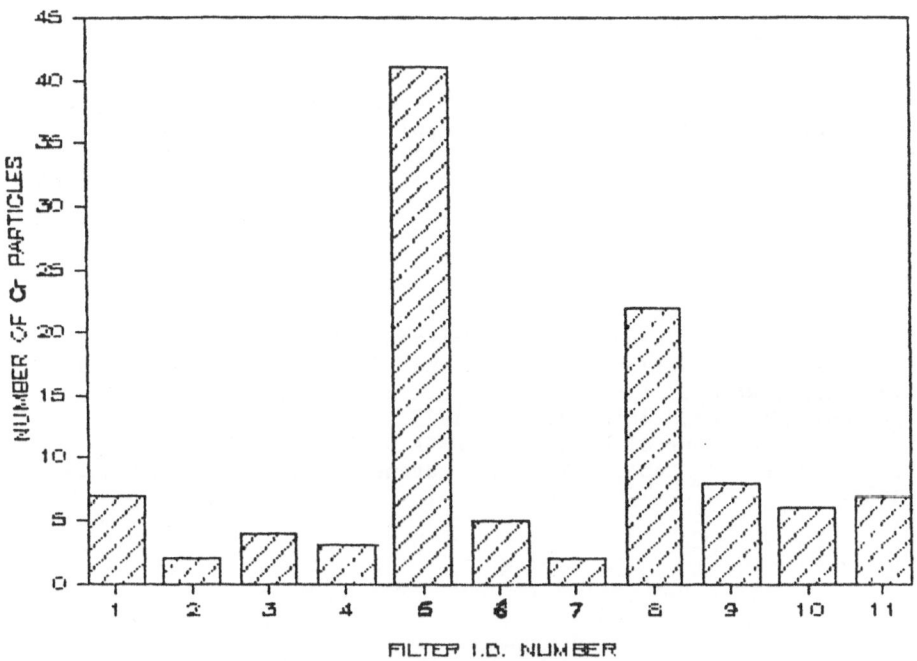

Figure 17. Number of Cr particles. Blanks (columns 1 through 4)
compared against samples collected ($>$ three months) from
disk files in the field (columns 5 through 11). Column 5
is a sample from one disk file that may have a chromium
contamination problem.

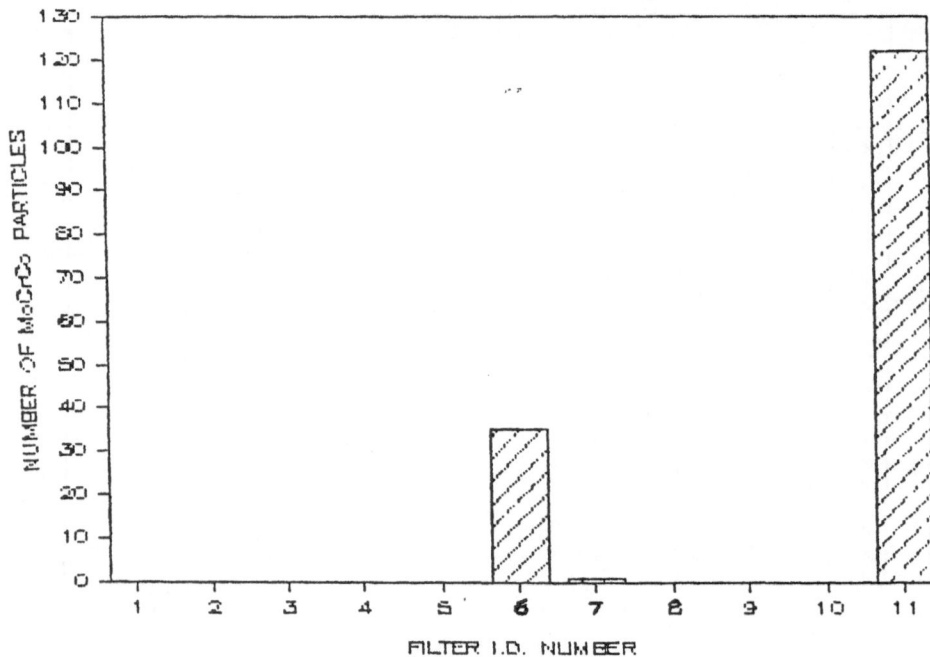

Figure 18. Number of Mo/Cr/Co particles. Mc/Cr/Co particles appear on only three of the continuous field samples (columns 5 through 11). The blanks are columns 1 through 4 and are empty because no particles were detected.

NOTE: Since this paper was written several different sets of filters have been analyzed. For the more recent polyester membranes the manufacturer has eliminated the source of the chromium contamination. I recommend that all laboratories with a supply of polyester membranes on hand should discard them and order a fresh shipment.

REFERENCES

1.   Nuclepore Corp., "Innovations in Membrane Filtration," 1984.
2.   M. Balazs and S. Walker, "Counting and Identifying Particles in High-Purity Water," Semiconductor International, April 1982.
3.   "Standard Methods," 14th edition, 1975.
4.   "Contamination Control Handbook," NASA SP-5076, 1969.
5.   R. M. Edwards, J. Lebiedzik, and G. Stone, Scanning, 8, 221-231 (1986).

PARTICLE SIZING AND COUNTING WITH THE INSPEX EX20/20

Charly Allemand

INSPEX, Inc.
40 Bear Hill Road
Waltham, MA 02254

Performances of the Inspex EX20/20 Wafer Inspection System are given. Precisely defined statistical expressions are used to describe measurement conditions close to the detection limit which is determined as 0.29µm for latex spheres. The counter's precision is given as well as the contamination it causes, its uniformity, its haze measurement capability of 0.4ppm, its calibration curve and considerations on missed particles and false counts.

1. INTRODUCTION

The objective of this paper is the presentation of the performances of the Inspex EX20/20 Wafer Inspection System. It will be preceded by the definition of statistical expressions taken from literature and well established by usage.

Unless specified otherwise, the distributions of the data are normal and the number of data is large enough to justify the omission of small sample corrections without adversely affecting the precision of the results. Because of their availability and uniformity, latex spheres are used as standard calibration particles and their detectability is assumed to be limited by the background noise. These simplifications allow for the examination of sizing and counting conditions close to the detection limit where the effect of both the background noise and the threshold setting on the counting accuracy is explained in detail.

The paper begins with a brief description of the EX20/20 and ends with the presentation of results obtained on it, which include the detection limit, the uniformity, the counter's precision, haze determination, and integrated distribution counting for different sized particles.

Several results are known for other commercially available instruments. Therefore, although no systematic comparison is possible at this time; comparable data are, nevertheless, mentioned when available.

2. INSTRUMENT DESCRIPTION

The Inspex EX20/20 is designed to detect contaminant particles on virgin wafers. It uses polychromatic light from a mercury-xenon arc lamp directed onto the wafer along its normal as a collimated and precisely limited beam. Light scattered by the contaminants is collected at an angle from the normal by a Silicon Intensified Target (SIT) vidicon camera. The target itself has 512x512 pixels. One image is read every 1/30 of a second; signals of consecutive readings are stored in a frame memory where up to 64 images can be accumulated. The integrated images can be displayed on a monitor and examined both visually and electronically, meaning that each particle signal can be interrogated separately or collectively, the answer being given in the form of a value, a table or of a histogram. When a particle signal covers more than one pixel, all the adjacent active pixels are counted together to form a single particle signal, the area of which is expressed in pixels.

More information can be obtained from the processor. For instance, if the contaminant produces an asymmetrical image, the ratio of the longest to the shortest chord is computed. It is then compared to a preset scratch-ratio to decide whether or not the defect is to be treated as a scratch. Furthermore, in order to estimate the amount of light scattered by the clean wafer surface (by its roughness or by much smaller particles) the total number of active pixels can be counted which is then compared to the number of pixels activated by a white surface under a dimmed illumination. This is known as a haze measurement.

3. DEFINITIONS

Definitions given in this section are used in section 5 to characterize the EX20/20.

3.1 The Uncertainty of Data[1]

The uncertainty of data is equal to one standard deviation, S, where $S^2 = \frac{1}{N-1} \Sigma (x_i - \bar{x})^2$, and x_i are the data, N their number, $\bar{x}$ their mean.

3.2 The Repeatability is the Relative Standard Deviation Times and Arbitrary Factor

In other words it is the standard deviation, S, divided by the sample mean, $\bar{x}$, times a factor K:

$$R = \frac{S}{\bar{x}} \cdot K$$

In experiments such as the determination of calibration curves, the factor K is 1, and the repeatability is equal to the relative standard deviation.

The term repeatability is also used to compare one of the measured data at the beginning of the experiment with the same data after a set working time. In that case the repeatability is again calculated as above with K = 1.

The factor K is introduced here to conserve the generality of the definition, because in some instances it is given a value different from one.

3.3 The Calibration Curve is the Relation between Known Standard Samples and the Output Signals They Produce

It may sometimes be derived from theoretical considerations, but it always has to be confirmed by experiment.

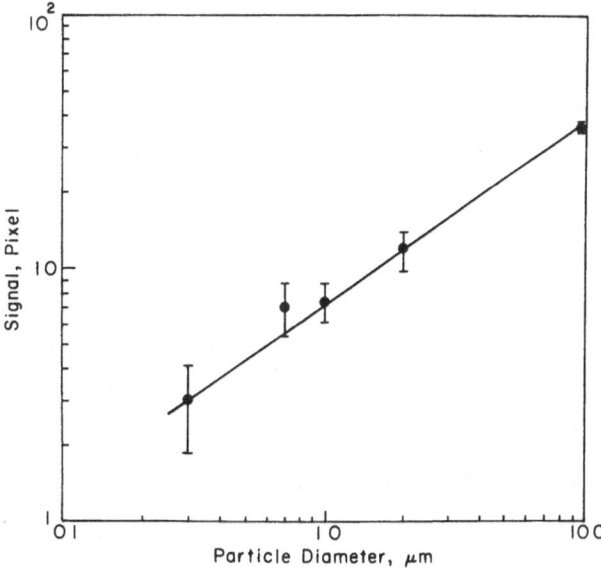

Figure 1   Typical calibration curve for particle-sizing limited by instrument noise. A 0.3μm particle produces a 3 pixels mean signal with an uncertainty of 1.5 pixels.

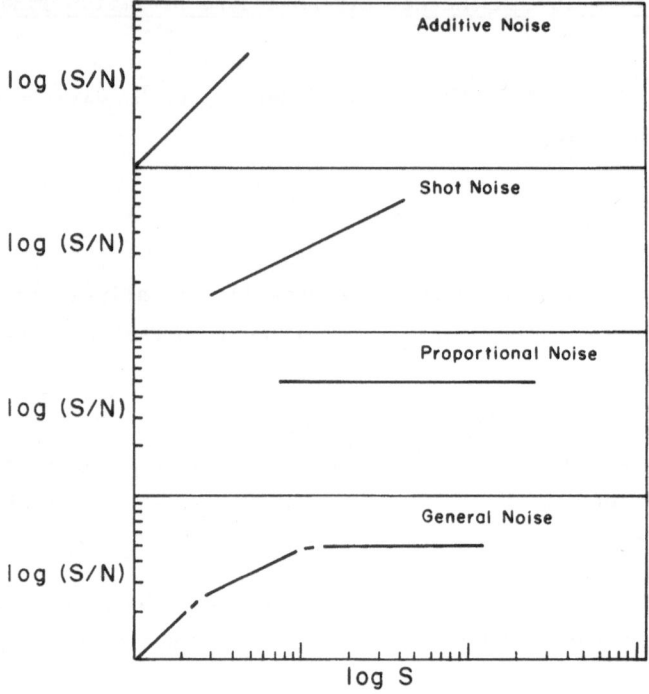

Figure 2   Signal-to-noise ratios for typical noises.

In the example shown here (Fig. 1), standard latex spheres of five different known diameters have been deposited on a clean wafer and the signal they produce has been recorded. For instance, a 0.3μm particle produces a 3-pixel signal.

3.4   The Noise is the Variation of the Measured Data

For instance, the determination of the above data has been repeated, and the uncertainty of the signal (its noise) is plotted on figure 1 as an error bar one standard deviation long on each side of the average. Because of the logarithmic scale used for the graph, the uncertainty bars appear to have different lengths. In fact, they all have the same 1.5-pixels length because, in this particular instrument (an Inspex 1515), the variations in the signals were caused by the input amplifier. This is a typical situation for an uncertainty, or noise, N, consisting only of an instrumental noise which is independent of the signal.

Such an instrumental noise is constant and additive to the signal, S; the signal-to-noise ratio increases linearly with the signal (Fig. 2, top).

$$\frac{S}{N} = C \cdot S$$

In general, the noise has several components such as background noise or the signal noise[2,4,5]. Three examples of noises are given in Figure 2. When the signal is statistical in nature, as the arrival of electrons for instance, the signal current mean-square fluctuation is $\overline{i^2} = eI\Delta f$, where e is the electron charge, I the signal current, and Δf the frequency band. Therefore, the noise is proportional to the square root of the signal. The signal-to-noise ratio is then

$$\frac{S}{N} = C \sqrt{S}$$

This situation is represented in Figure 2 as shot noise.

When the signal is large, its variations is often dominated by the 1/f noise, the power of which is given by $W = cI^{\alpha}/f^{\beta}$, where c is a proportionality factor, I the signal current, and f the frequency. α is very nearly equal to 2 and β can range from about 0.8 to 1.5. The noise is then proportional to the signal and

$$\frac{S}{N} = C \quad .$$

Many other reasons may cause the noise to be proportional to the signal. One of them is a response variation with location. When signals are collected over all locations, their population exhibits a variation that reflects the response variation and is, therefore, proportional to the signal. All three types of noise together cause the signal-to-noise ratio to vary with the signal as shown at the bottom of figure 2.

3.5 The Detection Limit is the Sample that Produces a Signal Equal to an Arbitrary Factor Times the Standard Deviations of the Background Variations

$$S_{DL} = C \cdot S_B$$

The constant C is often set equal to 1 (Refs. 3-5) or 2 (Refs. 6-9) or sometimes to 6 or 8 (Refs. 6 and 9). In this paper it will be set to 2.

The calibration curve of figure 1 had a noise of 1.5 pixels. The detection limit for that instrument is the particle that produces a signal of 3 pixels, namely a 0.3μm diameter latex sphere.

3.6 Missed Particles

Missed particles are those particles that the instrument fails to count because their signal is smaller than the set detection threshold.

The signals of particles of a particular size are distributed around an average value according to a Gaussian distribution

$$P_G = \frac{n}{\sigma\sqrt{2\pi}} \exp\left[-\tfrac{1}{2}\left(\frac{x-\mu}{\sigma}\right)^2\right]$$

where σ is the standard deviation and n the total number of particles, μ the mean, and x the value of a random observation. When a threshold signal T is set in the instrument, then signals smaller than the threshold are not allowed through the data processor. This means that particles that produce a signal smaller than T are not counted. Their number is given by

$$n_m = \int_{-\infty}^{T} P_G dx$$

Twice this value is tabulated in statistics textbooks[1]. Therefore, n_m can be determined when μ, σ and n are known as well as T expressed as a multiple of σ.

3.7 False Counts

False counts are the counts that are due to signals generated not by particles but by the background noise.

The background varies about its mean value and the distribution of the fluctuations is also normal. The detection threshold is usually set well above the average background level. Nevertheless, the background fluctuations that produce a peak signal extending above the threshold are detected and counted as a particle. Their number is given by

$$n_f = \int_{T}^{\infty} P_{GB} dx$$

where P_{GB} is the background distribution. This integral is found in the same above-mentioned tables.

3.8 The Counter's Precision is the Standard Deviation of the Particle Count

It is given as a function of the number of counted particles[10,11] for particles of different sizes. Since the number of missed particles and false counts depends on the size of the particles and the threshold setting, the counter's precision depends on the distribution of the particle sizes in the measured sample. More precise information is obtained when only one particle size is counted. In this paper only 0.36μm latex spheres were deposited on a clean wafer and counted.

3.9 Uniformity

Uniformity refers to the variation in the signal of a single standard particle deposited at different locations on the wafer. It is given as the relative standard deviation of the average signals in different locations.

3.10 Contamination is the Number of Particles Added per Wafer per Pass (PWP, ref. 11)

A cartridge of clean wafers is passed through the instrument and the number of particles larger than 0.3μm is counted. The wafers are placed in another cartridge in the same order and are passed through the instrument a number of times, and each time the number of particles is recorded. The result is one table per wafer, listing the number of particles, y, versus the number of passes, x. These data pairs are best-fitted to a straight line y = a + bx, its slope b is PWP and is related to the data by

$$PWP = b = \frac{n\Sigma x_i y_i - \Sigma x_i \Sigma y_i}{n\Sigma x_i^2 - (\Sigma x_i)^2}$$

The standard error of b, s_b, for statistical fluctuation (shot noise) is given by

$$s_b^2 = \frac{1}{\Delta} \Sigma \frac{1}{y_i} \quad , \text{ where } \Delta = \Sigma \frac{1}{y_i} \Sigma \frac{x_i^2}{y_i} - (\Sigma \frac{x_i}{y_i})^2$$

s_b is then treated like the standard deviation to determine the confidence limits of PWP.

3.11 Overlapping Signals

In vidicon particle counters, large particle-signals extending over several pixels are outputed as a group of several pixels. When a group of pixels from one particle touches a group of pixels from another particle, then the total area is counted as one particle only. Therefore, when the number of particles is large or when the particle-signal increases as the threshold is lowered, large number of particles will be undercounted.

The number of pixels in a Silicon Intensified Target (SIT) detector is 512 x 512. This is the maximum number of pixels, K, into which the observed field can be divided. If there are n particles of the same size randomly distributed on this wafer the probability of finding x particles within the same pixel is given by the binomial distribution[1]

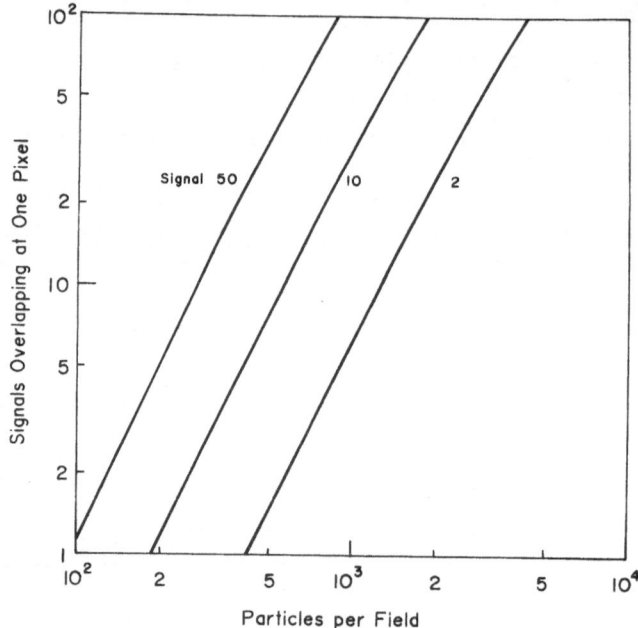

Figure 3   Probability to find particle-signals overlapping at one pixel
for signals of 2, 10, and 50 pixels as a function of particle-
density in the field of view in vidicon instrument.

$$P_B(x,n,p) = \frac{n!}{x!\,(n-x)!}\; p^x(1-p)^{n-x}$$

p (the probability for each particle to be observed) has to be given special attention. If the particle signal does not cover more than one pixel, then p = 1/K. On the other hand, if a particle produces a signal extending over s pixels, then K must be replaced by K/s. In other words,
the pixel size has been enlarged to equal the signal-area. The number
of signal - areas with x particles in them is equal to the probability
P_B given above, times the total number of signal - areas in the field.
When two s large particle-signals overlap on one pixel only, then the
total signal of 2s-1 pixels is counted as one particle. Therefore, in
order to find the number of particles that overlap on one pixel the signal area has to be increased to 2s-1, and p = (2s-1)/K.

   The order of magnitude of this effect is shown in Figure 3. For particles producing a signal of 50 pixels, for instance, the probability that
5 pairs of particles overlap on one pixel each, is roughly one in three

hundred. For particles producing a signal of 2 pixels, the probability of finding two particles overlapping at one pixel is about one in 600. In terms of todays cleanliness, a wafer with 600 or even 100 particles larger than 0.3µm on it is quite a dirty wafer. Therefore, the effect is usually not observed, except on dirty wafers or when the wafer is deliberately sprayed with a large number of particles.

One more line could be added to Figure 3: the line representing the probability of particles producing a one pixel signal overlapping on one pixel. This line would thus represent the probability for one pixel to be occupied by more than one particle. Since that probability is small, doubly occupied pixels are too rare to have a significant effect on particle-counting.

3.12 The Level of Confidence is the Relative Number of Determinations that Fall within Limits Set around the Experimental Average Value

This quantity if introduced when the number of measurements, n, cannot be considered as large. In that case, the data uncertainty is better defined as the confidence limits CF which are related to the standard deviation s, by

$$CF = st$$

where t is Student's normalized deviate. t is tabulated [12] and depends on the number of degrees of freedom, DF, which is the number of independent measurements, n, minus the number of calculated regression constants (like average and standard deviation, etc.), and of the level of significance, σ, which is related to the confidence level, CL, by

$$\sigma = 1 - CL$$

The confidence limits are set on each side of the experimental average and define a range within which a number CL of data will fall if the subsequent determinations are made many times. For instance, suppose that 12 measurements are made with a mean of 100 and a standard deviation of 10. Suppose further that a confidence level of 95% is desired. Then $\sigma = 1 - 0.95 = 0.05$, DF = 10, and t is tabulated as 2.23. The confidence limits are CF = ±22.3, and 95% of all future measurements will result in values between 77.7 and 122.3.

4. RELATION OF SIZING AND COUNTING CLOSE TO THE DETECTION LIMIT

The existence of missed particles and false counts mentioned earlier
will now be explained based on a simplified representation of particle
detection, and their effect will be described.

Each pixel of a vidicon detector carries an electrical charge that
is neutralized by an electron beam at each "reading". While the beam
neutralizes the pixel, it produces a current which constitutes the signal for that particular pixel. The charge on the pixel, and consequently
the signal current, depends on the amount of light received by the pixel;
but even in the absence of illumination, there will be a small charge and
therefore, a background signal afflicted by a certain variation, or noise.

Let us assume that each pixel has the same average dark current
(background), and let us consider one scan of fifty contiguous pixels.
The background, read in one scan, pixel after pixel, will vary around the
average background, just as if one single pixel were to be observed a
certain length of time. In Figure 4, the instantaneous background values
of each pixel are joined by a straight line to better separate the background from the particle-signal. The particle-signal is represented by
a heavy line on top of the background at every signal carrying pixel.
All these signal lines have the same length, consistent with the assumption that the signal-noise may be neglected, and that the signals are
all from identical particles. Furthermore, the length of the particle-
signal lines equals two standard deviations of the background-noise.

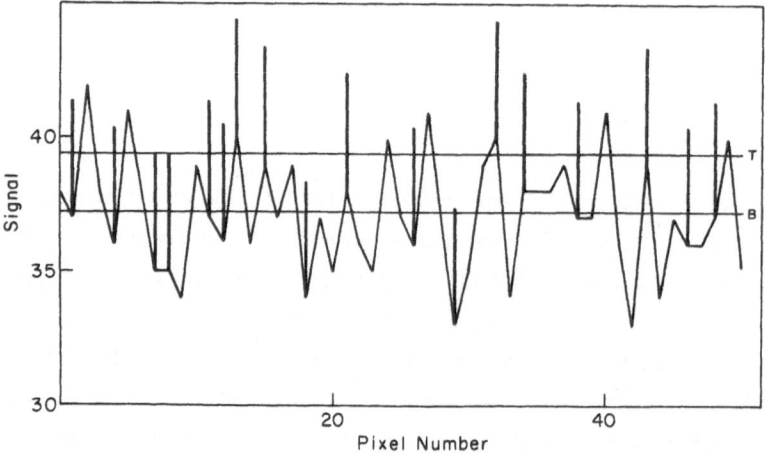

Figure 4   Particle-signals (vertical bars), and background-signal (values
connected by straight lines) for a single scan over 50 pixels
for particle-signals at detection limit.

316

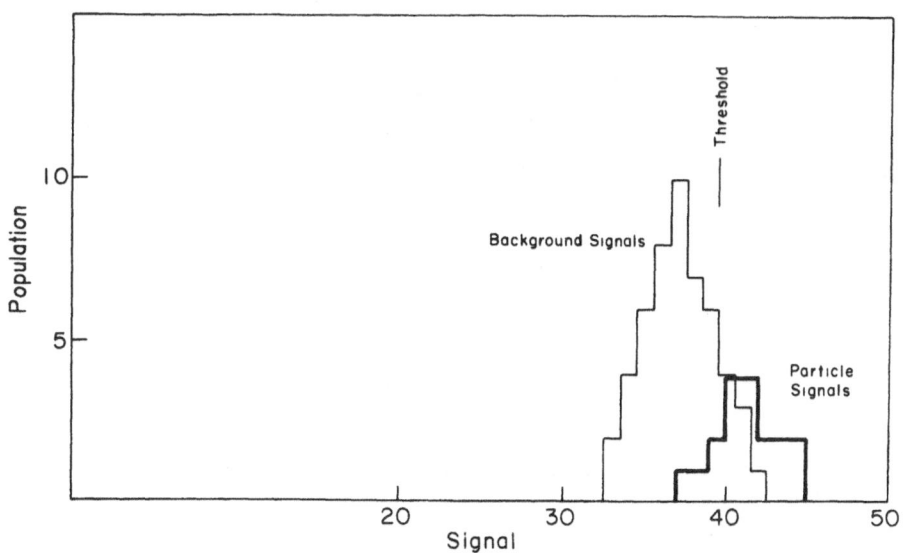

Figure 5   Histograms for particle- and background-signals for figure 4.

One horizontal line is drawn at the average background, another at one
standard deviation above it. These are the background-noise and particle-
signal conditions that prevail at the detection limit.

Only those signals are detected that are higher than the threshold.
With the threshold set at one standard deviation above the average background in the example of Figure 4, only 14 of the 18 particle-signals
will be detected, and 4 particles are missed. On the other hand, 6 pixels that do not carry a particle-signal have a background-signal larger
than the threshold and will, therefore, be counted as particles. Those
are 6 false counts. If we were to count particles at the level of the
detection-limit, this example shows that the count would be 20, but that
6 are false counts and that 4 true particle-counts are missed.

The signals of Figure 4 may be represented in the form of two histograms Fig. 5): one histogram for the signal carried by the 32 pixels
without a particle-signal, and one histogram for the 18 pixels with a
particle-signal. With the threshold at one standard deviation above the
average background, 4 particle-signals are below threshold and 8 background-signals are above threshold.

If the number of pixels were increased to infinity and if the width
of the histogram ranges were narrowed to zero, the histograms would become smooth distribution curves (Fig. 6). These curves are the parent
distributions from which the histograms are a small part.

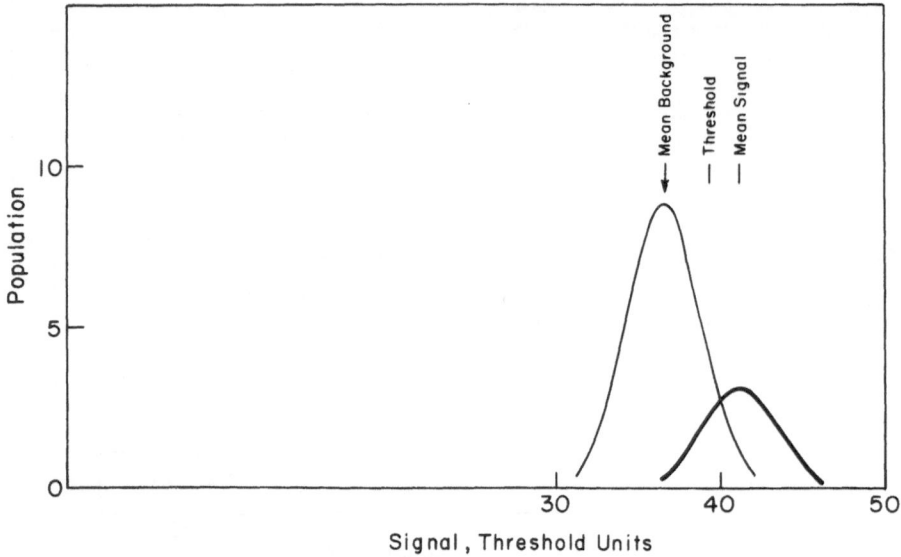

Figure 6   Parent distribution of histograms in Figure 5.

Of course, no single measurement can provide two neatly separated distributions, one for the background, the other for the particle-signals. What is obtained instead is the sum of both distributions (Fig. 7). Figure 7 still represents a distribution of signals at the detection-limit for 36% particle-signals on background-signals.

The presence of 36% particle-signals at the dection-limit level is noticeable only by a slight asymmetry of the distribution, but not by a secondary peak or by a marked shoulder.

Even the distribution of Figure 7 is not always directly accessible in a particle-counter, but the integrated distribution is (Fig. 8). If the threshold is first set very high, and then gradually lowered, the number of particles counted will increase as the integral of the particle size distribution. The background will contribute an amount represented

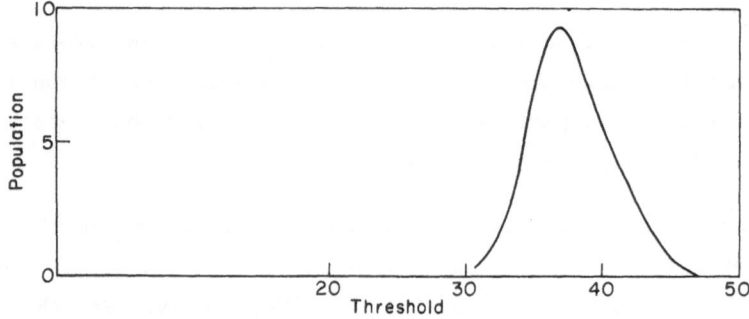

Figure 7   Sum of parent distributions of Figure 6.

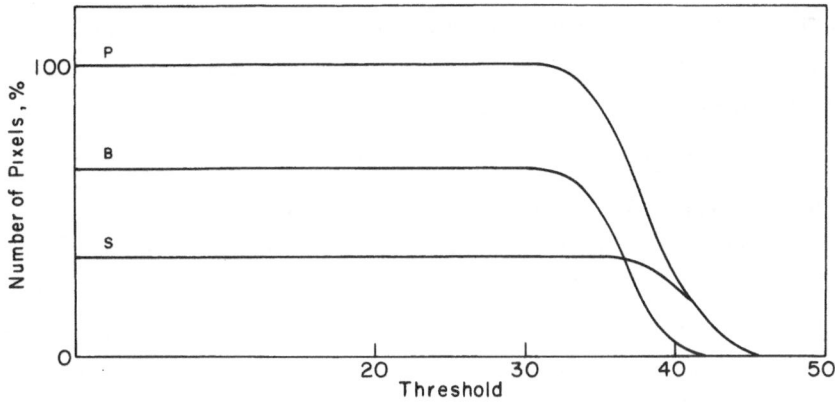

Figure 8 Integrated distributions of Figure 5 and 6. S is contributed
by the signal , B by the background and P is the measurable
sum of S and B.

Table I. Uniformity of a 0.36μm Latex Sphere Signal

	Left	Bottom	Right	Top	Center	All Positions
Average Signal	7.4	7.45	6.65	6.05	6.6	6.83
Standard Deviation	0.6	0.6	0.67	0.76	0.68	0.84
After 150 Additional Burning Hours						
Average Signal	7.25	6.6	5.95	6.55	7.2	6.71
Standard Deviation	0.79	0.94	0.76	0.69	0.77	0.91

by curve B, the particle-carrying pixels an amount represented by curve
S, and the sum is represented by curve P. Curve P is actually what can
be measured directly with most particle-counters.

5. RESULTS

5.1 Uniformity of Particles Sizing

Uniformity of particle sizing has been determined with 0.36μm latex
sphere located at 5 different positions on the wafer. The results for
each location are shown in Table I, The total uncertainty for all positions is 12%.

5.2 Repeatability

The one-particle measurements made for uniformity tests were repeated after 150 additional buring hours as read on the instrument's totali-

ser. The results are also given in Table I and are essentially unchanged.
Indeed, the difference is too small to justify a significant repeatability determination.

5.3 Calibration Curve

Figure 9 represents a calibration curve of the EX20/20. For particles larger than 1μm the signal is roughly proportional to the square root
of the diameter, and for particles smaller than 0.3μm the proportionality
is to a higher power of the diameter. The uncertainty in this case is
roughly proportional to the signal. The strong signal decrease for latex
spheres below 0.3μm could be confirmed in an experimental instrument for
particle diameters down to 0.9μm. The behavior of glass particles is
similar; and for metallic spheres, the down slope occurs at smaller diameters.

5.4 Haze Measurement

A clean wafer was introduced in the EX20/20 and the gain adjusted to
its highest value. 20.8% of the pixels were activated. Then another
wafer was introduced, covered with Kodak 6080 white reflectance coating.
The aperture of the collecting lens was closed down, and neutral density
filters were placed in front of the lens until the haze number would again
indicate that approximately 21% of the pixels were active. The ratio of
both light intensities gives the haze measurement which is also the ratio
of the light scattered from the sample to that scattered by a white wafer.
The ratio was

$$3.75 \cdot 10^{-7} \quad \text{or} \quad 0.4\text{ppm.}$$

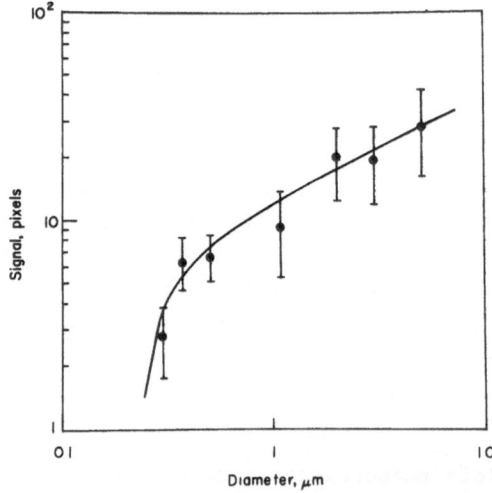

Figure 9   Calibration curve for latex spheres on an EX20/20.

5.5   Counter's Precision

Figure 10 is a plot of the standard deviation of counts as a function of average particle counts for 0.36μm particles. The standard deviation of the EX20/20 is approximately equal to the square root of the average count divided by 34. This value may be compared to a value and a statement made for another instrument[10] that a standard deviation equal to the square root of half the average count" is in approximate agreement with today's count for means of small samples."

This means that the EX20/20 is about 6 times more precise than the instrument considered by ref. 10. In fact, that comparison is unreasonaboe because in ref. 10, "random" particles are counted; their signal distribution, let alone their size and shape, is unknown. It can be shown that the counting precision is a function of the particle signal distribution close to the threshold. Particles that produce average sig-. nals below the threshold contribute to the total count and to the counting uncertainty. Therefore, counting precision determinations based on a mixture of particles should include a precise determination of the content of that mixture. Counting precisions determined on particles deposited "naturally" on the wafer exposed to the laboratory environment lack such a signal distribution determination and show a strong dependence on that environment. They are, therefore, inadequate to quality an instrument.

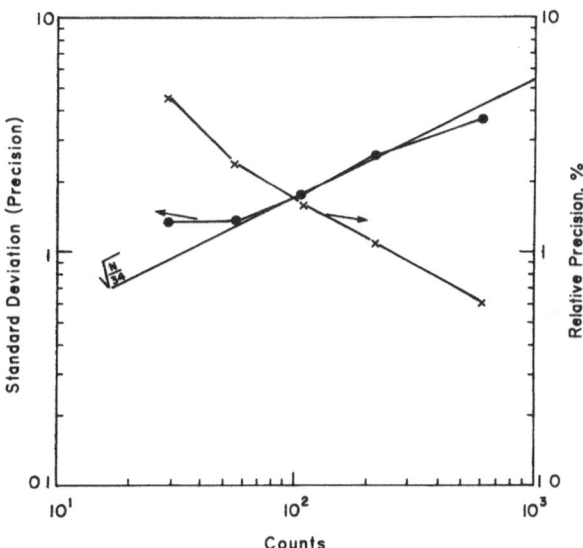

Figure 10   Counter's precision represented as the standard deviation of counts for variable counts of 0.36μm latex spheres. The straight line shows a proportionality to $\sqrt{N/34}$.

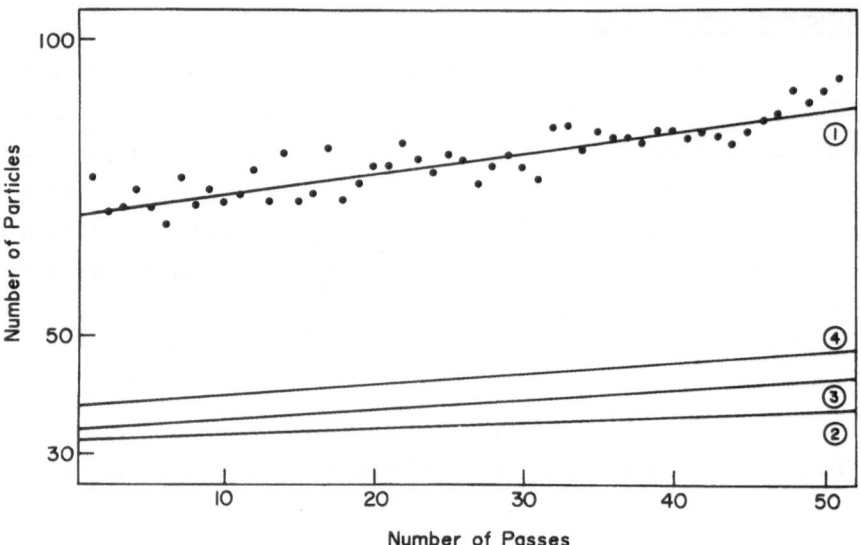

Figure 11   Contamination with particles larger than 0.3μm. Particle
counts for 52 consecutive passes through an EX20/20. All the
counts are represented for wafer 1 placed at the top of the
stack and linear regressions are shown for wafers 1 to 4.
The slope of the linear regression is the number of particles
added per pass.

5.6   Contamination

Four wafers were placed in a cassette and passed through the EX20/20.
The threshold was set to detect 0.3μm particles. The four wafers were
passed 52 times through the instrument and their order in the cassette
was not changed. Wafer number 1 was at the top. In Figure 11 the number
of particles that produced a signal of two pixels or more are plotted
versus the number of passes. All the counts are plotted for wafer number
1, together with their linear regression line. For wafer #2,3 and 4, only
the linear regression is shown.

Then the cassette was left next to the instrument in a class 100
clean room for one hour, and one more particle count was performed. During that one hour, no significant change could be registered in the particle count. Therefore, the slopes of the straight lines are the PWP's.
They are shown in Table II. The confidence limits ±CF have been computed
for a desired level of confidence of 95% for our sample with 50 degrees
of freedom.

In another PWP determination made in a cleaner environment, the results were significantly lower and could mean that despite the one hour

322

Table II. Contamination by Particles 0.3μm or Larger

Wafer	PWP	Confidence Limit
1	0.37	± 0.053
2	0.10	± 0.045
3	0.16	± 0.045
4	0.17	± 0.054

control test, our clean room is responsible for most of the added particles.

5.7 Integrated Distribution

Curves shown in Figure 12 were obtained from particle-counting on
part of a clean wafer sprayed with a large number (about 150 to 200) of
latex spheres of a single diameter. The count was recorded as the threshold was lowered. The dots of the 1μm curve represent averages of 10 to
30 counts. The upper end of the solid line represents a normal distribution computed from the integral of Gaussian function:

$$P_G = \frac{100}{\sigma\sqrt{2\pi}} \; \exp \; [-\tfrac{1}{2}(\frac{x-\mu}{\sigma})^2] \, dx$$

$$\text{with} \quad \mu = 233$$

$$\text{and} \quad \sigma = 27.$$

The excellent agreement of experimental value with a computed Gaussian
function confirms that the signal distribution is indeed close to normal.

Still on the 1μm curve, a depression is noticeable for threshold
values below 100, because of the high particle-density used. This is also
in agreement with our consideration on overlapping signals. At the lower
end, the background contribution is apparent and increases the particle-
count until overlapping signals saturate the pixels and decrease the count.

The curve for 0.5μm illustrates the effect of both the presence of
larger particles that produce a tail at the upper end of the curve, and
the presence of smaller particles that increase the count at lower tres-
hold values.

The curve for 0.3μm presents a strong shoulder, definitely more pro-
nounced than that on the integrated distribution of Figure 8 computed at
detection-limit. This indicates that the detection limit of this instrument is well below 0.3μm.

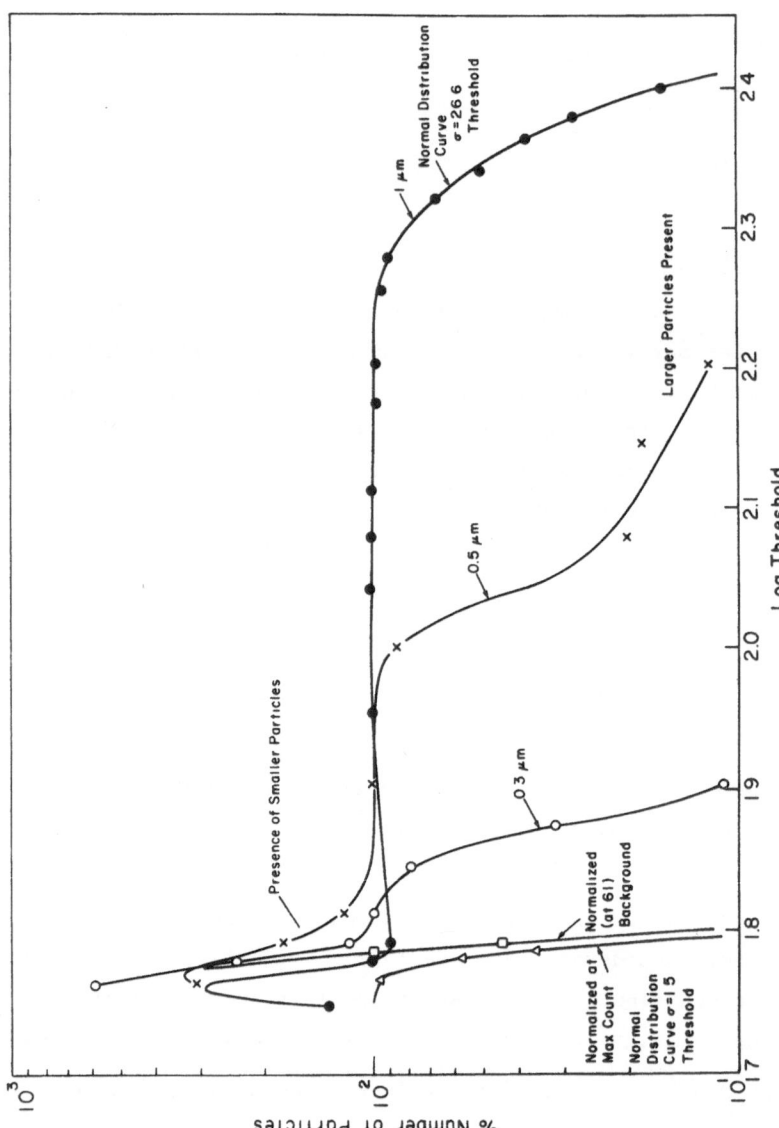

Figure 12 Experimentally determined integrated particle-distribution in an Inspex EX20/20 particle sizing and counting instrument. The upper end of the 1μm curve is a theoretical curve computed from a Gaussian function. The dots are experimental results. The 0.3μm curve presents a shoulder, much larger than that of Figure 7 computed at the detection limit. It shows that the detection limit of the Inspex EX20/20 is well below 0.3μm for the standard samples used here.

5.8 Counting Confidence Limits at 95% Confidence Level

The counting confidence limits have been determined for a 95% confidence level and 10 degrees of freedom. They are reproduced in Table III. They mean, for instance, that if one micrometer particles are counted many times, 95% of the results will fall within ± 0.4% of the experimental mean of 12 determinations used to establish these limits. In other words, 95% of the count will lie between 99.6 and 100.4% of that experimental mean.

Table III. Confidence Limits for Particle Counting at a Confidence Level of 95%

Particle Diameter	Confidence Limit	Count Range
1.00μm	0.4%	99.6 - 100.4%
0.50μm	2%	98 - 102%
0.29μm	8%	92 - 108%

6. CONCLUSION

The results obtained on the Inspex EX20/20 demonstrate that the instrument is capable of a detection limit smaller than 0.3μm and that statistical computations based on normal distributions can be verified with this instrument. Consequently, once the calibration curve and the noise have been determined experimentally, the counting precision for any particle size may be obtained from statistical computation. Thus determination of false counts and of missed particles based on normal distributions are justified.

7. ACKNOWLEDGEMENTS

I gratefully acknowledge the support given by Mario Maldari and Hitoshi Iida whose many suggestions helped realize the EX20/20. Robert Wilson prepared most of the wafer samples and performed many of the time consuming experiments. Karen Yilmaz, Francine Phelan and Wilma Finn patiently typed, corrected and edited the manuscript. All deserve my lasting gratitude.

REFERENCES

1. P.R. Bevington, "Data Reduction and Error Analysis for the Physical Sciences", McGraw-Hill, New York, 1969.

2. R.D. Hudson, Jr., "Infrared System Engineering", Wiley-Interscience, New York, 1969.

3. R.C. Jones, Advan. Electron, 5, 68 (1954).

4. M. Garbuny, "Optical Physics", p. 452, Academic Press, New York, 1965.

5. M. Schwartz, "Information Transmission, Modulation, and Noise", p. 260, McGraw-Hill, New York, 1959.

6. V.A. Fassel and R.N.Kniseley, Anal. Chem., 46, No. 13, 7003 (1974).

7. D.R. Demers and C.D. Allemand, Anal. Chem., 53, 1915 (1981).

8. V.A. Fassel, C.P. Peterson, F.N. Abercrombie and R.N. Kniseley, Anal. Chem., 48, No. 3, 516 (1976).

9. V.A. Fassel, "Electrical Flame Spectroscopy", XVI Colloquium Spectroscopicum Internationale, Heidelberg, Germany, Oct. 9, 1971.

10. B.J. Tullis, Microcontamination, 67, November 1985.

11. B.J. Tullis, Microcontamination, 15, December 1985.

12. Lothar Sachs, "Applied Statistics", Springer, New York, 1984.

PART IV. PARTICLE REMOVAL

METHODS FOR SURFACE PARTICLE REMOVAL: A COMPARATIVE STUDY

Juan Bardina

Atcor Corporation
2350 Charleston Road
Mountain View, CA 94043

Precision cleaning techniques and their associated chemicals are as varied as the contaminants and surfaces to be cleaned. The objective of this paper is to evaluate the effectiveness of different methods used to remove particles in the size range of 0.1 to 10 micrometers (μm) and improve the cleanliness level of a surface from a density of 100 particles per square inch to one particle per square inch or better. Surfaces considered include plastics, metals, silicon, and quartz.

The research was conducted in three phases. An extensive survey was made of all available literature on the subject. Then, the forces and processes involved in cleaning were analyzed from a theoretical perspective. Finally, these findings were compared with current industry practice.

The subject concerns four main topics: methods, chemicals, contaminants, and surfaces. This paper looks at how effective different chemicals and methods are at removing various contaminants from a wide range of surfaces. It also notes any incompatibilities between these topics.

Results have been classified into three categories of cleaning techniques: (1)those that do not work because of incompatibilities between methods, chemicals, contaminants, and surfaces; (2)those cases where compatibility exists and is confirmed by experimentation or industry practice; and (3)most significantly, those cases where all factors combine favorably to present distinct new possibilities for precision cleaning techniques not known to be in current use.

INTRODUCTION

Precision cleaning methods are often chosen in a hit-or-miss fashion. Potential users will see what is already being used and either use it or modify it to fit their needs. This often leads to inadequate or inefficient cleaning systems which end up costing more in the long run.

In order to intelligently choose a cleaning system, one must be aware of all available options and the high points and drawbacks of each. This paper provides such an evaluation. It gleans information from literature, theoretical analysis, and industry and summarizes and evaluates each method. Because it draws from such varied sources, this paper provides a good overview of current cleaning systems.

The first step in discussing precision cleaning methods is to narrow the field of potential techniques. With over 25 possible methods to discuss, the criteria used to judge them must be strict. This paper concentrates on techniques that the literature claims will remove particles that are less than 10 μm in diameter. This limit reflects the current needs of the semiconductor industry. The methods which will be covered include: ultrasonics, megasonics, wiping, brush scrubbing, low-pressure surfactant spraying, high-pressure jet spraying, etching, and centrifugal spraying. Methods which will not be considered include: air and gas jets, mechanical abrasion, chemical displacement and emulsion, electrostatic elimination, rinsing, UV/ozone treatment, plasma etching, ion milling, immersion, aqueous, vapor degreasing, strippable adhesive coatings, polishing and buffing, laser cleaning, heat treatment, ion bombardment and combustion, and leaching. These methods either do not remove small particles or are grossly inefficient.

The surfaces which will be considered are silicon wafers and the plastic containers which hold them. The main composition of the particles considered in this study is silicon. These choices also reflect the needs of the semiconductor industry.

PARTICLE ADHESION

Before removing a particle from a surface, one must know the strength of the adhesion force holding the particle. The main components of the adhesion force are the van der Waals force, the electrical double layer force, the electrostatic image force, and the capillary force.[1] The van der Waals force is the attraction any molecule or atom has for another

Table I. Various Adhesion Forces Between a Particle and a Surface
(See Ref. 3 for Similar Equations)

Force	Equation	Reduces to:
van der Waals	$\dfrac{\hbar w}{16\pi z^2}$	$1.43 \times 10^2 d$
Electrostatic double layer	$\pi \epsilon_o \dfrac{(\Delta\phi)^2 d}{2z}$	$34.8 d$
Electrostatic image	$\dfrac{Q^2}{\epsilon_o \epsilon d^2}$	$5 \times 10^4 d^2$
Capillary	$4\pi r \gamma$	$4.52 \times 10^2 d$

WHERE: the particle is spherical
d = particle diameter in cm.
$\hbar w$ = van der Waals constant = 7.2eV
$\Delta\phi$ = difference in work function $\approx$ 1V
ϵ = dielectric constant = 1 for air
ϵ_o = permittivity of free space = 8.86 pF/m
z = adhesion distance = 4Å
Q = particulate charge = 10 - 16 C/one μm size
   particle capillary liquid is water, so γ = 73 dyn/cm

molecule or atom.[2] Electrical double layer forces are caused by electrostatic contact potentials due to differences in local energy states and electron work functions between two materials, and electrostatic image forces are caused by bulk excess charges present on the surface which produce a coulombic attraction.[3] The last component, capillary forces, exists in the presence of humidity. Equations which express these forces are shown in Table I.

When a cleaning solvent is used, only the van der Waals force needs to be taken into account. The best cleaning solutions available will quickly dissipate charges so that the static forces are overcome. When the solvent moves over the surface and the gas/solid interface disappears, the capillary force will also disappear.[4]

PARTICLE REMOVAL

Sonic Cleaning

Sonic cleaning encompasses two popular cleaning methods: ultrasonics and megasonics. Although both are based on high frequency sound waves, they operate quite differently.

In ultrasonic cleaning a part is immersed in a suitable liquid medium and sonicated or agitated at a high frequency (18-120 kHz[5] or 20-80kHz[6]). This usually lasts for several minutes, and then the part is rinsed and dried.[5] Cavitation occurs when microscopic bubbles form in the liquid medium and then violently collapse or implode, mechanically scouring the part to be cleaned and displacing or loosening the contamination on it.[5] Besides scouring the surface, this also promotes a chemical reaction by raising the liquid temperature.

Because a reliable means of quantifying cavitation has yet to be determined, performance assessments are usually based on the level of cleanliness that is achieved. It is known that cavitation intensity depends on three variables: (1) rheological properties of the medium including laminar and turbulent flow and static conditions, (2) colligative properties of the liquid such as viscosity, density, surface tension, and vapor pressure, and (3) the amplitude and frequency of the radiating wave.[5] It has also been found that the ultrasonic force required to remove a particle decreases approximately linearly with the diameter.[7]

The ultrasonic cleaning process is thorough because the bubbles can penetrate wherever the liquid does. It is extremely effective for removing tenacious deposits such as rust, scale, and tarnish.[8] Some say its most successful application is in the removal of insoluble contaminants. It also removes films and small particulate matter such as solder flux, greases, light oils, fingerprints, and microbial contaminants.[9] Ultrasonics is particularly useful in the manufacturing of silicon wafers. It can even erode such hard surfaces as quartz, silicon, and alumina, but experts warn that unless cavitation forces are controlled, erosive "cavitation burn" can occur. In fact, ultrasonics is too harsh for some sensitive, delicate components.[5]

Some critics say that ultrasonics is just plain ineffective for particle removal, even when used at power densities high enough to produce crystal damage.[6] Others have recommended ultrasonics for gross and not precision cleaning of optics,[4] and some point out that ultrasonics can not adequately remove particles less than about 25 μm.[10] Even so, most acknowledge that ultrasonics is a useful tool for increasing the effectiveness of immersion cleaning and solvent extraction.[4]

In order to completely remove all contamination, an ultrasonic system must have continuous filtration, overflow, or both. A filtration system would remove soils as they are displaced and overflow consists of at least two consecutive tanks in series. Even with these, only 65 percent of the contaminants are removed in a single cycle, and three cycles are necessary to achieve 90 percent removal. [5]

There are many advantages to ultrasonic cleaning. It is fast, effective, and safe to use. It requires less heat than other cleaning methods, and when used properly it can vigorously clean delicate parts without harming surface finishes. Besides these, there is no need to dismantle assemblies. [9]

Disadvantages of ultrasonic cleaning include its complexity and the noise it makes. In fact, ultrasonic cleaning makes so much noise at frequencies lower than 20kHz that a 40kHz frequency is recommended even though it is less efficient. [8] Ultrasonic cleaning also has much more impact on the bottom of the part being cleaned since the bubbles float up. The top only receives 10-25 percent of the energy the bottom does. [11] Also, care must be taken to choose the proper cleaning solution otherwise damage may occur to circuit boards and transistors. [9]

Megasonic cleaning is another cleaning method which consists of basically the same steps as ultrasonic cleaning: immersion, agitation or sonication, rinsing, and drying. The major difference between the two is that while ultrasonic frequencies range from 18-120 kHz[5], megasonic frequencies are in the range of 0.8-1.0 MHz[6] with input power densities from 5-10 W/cm. [12] Whereas the cleaning action in ultrasonic cleaning comes from cavitation, the cleaning action in megasonic cleaning comes from high pressure waves pushing and tugging at contaminants lodged on a part's surface.[13] Megasonic cleaners are also able to use more chemically active cleaning solutions such as hydrogen peroxide and ammonium hydroxide[14] but they can not handle such cleaning solutions as strong hydrofluoric acid (HF). [6]

Quantification of the forces megasonic cleaning can impart is difficult, but it is important to note that the frequencies it uses are very nearly the natural frequencies of oscillation of the particles it is removing. Therefore, large oscillations develop which allow the particle to move far enough from the surface for wetting to occur beneath it. One model predicts that the frequencies will need to increase as the particle size decreases, and that the frequency is proportional to the inverse three-half power of the particle size.[7] Megasonic cleaning has been found capable of removing particles as small as 0.2 μm[14] to 0.3 μm from all sides of a surface. [10]

Megasonic cleaning is commonly used on silicon wafers, but it can also clean ceramics and photomasks. Besides particles, it can remove thin organic films, ionic impurities, photoresist, wax, and grease.[6] The time it takes to do this is quite low: it takes 15 minutes to clean a wafer, and at least 100 can be cleaned at the same time.[6]

There are many advantages to megasonic cleaning. It causes almost no scratches, breakage, or chipping since substrates are not transferred or subjected to any mechanical stress. It is three to four times more productive in terms of throughput than scrubbing or chemical cleaning at an equal or lower investment cost and produces superior wafer cleanliness. Megasonic cleaning uses about one-eighth the amount of chemicals as chemical cleaning does, and megasonic cleaners have low maintenance and are simple to automate.[6] Megasonic cleaners are also able to use more chemically active cleaning solutions such as hydrogen peroxide and ammonium hydroxide. [14]

Megasonic cleaning also has some disadvantages. For example, the solvent system must be adopted to the contaminant-substrate combination and the transducer matrix is not a commercial item. Cleaning solutions such as strong hydrofluoric acid (HF) can not be used, and the substrate containers must be designed to minimize obstruction to the megasonic beam. [6]

Wiping

Although it may not be as efficient as sonic cleaning, wiping is another successful cleaning method. It is an inefficient but effective method of particle removal [10] that is commonly used to clean optical surfaces. [4]

The typical wiping procedure consists of rubbing a surface with a solvent saturated lens tissue. This takes advantage of solvent extraction [4] and applies a liquid shear force on particles with respect to the surface. [10] Wiping leaves only 2-40 particles per square centimeter larger than 5 μm in size. [4] Some point out that this can be inefficient because it takes 8 man-hours per square foot to achieve a cleanliness level of less than one 10 μm size particle per square foot. In fact, cleanliness verification and equipment preparation bring this total to more than 80 man-hours per square foot. [10]

Besides the amount of time it takes, a major drawback to wiping is that particles can be deposited from the lens tissue or the solvent being used. [4] Wiping is also unable to reach irregular surface geometries [10] and the results depend on the wiper's skill and attention to detail. [4]

Despite these drawbacks, wiping is still one of the simplest and most effective methods in use today. It should be seriously considered if other methods are impractical or time is not a major concern.

Brush Scrubbing

No cleaning method has more contradictory material written on it than brush scrubbing. Some say the method has been all but abandoned [15] and others say it is one of the more common methods of cleaning. [13] Some say it is highly effective [13,15] and others claim that it is ineffective. [6] The real truth is a combination all of these.

Brush scrubbing is most effective when the brush, which is made of camel's hair, [16] nylon, polypropylene, or mohair, [13] is first sprayed with the scrubbing solution. [15] The scrubbing solution is usually water-based and will only remove contaminants that are soluble in it. The brush shape is usually either a cylinder or a cup [13] or a sponge material [16] instead of bristles. The brush never actually touches the surface being cleaned due to the brush's hydrophilic nature. There is always a film of the scrubbing solution between the brush and the surface. [13] The hydrophilic brush will only remove contaminants from hydrophobic surfaces. Surfaces that are hydrophilic are more difficult to clean because suspended contaminants can precipitate onto them. [13] To increase its power, brush scrubbing is often used in conjunction with high pressure jets. [16]

The literature on brush scrubbing is contradictory in regard to whether or not it should be used with high pressure scrubbing. Two sources say that only one is used, depending on the pattern of the surface being cleaned. [13,17] A third article states that high pressure jet scrubbing is almost always used with brush scrubbing. [16] It says that the jet spray will remove microscopic debris that the brush can not.

There are several things which can make brush scrubbing ineffective. First, the aqueous neutral detergent solutions that scrubbers are usually used with can leave behind thin nylon films and corrode rapidly when used with chemically active cleaning solutions. If this is true, then chemical cleaning is also required, and this almost always introduces more particles. Another problem can arise when brushes become infested with dirt particles and debris from the surface breaking or chipping. When this happens the brushes themselves can become sources of contamination and scratches. Finally, critics of brush scrubbing point out that scrubbers are sequential in operation and can only clean one part, and often only one side at a time. [6]

Low-Pressure Surfactant Spraying

Unlike brush scrubbing, low-pressure surfactant spraying relies on chemical means to remove particles. Its success depends on the effectiveness of the detergent that it sprays. The pressure of the jet itself, which people surveyed fix anywhere from 5-40 psi[10] to 60-80 psi, is not nearly enough by itself to remove particles. One need only insert lower velocities into the equations used to evaluate high-pressure spraying to see this. Therefore, the compatibility of the detergent with the contaminant and the surface is crucial.

Accurately quantifying a detergent's power is difficult, but there are some basic considerations which should be kept in mind when choosing one. Detergents are surface-active agents (surfactants) which reduce the surface tension of water and remove soils through emulsification and by concentrating at interfaces.[18] Surfactants are made up of two parts: a head group and a chain. The hydrocarbon chain has a very weak interaction with water. It is classified as hydrophobic because the strong interactions between water molecules due to hydrogen bonding and dispersion forces squeeze the hydrocarbon out of the water.[19] The head group is hydrophilic because it is solvated by interacting strongly with water through ion-dipole or dipole-dipole interactions. [19] Whether a detergent is anionic, cationic, or nonionic depends on the charge of its head group and its behavior in emulsification. [20]

At high concentrations, detergent solutions form micelles. Micelles occur when the chains form a low polarity region that is stabilized by having the polar ends in contact with the water. The cleansing action occurs because the lowered water surface tension allows the detergent to penetrate and the micelles to dissolve greases and oils by taking them into the hydrocarbon regions. [21,22]

High-Pressure Jet Scrubbing

High-pressure jet cleaning operates under very different principles than surfactant spraying. It has been referred to as shear stress cleaning[10] because it works when the shear force it exerts is greater than the adhesion force holding a particle to a surface.[4] The method consists of a high velocity [10] jet of liquid sweeping across a surface at pressures of 100-4000 psi. (Different people surveyed quoted different pressures: 2000-3000 psi[16] or 300-4000 psi[13] or 100-2200 psi[10] or 300-3000 psi.[23]) The liquid used is often static-free DI water,[16] though solvents such as Freon TF are also used.[10] Denser and more viscous fluids impart more momentum to attached particles.[4]

The main advantage of high-pressure jet scrubbing is that it is able to remove microscopic debris from difficult surface geometries such as depressions and circuit corners.[16] There has been some concern that the jet action will scratch or mar the surface[12] but some claim it will not.[23] At least one proponent of the method claims that it is able to remove particles as small as 0.1 μm. [10]

334

An example of the effectiveness of high-pressure water cleaning is one where DI water at 2500 psi was directed onto a mask spinning at 3000 rpm. Whereas the mask had 75% good chip sites before cleaning, it had 95% after. It was also found that an acid soak prior to high-pressure water cleaning greatly improved cleaning effectiveness.[17]

Features to consider in designing a high-pressure jet cleaning system include the position of the spray nozzle and the ability to change the angle and speed of the water.[13] The nozzle should be as close to the surface as possible in order to maintain fluid momentum and the angle and the speed should be able to adapt to different sorts of surfaces. Having a nozzle close to a surface or adding CO_2 bubbles or isopropanol or methanol to the fluid can also serve to reduce static contamination.[23]

The shear stress that high pressure jets must impart to remove particles is determined by Equation (1). The distance in the fluid from the surface affects the speed of the laminar or turbulent flow.

$$F = \frac{C \rho V^2 A}{2} \qquad (1)$$

WHERE: F = drag force on particle
         C = coefficient of drag
         ρ = density of fluid
         V = velocity of fluid
         A = projected frontal area of particle

Etching

Etching is one of the most common methods discussed. It is a chemical cleaning method that consists of dissolving unwanted substances on a surface and is not as severe to the surface as mechanical means of surface treatment.[4] Etching is closely related to acid cleaning and is one of the more important procedures in microelectronic device fabrication.[24]

Liquid etchants are used for dissolving low molecular weight or unwanted surface contamination and for chemical modification. In order to uniformly and efficiently remove contamination, the liquid must be able to spread and wet the contaminated surface. Spreading occurs when the surface tension of the liquid is lower than the contaminated surface's critical surface tension. In other words, the contact angle of the liquid on the surface must vanish. Lower viscosity liquids accomplish this more quickly.[10]

Etching is commonly employed to clean silicon, high alumina ceramics, Pyrex, quartz and component parts such as lead frames, cans and headers. Alkaline solutions with complexing agents such as KOH or n-butanol are often used to etch silicon. Acids and acid mixtures are used to clean product handling equipment such as Pyrex and quartz. The more commonly used acids are glacial acetic acid, nitric acid, hydrofluoric acid, and hydrochloric acid.[24] The most common combination of chemicals used to etch surfaces is the RCA cleaning process.[15]

The RCA cleaning process is still the dominant cleaning technique in use today, and many other techniques are actually simple variations of it. Sometimes referred to as the "RCA immersion technique"[15] or the "RCA chemical clean recipe,"[25] this technique has been called "the most recent leading-edge industry work in cleaning technology," despite the fact that it was developed in the late 1950's.[15]

The RCA technique is actually more of a cleaning recipe than a cleaning method. It is a safe and sequential cleaning process "based on oxidation and dissolution of residual organic impurities and certain metal contaminants in a mixture of $H_2O-H_2O_2-NH_4OH$ ("SC-1" for a standard clean, solution 1) at 75-80 C."[25] This solution is generally mixed in parts from 5-1-1 to 7-2-1.[26] The next step consists of dilute HF (or Buffered Oxide Etch) and H_2O_2-HCl.[27] The proportions in the latter are usually 6-1-1 to 8-2-1 by volume. This solution serves to "remove heavy metals and to prevent displacement replating from solution by forming soluble complexes with the resulting ions". These solutions were chosen because they are completely volatile. They are also less hazardous than other possible cleaning mixtures and present no disposal problems.[26]

This recipe works because at a high pH hydrogen peroxide (H_2O_2) solutions are effective in removing organic contaminants by oxidation, and at a low pH they are effective at desorbing metal contaminants by complexing. The technique is particularly useful in cleaning silicon device wafers, quartz tubes, and parts used in semiconductor processing.[26]

Centrifugal Spray Cleaning

Centrifugal spray cleaning is an effective alternative to chemical immersion processes.[28] It is commonly used for cleaning wafers. The wafers are enclosed in a sealed chamber purged with nitrogen. As the wafers spin they are subjected to a series of continuous fine sprays of reagent solutions, including a hot aqueous solution of hydrogen peroxide and ammonium hydroxide (SC-1), an aqueous solution of hydrochloric acid and hydrogen peroxide (SC-2), and high purity water.[25] Recontamination is prevented by arranging the sprays so that each wafer is continuously exposed to fresh solutions.[13] A commercially available centrifugal cleaner sprays wafers at 2500 psi (17 MPa).[4] After the wafers have been sprayed they are dried with N_2.[25]

Therefore, centrifugal spray cleaning relies on centrifugal force, shear force, and solvency. The centrifugal force is expressed by Equation (2). The shear force has been discussed with regard to high pressure spraying, and solvency was dealt with in regard to etching.

$$F = \frac{\pi \, d_p^3 \, \rho_p \, w^2 \, R}{6} \tag{2}$$

WHERE:
  F = removal force in dynes
  w = angular speed of rotation in radians/sec.
  d_p = particle diameter in cm.
  R = distance from the centrifuge axis of rotation to the substrate in cm.
  ρ_p = particle density in g/cm^3.

Comparison of Particle Removal Methods

The following is a summary of the lowest particle size that each method can remove.

Size (μm):	Methods:
0.2	- Megasonic Cleaning - Low-Pressure Surfactant Spraying - High-Pressure Jet Spraying (15,000 psi)
0.3	- High-Pressure Jet Spraying (10,000 psi)

Size (µm):	Methods:
0.5	- Brush Scrubbing - Centrifugal Spray Cleaning - High-Pressure Jet Spraying (4,000 psi) - Etching
5.0	- Wiping
25.0	- Ultrasonic Cleaning

CONCLUSIONS

This paper should be used as a guideline to potential cleaning methods. It is by no means exhaustive, as it does not include all possible surfaces, contaminants, or cleaning methods. The reader must also decide which sources he values most, as relevant literature, experimentation, and industry practice do not always agree.

It is important to note that not all methods have been evaluated quantitatively. In some cases this is because no means of doing so has been discovered yet (e.g., ultrasonics, megasonics) and in others it is because it is extremely complicated and potentially inaccurate (e.g., surfactants, brush scrubbing). In these cases the best sources of information are literature and industry experience.

This paper also does not deal with the factors one must consider when choosing a cleaning system. These include cost, process time, space, and flexibility. Any of these could take higher priority than the apparent effectiveness of a given system.

Hopefully this paper will help people to have a fundamental basis for choosing a cleaning system. Whereas factors such as cost and process time are easily quantifiable, the effectiveness of a system often is not. It is important to understand this factor as it is often the most crucial of all.

REFERENCES

1. S. Bhattacharya and K.L. Mittal, Mechanics of removing glass particulates from a solid surface, Surface Technol., 7, 413 (1978).
2. M. Corn, The adhesion of solid particles to solid surfaces, I: A review, J. Air Pollution Control Assoc., 523 (November 1961).
3. R. Allen Bowling, An analysis of particle adhesion on semiconductor surfaces, J. Electrochem. Soc., 132, 2208 (September 1985).
4. I.F. Stowers and H. G. Patton, in "Surface Contamination: Genesis, Detection, and Control", K. L. Mittal, editor, Vol. 1 , pp. 341-349, Plenum Press, New York, 1979.
5. M. O'Donoghue, The ultrasonic cleaning process, Microcontamination, 63 (October/November 1984).
6. A. Mayer and S. Shwartzman, Megasonic cleaning: A new cleaning and drying system for use in semiconductor processing, J. Electronic Mater., 8 (6), 855 (1979).
7. Personal communication with Prof. Joel Ferziger, Dept. of Mechanical Engineering, Stanford University, Stanford, CA, June 1986.
8. B. H. Armstead, P. F. Ostwald and M. L. Begeman, "Manufacturing Processes", 7th Edition, pp. 658-665, John Wiley & Sons, New York, 1977.

9.   P.W. Morrison, editor, "Environmental Control in Electronic Manufacturing", Van Nostrand Reinhold Company, New York, 1973.

10. R.P. Musselman and T.W. Yarbrough, Shear stress removal of submicron particles from surfaces, J. Environmental Sci., 51 (January/February 1987).

11. J. Tuck, Ultrasonic cleaning, Circuits Manufacturing, 50 (January 1982).

12. S. Shwartzman, A. Mayer, and W. Kern, Megasonic particle removal from solid-state wafers, RCA Review, 46, 81 (March 1985).

13. A. D. Weiss, Wafer cleaning update, Semiconductor Intl.,82 (April 1984).

14. P. S. Burggraaf, Wafer cleaning: sonic scrubbing, Semiconductor Intl., 103 (July 1981).

15. G. D. Hutcheson, Recent trends in clean technology, Microcontamination, 8 (January 1986).

16. G. Fong and M. Daszko, Wafer cleaning technology for the eighties, Microelectronic Manufacturing & Testing, 1 (May 1985).

17. Cleaning photomasks with scrubbers and high-pressure water, Circuits Manufacturing, 16 (May 1978).

18. G. G. Hawley, "The Condensed Chemical Dictionary", Van Nostrand Reinhold Co., San Francisco, 1981.

19. Th. F. Tadros, Editor, "Surfactants", Academic Press, New York, 1984.

20. C. A. Hampel, G. A. Hawley, "Glossary of Chemical Terms", 2nd ed., Van Nostrand Reinhold Co., New York, 1982.

21. K. L. Mittal, Editor, "Micellization, Solubilization and Micro-emulsions", Vol. 1 and 2, Plenum Press, New York, 1977.

22. K. L. Mittal and B. Lindman, editors, "Surfactants in Solution", Vols. 1-3, Plenum Press, New York, 1984.

23. P. S. Burggraaf, Wafer cleaning: Brush and high-pressure scrubbers, Semiconductor Intl.,49 (July 1981).

24. S. Iwamatsu, Effects of plasma cleaning on the dielectric breakdown in SiO_2 film on Si, J. Electrochem. Soc. , 129, 224 (1982).

25. W. Kern, Purifying Si and SiO_2 surfaces with hydrogen peroxide, Semiconductor Intl.,94 (April 1984).

26. W. Kern and D. A. Puotinen, Cleaning solutions based on hydrogen peroxide for use in silicon semiconductor technology, RCA Review, 187 (June 1970).

27. "Search for an Optimum Cleaning Procedure for Silicon Semiconductor Wafers Prior to High Temperature Operations," TR117, FSI Corp., August 1979.

28. P. S. Burggraaf, Wafer cleaning: State-of-the-art chemical technology, Semiconductor Int., 91 (July 1981).

29. C.N. Davies, "Aerosol Science", Academic Press, New York, 1966.

ELECTROSTATIC REMOVAL OF PARTICLES FROM SURFACES

D. W. Cooper, H. L. Wolfe, and R. J. Miller

IBM Research Division
T.J. Watson Research Center
Yorktown Heights, NY 10598

We have achieved electrostatic removal of micron-size particles adhering to conductive surfaces. Electric fields much
greater than the breakdown field for air (30 kV/cm) are needed
for removing particles having diameters of the order of
microns (μm), so these measurements were done in vacuum. We
have also measured the fraction of particles removed versus
applied electric field. For spherical nickel particles, the
field needed to remove 50% of the particles of a given size
was approximately inversely proportional to particle size.
Removal of 95% of the particles tested (nickel, SiO2,
polystyrene latex) typically required about twice the field,
about four times the force, that removal of 50% of the particles required.

INTRODUCTION

Particulate contamination is implicated in a large fraction of yield
losses in the manufacturing of semiconductor chips and in yield and reliability losses in high-density magnetic storage devices. Effective
cleaning should help reduce such losses. Wet cleaning can be effective
in many instances, but there are situations where this is not advanta-
geous:
1. wet cleaning may leave harmful residues;
2. wetting the material may damage it;
3. the drying times may be prohibitive;
in such cases, dry cleaning is to be preferred. One common method of dry
cleaning, using jets of high-velocity gas, is often used and sometimes
effective, but its effectiveness diminishes as particle size decreases,
and it becomes ineffective on particles of diameters of a few microns and
smaller.[1] Thus, there would be advantages to developing a dry method for
removing particles a few microns and smaller in diameter. We decided that
electrostatic methods offered the potential for such removal. We investigated applying high electrostatic fields to remove particles from
conductive surfaces by using a vacuum to insulate the conductive surface
to be cleaned (the source) from a closely-spaced counter electrode, while
applying a high voltage difference to them.

The use of electric fields to remove particles can also be employed
as a method of determining the strength of adhesion of these particles.

Others have recognized the possible use of electric fields to remove charged particles from surfaces. A patent was obtained by DeGeest[2], theory was developed by Lebedev and Skal'skaya[3] and Cho[4] and others[1], and measurements were made by Myazdrikov and Puzanov[5] and Cho[4] and others[1] Hays[6] measured the fields needed to detach toner particles roughly 13μm in diameter from nickel carrier spheres (250μm in diameter), and he measured the charge on the particles that were detached, so that he was able to determine the force needed to detach them. We have not seen in the literature, however, data on fraction removed versus applied field and particle size for particles as small as those we studied.

BACKGROUND: ADHESION AND ELECTROSTATICS

Theory[2] predicts that the particles on a conductive surface will acquire a charge proportional to the particle area and the electric field. Further, theory predicts that the particles will experience a net force in the field proportional to the field times the charge. Thus, the force (F) on a conducting sphere is proportional to the square of the sphere diameter (d) and the square of the electric field (E):

$$F = k\, E^2\, d^2$$

where $k = 3.8 \times 10^{-6}$ dyn/V^2 when the field is in volts/cm and the particle diameter is in cm, giving a force in dynes. The force was derived[3] from the electrostatic field equations for the geometry of the sphere and the plane, using bi-spherical coordinate expansions. Good approximations of this force can be obtained by using the predicted charge, the field at the planar surface, and a correction for image-charge attraction of the sphere to the plane.

Clearly, high fields are desirable. The dielectric breakdown of gas depends on geometry and gas density and the gas atomic or molecular characteristics, but for air at normal temperature and pressure in a gap of the order of 1 cm, the breakdown occurs at 30 kV/cm. To get higher fields, one needs higher dielectric breakdown strengths; otherwise, arcing occurs.

Much of the previous work was done in air at NTP and was thus limited to about 30kV/cm, although Myazdrikov and Puzanov[5] did go to higher fields by using high-pressure gas. Air has its minimum dielectric strength at a fraction of an atmosphere. High pressures and gases other than air (such as sulfur hexafluoride) are approaches to extending the field strength range. One of the approaches that we are investigating is the use of vacuum (< one-millionth torr).

As for adhesion, much research has been done on this complicated topic, and excellent reviews of the literature on adhesion are available.[1][7][8][9]

To remove particles from a surface one must diminish the forces of adhesion of the particles or overcome them or both. Liquid cleaning with surface-active agents does both.[8] Dry cleaning primarily or exclusively depends on applying forces to the particles that overcome the adhesion forces.

Contributing to adhesion are the van der Waals force from interacting dipole moments of the molecules, capillary force from the surface tension of any adsorbed liquids, and electrostatic forces from any charges on the particles and from fields set up by the particles and the surfaces and charges external to both. Van der Waals force and the capillary force are predicted to vary linearly with particle diameter, and adhesion of particles has been found empirically to follow approximately the following relationship:[10]

F = (75 + 0.68(%RH)) d

in which %RH is the percentage relative humidity and d is the particle diameter in cm, giving a force in dynes. This equation is for the adhesion forces for "hard materials and clean surfaces " in air.

Adhesion decreases approximately proportionately to particle diameter as particles get smaller, but the forces we can apply to particles to remove them tend to decrease even more strongly with particle diameter, which is why the smaller the particle, the harder it is to remove, generally.

In Figure 1 we have plotted the lines along which the electrostatic force on a particle on the surface of a conductor just balances either the force of gravity (assuming density is 1 g/cm³) or the force of adhesion (0%RH). Moving to the right, toward higher electric fields, the electrostatic force becomes stronger, finally exceeding that of gravity or adhesion. Moving upward, toward larger particle sizes, the gravitational force becomes stronger, finally exceeding that of adhesion or electrostatics. Moving downward, toward smaller particle sizes, the force of adhesion diminishes less rapidly than that of electrostatics or gravitation, finally exceeding both. For micron-size particles and smaller, for fields of kV/cm and larger, the particle removal is determined by the adhesion. Fields of hundreds of kV/cm should suffice to

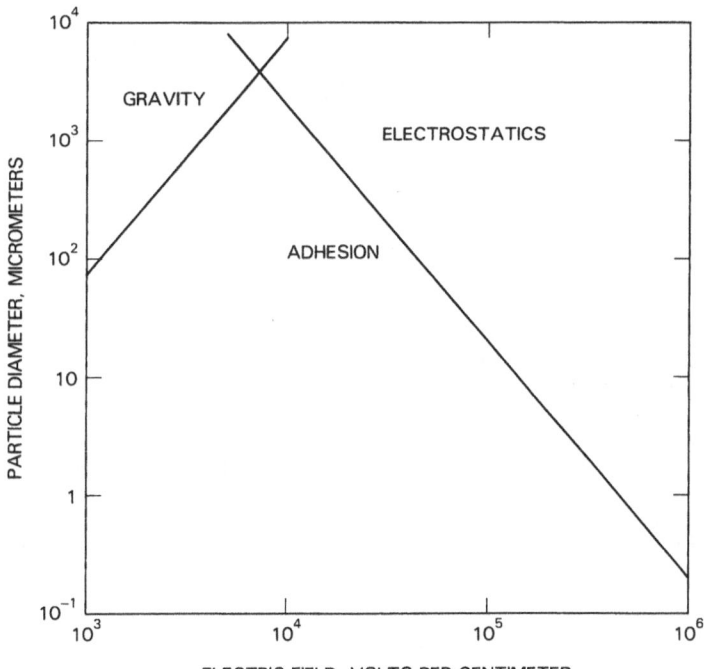

Figure 1. Particle sizes for which electrostatic force just equals that of adhesion or of gravitation.

remove particles a few microns and larger. In fact, Shelkunova and Gegin[11] reported that particles having diameters of a micron to tens of microns and being made of refractory metals (W, Ta, Mo, tungsten carbide) were removed at room temperature and pressure in air from metallic surfaces at electric field intensities of about 25 kV/cm. The material of which the particles are made will influence adhesion through the particles' surface properties and through their deformability, however; Hays needed fields of 100 kV/cm in vacuum to remove plastic particles 13μm in diameter. We expect that removing hard submicron particles will require fields of 1 MV/cm and perhaps greater.

EXPERIMENTAL

Equipment

Figure 2 shows a schematic of the experimental equipment. Inside a vacuum chamber (at pressure < 10^{-6} torr) a highly-polished conductive surface at the top of a pedestal was spaced 0.36 or 0.40 mm away from the conductive top plate of the chamber, used generally as the counter-electrode. One pedestal top was 25.4 mm in diameter; a second was 6.3 mm in diameter. The counter-electrode was usually coated with a thin film (< 10μm thick) of insulating material (acrylic). Sometimes a top plate made of conductively-coated quartz was used, to allow viewing the surfaces and particles through a microscope mounted outside and atop the vacuum chamber. The pedestal was connected to a high-voltage power supply and the outer surface of the chamber, including the top plate, was grounded.

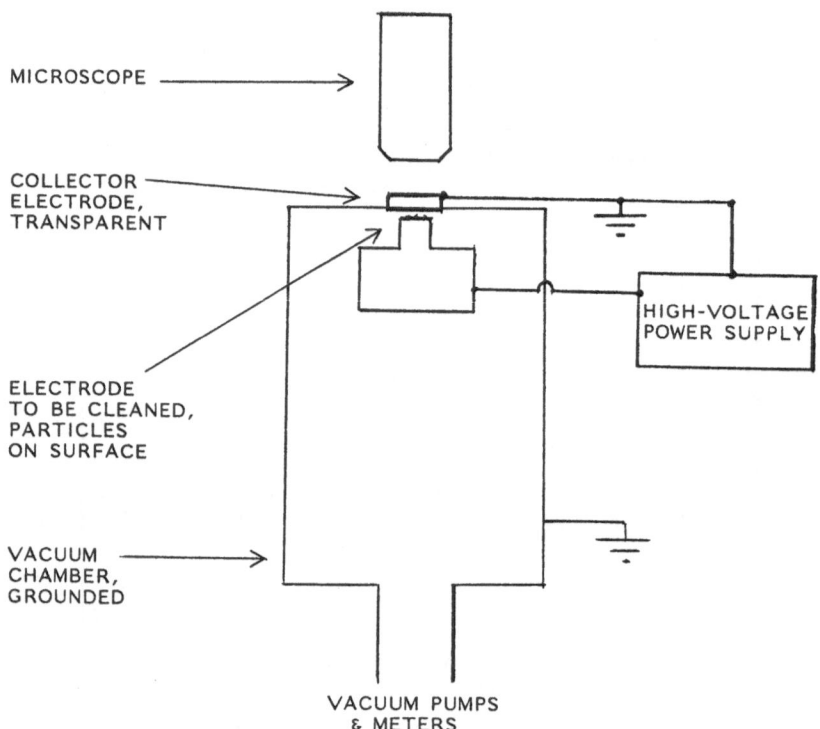

Figure 2. Schematic of equipment for study of electrostatic removal of particles in a vacuum.

Procedure

Fine particles were put on the pedestal in one of the following ways: (a) by exposing the pedestal to an aerosol produced by atomization or (b) by pouring a small amount of powder on the surface and tapping and blowing to remove the excess or (c) by using larger spheres coated with particles to transfer the particles to the pedestal surface. The particles discussed in this report are spheres of nickel, spheres of glass, and spheres of polystyrene latex (PSL). After putting particles on the pedestal, we put the pedestal back in the chamber and turned on the pumping system. In somewhat less than an hour, the system was below one-millionth torr. (This low pressure would correspond to 0% RH and the presence of little or no volatile material on the surfaces.) The pre-selected voltage was applied for five minutes, then turned off. The system was allowed to return to atmospheric pressure (with dry nitrogen), and the pedestal was removed. Exactly the same areas on the pedestal on which particles had originally been sized and counted before the application of voltage were recounted after the application of different voltages. From the counts we determined the fraction of particles that were initially singlets that remained after the voltage had been applied.

Analysis

Figure 3 shows an example of the data obtained and the analysis process, for data from a single run. Plotted are the fraction of particles remaining at the indicated field strength versus field strength (kV/cm). The dots represent the best-fit cumulative normal distribution through the data, as determined by the SAS (TM, SAS Institute, Inc. Cary, NC) PROBIT computer program. (In some cases, a chi-square analysis indicated that the fit to a cumulative lognormal distribution was better, but for consistency the normal distribution was used throughout.)

Adhesion has generally been found to have substantial variability for particles of the same size on a given surface in any single experiment. That variability increases when several experiments are done and where the particles are not of a single size. The analysis method we have chosen presents the data succinctly and allows determination of a characteristic field strength (the fiftieth percentile field strength) as well as the spread in field strengths associated with particle removal.

RESULTS

Experiments with the transparent windows showed that at a given voltage, all the particles 15 μm and smaller that were going to leave the pedestal did so within half a minute, usually within a second or less. Repeated experiments without applied voltage showed that removal due to gas motion during evacuation of the chamber was negligible.

Figure 4 shows the results for Ni spheres with diameters between 3 and 5μm, as an example of the results of two runs with Ni. Figure 5 shows percentage removed versus electric field for glass spheres 9 - 15μm in diameter, for four runs, demonstrating the inter-run and intra-run variability.

Table 1 shows the 5th, 50th, and 95th percentile values for removal versus electric field for the particle sizes studied. (Where all three values are shown, these were obtained from the PROBIT analysis.) The differences between these values are caused partly because adhesion depends on particle orientation and perhaps other very local phenomena, showing substantial variability, and partly because the particles were

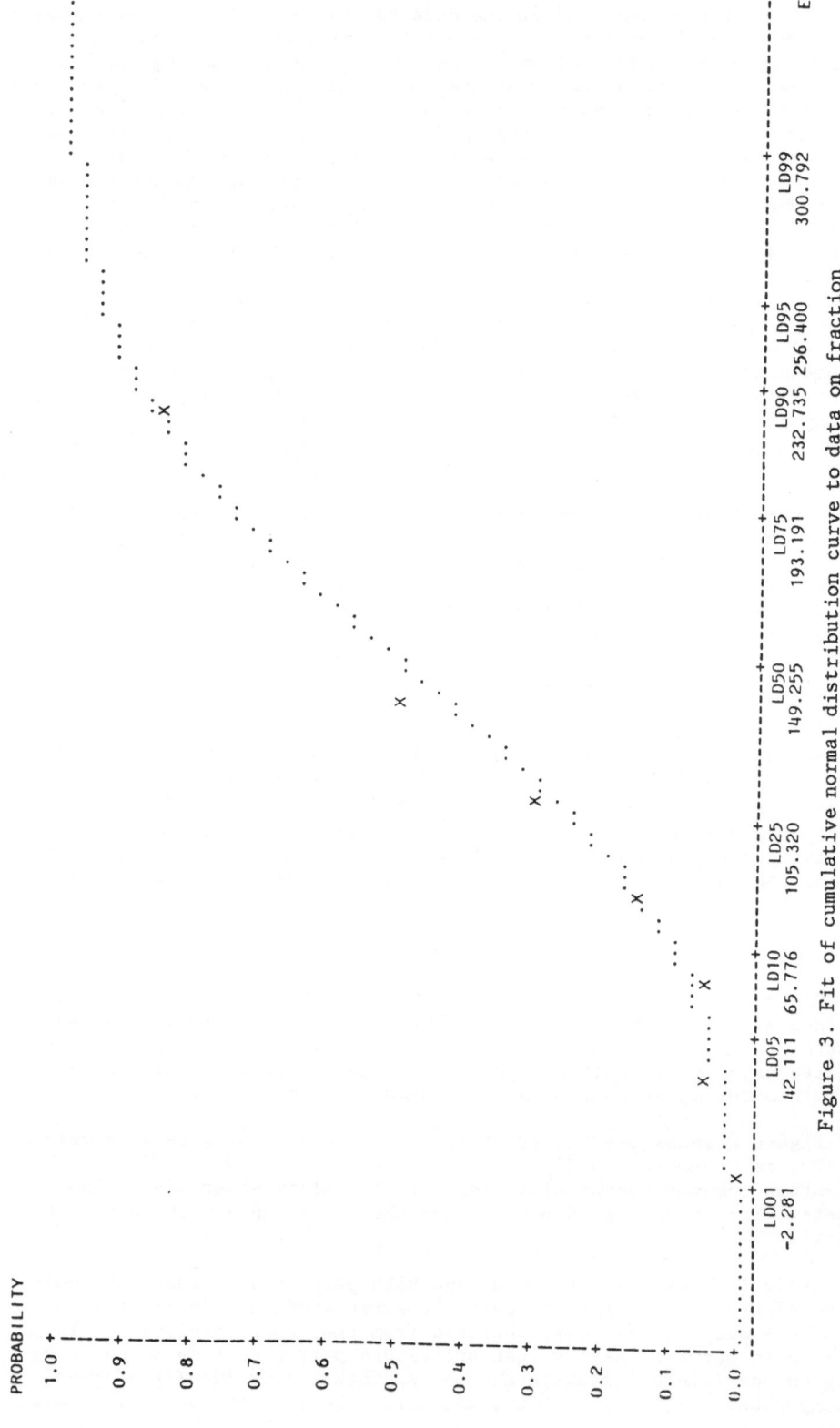

Figure 3. Fit of cumulative normal distribution curve to data on fraction of Ni particles 3 to 5μm in diameter removed by electric fields of various strengths.

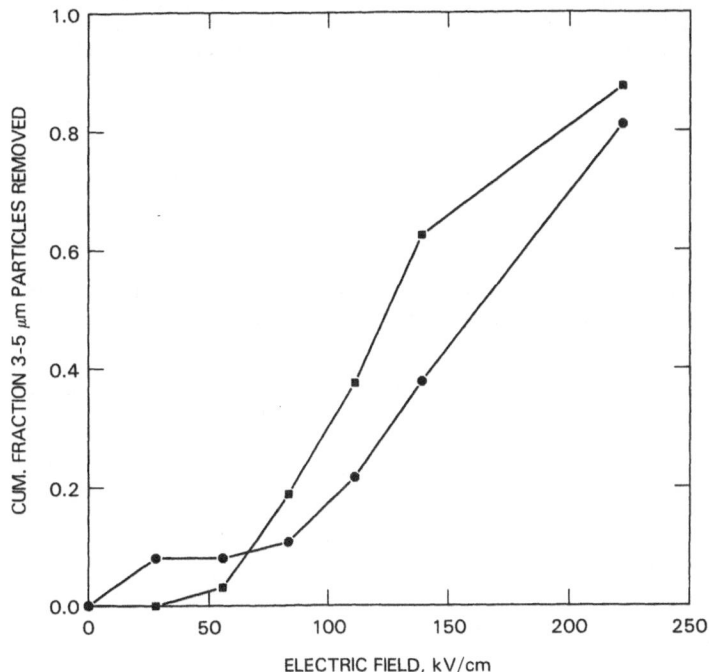

Figure 4. Fraction of particles removed in vacuum versus electric field for Ni particles of diameters 3-5μm. Symbols: two different runs.

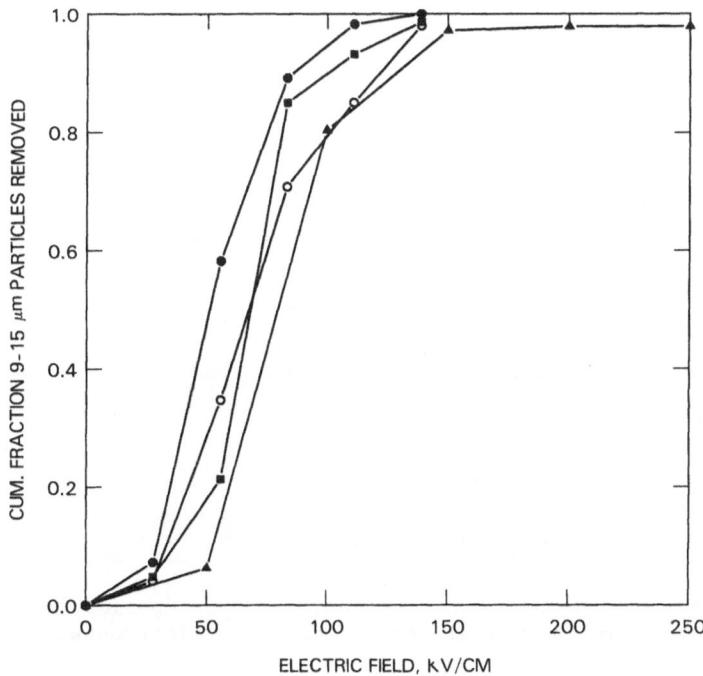

Figure 5. Fraction of particles removed in vacuum versus electric field for glass particles of diameters 9-15μm. Symbols: four different runs.

345

not of a single size in each group. Thus, the 95th percentile includes
particles somewhat smaller than average in the group as well as locations
and orientation somewhat less favorable for removal than the average, we
believe.

Table 1. Electric Fields Needed to Remove Particles of the Indicated Size
and Composition from Stainless Steel in Vacuum.

Particle Material	Particle Diameter (μm)	Field (kV/cm) Needed to Remove the Percentage of Particles Indicated		
		5%	50%	95%
glass	9 - 15	20	55	91
		29	69	109
		22	72	123
		20	88	156
	0.5	> 525		
nickel	8 - 10	18	87	155
		<0	77	163
	5 - 8	27	100	174
		12	78	144
	3 - 5	29	159	289
		46	137	227
	1 - 2	600 (a)		
			600 (b)	
polystyrene	10 (c)	41	243	446
		70	367	664
	1.1	> 550 (d)		

Notes:
 (a) 2 of 35 removed
 (b) 23 of 43 removed
 (c) only 25 particles total, both runs
 (d) 1 of 109 removed

Figure 6 compares the 50th percentile values superimposed on the
lines shown in Figure 1, the diameters where the electrostatic force of
removal should equal the dry adhesion force. (The point for nickel
spheres 1 to 2μm in diameter is for 46% removal.) The latex particles
were more difficult to remove than were the glass particles, and the
nickel particles were intermediate in difficulty of removal.

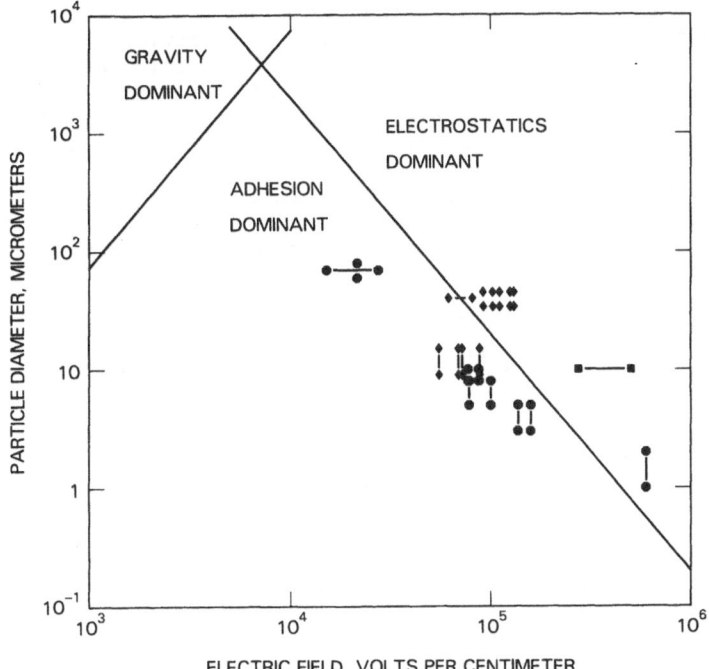

Figure 6. Particle sizes and fields needed for 50% removal (from our measurements) superimposed on Figure 1. Symbols: filled circles = nickel; filled diamonds = glass; filled squares = polystyrene.

DISCUSSION

It is necessary to coat the collecting electrode with an insulator or to make it highly adhesive; otherwise, particles will exchange charge, become repelled, and redeposit on the source electrode. We spray-coated our collector with a film of acrylic finish, which we found to be adequate to prevent redeposition. Operating without an insulator on the collector, we could identify redeposited particles as particles appearing in new locations in the photographs of the counted areas. The data reported here have no redeposited particles counted.

The behavior of doublets, triplets, etc. may be quite different from those of singlets, so only particles that were initially single were counted for these evaluations. In the case of half-micron SiO2 particles, we found particles on the collector even when the removal of singlets from the other electrode was zero, showing that some of the particles that were in agglomerates were more readily detached than any in direct contact with the stainless steel.

There were instances where lateral forces developed between particles on the surface, as evidenced by lateral spreading of some particles near each other on the pedestal. Further, particles deposited on the collector electrode tended to be more uniformly distributed than those on the source electrode, even after the source electrode had all its particles removed. These observations indicate that appreciable inter-particle repulsive forces were operating while the particles were on the source electrode surface and during the flight across the gap between the electrodes.

The experiments showed the trend expected: smaller particles required larger fields for successful removal. Sub-micron particles are expected to require fields of MV/cm magnitudes. This can be achieved in various ways, including the use of higher voltages, smaller spacings, and alternative dielectrics. The fields achievable are limited by the electrical or mechanical breakdown of the conductors and insulators used.

The material of which the particles were made clearly influenced the difficulty with which they were removed. Particle deformability and particle bulk and surface conductivity may have been important factors. Particle deformation may have played a role in the difficulty in removing the deformable polystyrene latex particles. Hays[6] found that fields of 10 V/μm = 100 kV/cm were needed to detach 13 μm plastic toner particles from Ni. Hays did not report what fraction of the particles were removed (and may not have been able to determine that). We found that 100 kV/cm would remove more than 5% but less than 50% of the polystyrene latex spheres 10 μm in diameter from stainless steel in a vacuum. We do not think the differences in removal difficulty were due to differences in resistivity. The conductivity of polystyrene is quite low, 10^{-17} mho/m, about 4 orders of magnitude less conductive than glass[12], and of course Ni is a conductor, perhaps coated with a thin insulating oxide. Generally, micron-size particles behave electrically much the same whether made of insulating or conductive materials, because the times involved in charge movement within and on the surface of the particles are usually fractions of seconds in either case.[13] Surface conductivity may be more important than bulk conductivity for micron-size particles, having large ratios of surface to volume. If the particles we tested all acted as conductors over the time scale of the tests, then the differences in electric fields needed to remove particles of the same size would seem to be due to differences in adhesion of the particles to the stainless steel.

We are continuing this investigation, toward the removal of submicron particles. We are also studying whether techniques that remove particles from conductor surfaces can be adapted successfully to semiconductor and insulator surfaces. Electrostatic removal in a vacuum would provide a dry cleaning technique of considerable utility.

The vacuum electrostatic removal technique also has promise as a tool for studying adhesion.

CONCLUSIONS

Electrostatic forces in a vacuum can be used as a dry non-contact method to remove micron-size particles from a conductive surface and perhaps to remove even submicron particles. Submicron particles are expected to require fields on the order of MV/cm for their removal. Removal of 95% of the particles of the size and material tested required about twice the field, thus four times the force, as did removal of 50% of the particles.

REFERENCES

1. A.D. Zimon, "Adhesion of Dust and Powder," 2nd Edn., Consultants Bureau (translation), New York, 1982.

2. W.F. DeGeest, "Electrostatic Cleaner and Method," U.S. Patent #3,536,528, 1970.

3. N.N. Lebedev and I.P. Skal'skaya, Zh. Tekh. Fiz. 32(3), 375-378 (1962).

4. A.Y.H. Cho, J. Appl. Phys. 35, 2561 (1964).

5. O.A. Myazdrikov and V.N. Puzanov, Zavod. Lab. (USSR) 35(10), 1265-1267 (1969).

6. D.A. Hays, Photog. Sci. & Engg. 22(4), 232-235 (1978).

7. R.A. Bowling, J. Electrochem. Soc. 132 (9), 2208-2214 (1985).

8. J.N. Israelachvili, "Intermolecular and Surface Forces," Academic Press, New York, 1985.

9. H. Krupp, Advan. Colloid Interface Sci. 1, 111-239 (1967).

10. W.C. Hinds, "Aerosol Technology," Wiley, New York, 1982.

11. Z.V. Shelkunova. and S.V. Gegin, Zavod. Lab. (USSR) 6, 708-709, June 1976.

12. A.D. Moore, editor, "Electrostatics and Its Applications," Wiley, New York, 1973.

13. N.A. Fuchs, "Mechanics of Aerosols," Pergamon, New York, 1964.

14. S. Bhattacharya and K. L. Mittal, Surface Technol. 7, 413-425 (1978).

ELECTRIC FIELD DETACHMENT OF CHARGED PARTICLES

Dan A. Hays

Xerox Corporation
Webster Research Center
Rochester, N.Y. 14644

Electric field detachment of charged toner particles in the xerographic process is of central importance in the image development and transfer steps. The theoretical concepts for electric field detachment of single charged particles and particle layers are reviewed. Limitations on the applied force imposed by dielectric polarization and air electrical breakdown are considered. Measurements on the electric field detachment of toner layers in vacuum with different surface coverage are presented. The results for 21 μm toner suggest that the adhesion is described by an electrostatic image force model due to a nonuniform charge distribution on irregularly shaped particles.

INTRODUCTION

Particle adhesion to surfaces is usually considered undesirable due to the problems it causes. For example, particle contamination in the fabrication of devices in the microelectronics industry represents an important current problem.[1] On the other hand, the adhesion of <u>charged</u> particles is beneficial in many industrial processes including the separation of materials, electrostatic precipitation and electrophotography.[2] In the xerographic process, the force balance between an applied electric field acting on charged toner particles and their adhesion to various surfaces controls the image development and transfer steps.[3]

Adhesion measurements on xerographic toner using the centrifuge technique have been reported in a number of studies with conflicting conclusions.[4-7] Krupp[4] was first to demonstrate the dominant role of short-range van der Waals forces in extensive studies. Donald[5] observed a significant electrostatic image force contribution, but Mastrangelo[6] concluded that van der Waals forces are dominant. More recently, Lee and Ayala[7] have described centrifuge adhesion data that show a strong dependence on toner charge. The conflicting data tend to suggest that both the van der Waals and electrostatic image forces can be significant depending on the material, particle size and measurement conditions.

The detachment of charged toner particles in the xerographic image development and transfer steps is obtained via an electric field. Studies of toner adhesion by electric field detachment have been limited since the detachment force is more complicated than the centrifugal force due to its strong dependence on particle geometry, texture and charge distribution. Goel and Spencer[8] used electric field and centrifugal force adhesion measurements to conclude that for large toner particles the image force is dominant while for small toner, the van der Waals forces are more important. Hays[9] obtained simultaneous measurements of the particle charge and adhesion to conclude that an electrostatic image force model for a uniformly charged spherical particle is not an adequate description of the adhesion. The same conclusion was also obtained for adhesion energy measurements of charged particle attachment to a surface.[10]

The purpose of this paper is to describe electric field detachment measurements on toner layers in vacuum with different surface coverage. Theoretical considerations are discussed first followed by a description of the detachment measurements.

THEORETICAL CONCEPTS

Applied Force due to an Electric Field

Consider an idealized particle that is a spherical dielectric with a spherically symmetric surface charge density. One can calculate the electrostatic forces acting on the particle resting against a conducting substrate in the presence of an applied electric field.[11] The applied force, F_A, consists of two terms: an electrostatic force proportional to the particle charge, Q, and the applied electric field, E_A, and a polarization force proportional to the square of both the particle diameter, D, and E_A. The expression for the applied force in SI units is

$$F_A = - \beta Q E_A - \gamma \pi \varepsilon_0 D^2 E_A^2, \qquad (1)$$

where β and γ are coefficients that depend on the dielectric constant of the particle and ε_0 is the permittivity of free space. If the dielectric constant of the particle happened to be 1, β and γ would equal 1 and 0, respectively. For a typical dielectric constant of 4, β and γ are equal to 1.6 and 0.25, respectively. Since the polarization force is always directed towards the substrate, the electric field that produces the maximum detachment force occurs when

$$E_A = - \beta Q / (2 \gamma \pi \varepsilon_0 D^2). \qquad (2)$$

A typical xerographic toner size and charge is 10 μm and 20 μC/gm, respectively. (Toner charge is usually expressed as a charge-to-mass ratio, Q/M, where the particle density is typically 1 gm/cm^3). For the data described later, the average toner size and charge are 21 μm and −5 μC/gm, respectively. For a toner dielectric constant of 4, we expect a maximum in the applied electrostatic force for an electric field of 6 V/μm. The theoretical maximum applied force is 12 mdyn for our toner particles.

When an electric field is applied in an air ambient, air breakdown commonly limits the field strength. The maximum electric field that can be applied across an air gap between two electrodes depends on the electrode gap, G.[12] For gaps on the order of 1000 μm, the air breakdown field in units of V/μm is

$$E_B = 3.0 + 1350/G. \qquad (3)$$

For G = 400 μm, E_B = 6.4 V/μm. Although higher electric fields are obtained in vacuum, an upper limit to the field is set by ionization of material on the electrodes.

Adhesion due to Electrostatic Image and Short-Range Forces

The electrostatic image force, F_I, for a spherically symmetric charge distribution on a spherical dielectric particle in contact with a conducting substrate is (in SI units)[11]

$$F_I = \alpha Q^2/(4\pi\varepsilon_o D^2), \tag{4}$$

where α is a coefficient that depends on the dielectric constant, uniformity of the surface charge density and separation between the particle and substrate. For a uniformly charged sphere in contact with the substrate, α is equal to 1 and 1.9 for dielectric constants of 1 and 4, respectively. For D = 21 μm and Q/M = -5 μC/gm, F_I = 2 mdyn.

In addition to the electrostatic image force, the particle adhesion includes contributions from short-range forces, F_{SR}, due to van der Waals interactions, F_{vdW}, and a possible electrostatic force contribution from a double layer (charge exchange) at the interface. The force due to van der Waals interactions for an undeformed spherical particle is[8]

$$F_{vdW} = AD/(12Z^2), \tag{5}$$

where A is the Hamaker constant and Z is an atomic separation. With A = 1 X 10^{-12} erg and Z = 4 Å, F_{vdW} = 110 mdyn for D = 21 μm. If plastic deformation occurs at the contact, F_{vdW} will be enhanced. It would appear that van der Waals interactions dominate the adhesion of our particles and prevent electric field detachment. However, since the data described later indicates that electric field detachment is obtained for high electric fields, we assume that the irregular shape of the particles and the surface morphology due to a crazed layer formed during fracture in the manufacturing process must reduce the effectiveness of the van der Waals forces. For purposes of discussion, we assume F_{vdW} is a constant. The total adhesion force is

$$F_{AD} = \alpha Q^2/(4\pi\varepsilon_o D^2) + F_{SR}. \tag{6}$$

Electric Field Detachment

When the applied force of Equation (1) is equal to the adhesion force of Equation (6), particle detachment is obtained. If we assume that F_{SR} is negligible compared to F_I, the electric field for detachment, E_D, is 0.5 V/μm for a particle size of 21 μm and Q/M of -5 μC/gm. (The polarization force is weak for this field strength.) For electric fields such that the polarization force is small compared to the electrostatic force, the Q/M of the detached toner should be proportional to E_D. If F_{SR} is much larger than F_I, Q/M of the detached toner should be inversely proportional to E_D.

Nonuniform Surface Charge

Toner particles are usually not spherical but rather highly irregular. Since charging of the insulating particles occurs by triboelectricity, one can expect a nonuniform charge distribution on the surface. The magnitude of the electrostatic image force is critically dependent on the distribution of the surface charge. For a nonuniform distribution, the particle will tend to orient with the highest charged regions near the substrate. Under these conditions, the effective α in

Equation (4) is much greater than 1. Goel and Spencer[8], Hays[9] and Lee and Ayala[7] have discussed the enhancement of the electrostatic image force due to a nonuniform surface charge.

Electrostatic Force due to Neighboring Charged Particles

A charged particle on a conducting substrate produces a fringe electric field in the surrounding region. A summation of the fringe electric fields acting on one particle due to neighboring charged particles increases the electrostatic force of adhesion. Goel and Spencer[8] have discussed the dependence of this force on surface coverage. For a monolayer of close packed spheres, the electrostatic force due to neighboring charged particles is 6.95 times greater than the electrostatic image force. This implies that as the surface coverage increases, the particle adhesion should increase significantly if the electrostatic image force is important. Furthermore, the increased adhesion due to neighboring particles suggests that when the applied force is near the threshold for detachment, a detachment of one particle will cause neighboring particles to be released in a chain reaction.

TONER LAYER DETACHMENT

Figure 1 shows an electrode configuration for measuring the electric field detachment of toner. A DC voltage up to 6 kV is applied between the polished aluminum electrodes spaced by 400 μm in vacuum. Toner with a volume median of 21 μm and a geometric standard deviation of 1.24 was mixed with 250 μm spherical carrier beads to form a developer mix with a surface coverage of 0.3 mg/cm^2. The toner material was a standard composite consisting of carbon black dispersed in a styrene/acrylic copolymer. The particle shape was highly irregular since the particles were formed by fracture. To load the grounded electrode with toner, the developer mix was cascaded between the grounded 5.4 cm X 7.6 cm electrode and a biased development electrode. The toner deposition was varied by adjusting the development time. The toner surface coverage, M/A, was calculated from a weighing of the toned electrode.

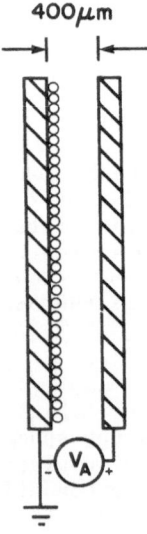

Figure 1. Electrode configuration for measuring the electric field detachment of toner layers in vacuum.

The procedure for measuring the adhesion distribution was as follows. After the electrode assembly was placed in a vacuum of 10^{-4} torr, an incremental DC voltage was applied across the gap. The electrode assembly was then returned to atmospheric pressure and disassembled. After the receiver electrode was weighed, the electrode was attached to an electrometer and the transferred toner removed with an air jet to measure the toner charge and mass. The fractional toner mass transferred per unit incremental electric field was calculated as well as the toner charge-to-mass ratio. The electrode assembly was then replaced in the vacuum system and the DC voltage increased to a magnitude that was an increment above the previous value. The toner detachment was independent of the time the voltage was on between 2 and 60 secs. (If a repeat measurement was made without a voltage increase, no additional toner was detached.) By repeatedly making a sequence of measurements with incremental increases in the applied voltage, one can obtain curves for the fraction of toner detached per unit incremental electric field and the charge-to-mass ratio as a function of the applied electric field.

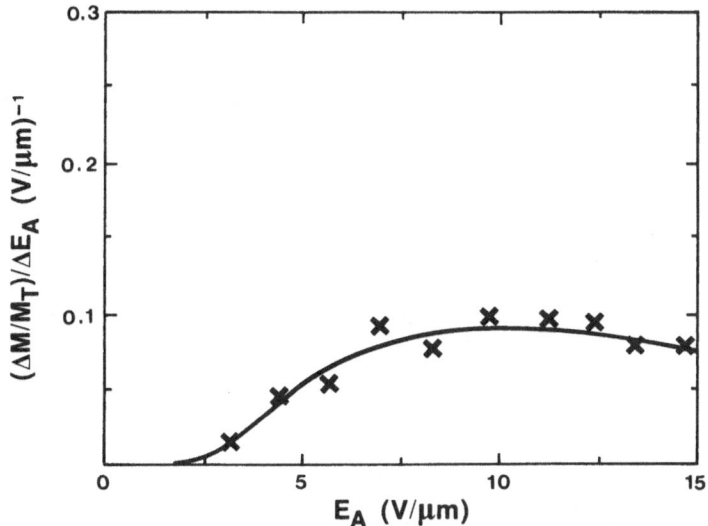

Figure 2. Fractional toner mass transferred per unit electric field as a function of the applied electric field for 21 μm toner loaded on the donor electrode with a M/A = 0.32 mg/cm^2. The residual toner on the donor is 15%.

Figure 2 shows the distribution in the fraction of toner detached per unit electric field as a function of the applied electric field for a low toner coverage of M/A = 0.32 mg/cm^2 corresponding to 0.25 of a monolayer. The residual toner on the donor after an applied field of 15 V/μm was 15%. It is clear that an applied electric field much greater than the theoretical 0.5 V/μm is required to detach an appreciable fraction of the toner layer. Furthermore, it is clear that the applied electrostatic force continues to increase for fields up to 15 V/μm, well in excess of the theoretical limit imposed by the opposing polarization force.

For a toner coverage approaching a monolayer, one would expect the peak in the distribution function to be shifted to a higher field as a consequence of the dipole force from neighboring particles. Figure 3 shows the distribution function for M/A = 0.85 mg/cm^2 corresponding to 0.67 of a monolayer. The residual toner was 5%. The peak shift to lower fields is likely due to the lower adhesion of some toners residing on top of other toner. This interpretation is supported by the observation that the peak field is 4.5 V/μm for a M/A = 1.77 mg/cm^2, a surface coverage in excess of a monolayer. Evidently the dipole force due to neighboring toner particles does not increase toner adhesion. This conclusion is supported by optical microscope observations that do not show simultaneous detachment of closely spaced toner.

The electric field dependence of the toner charge-to-mass ratio for the toner layer with a M/A = 0.85 mg/cm^2 is shown in Fig. 4. The Q/M of the toner removed for each field increment is independent of the applied electric field. Assuming there is a distribution in toner charge, this is contrary to expectation if the particle adhesion is dominated by either short-range forces or an electrostatic image force for a uniformly charged spherical particle.

If approximately half of the toner on the donor electrode is transferred to the receiver electrode by an electric field in one step, one would expect to transfer all of the toner back to the original donor for the same electric field if the adhesion of each toner particle is unique. To transfer toner to the receiving electrode, an electric field of 8 V/μm was applied to the donor electrode toned to M/A = 0.85 mg/cm^2. Since 60% of the toner was detached, the surface coverage on the receiver electrode was 0.51 mg/cm^2. The toner remaining on the donor electrode was removed and then the distribution function of the transferred toner was measured, as shown in Fig. 5. It is clear that electric field strengths greater than 8 V/μm are required to detach most of the particles. The distribution shown in Fig. 5 is equivalent to the distribution obtained when the donor electrode is toned to M/A = 0.5 mg/cm^2 by cascade development.

DISCUSSION

The results displayed in Fig. 2 show that the adhesion of 21 μm toner to a conducting surface for a low surface coverage is broadly distributed. If the irregularly shaped toner is modeled as a sphere with a uniform surface charge density, an electrostatic image force calculation predicts a detachment electric field of only 0.5 V/μm compared to the observed range of 2 to > 15 V/μm. The enhanced adhesion could be due to van der Waals forces since the calculation for a sphere predicts a force considerably greater than the measured values. Apparently the van der Waals forces are considerably reduced for an irregularly shaped particle. One could nevertheless argue that the enhanced toner adhesion is due to van der Waals forces. But if the van der Waals forces dominate the adhesion, the charge-to-mass ratio of the detached toner should be inversely proportional to the detachment electric field. Figure 4 shows that this is not observed. Instead, the charge-to-mass ratio is independent of the applied electric field which implies the toner adhesion is proportional to the average charge-to-mass ratio. The results of Fig. 5 suggest that the toner adhesion of a particle is not unique but can assume widely different values depending on the particle orientation on the substrate.

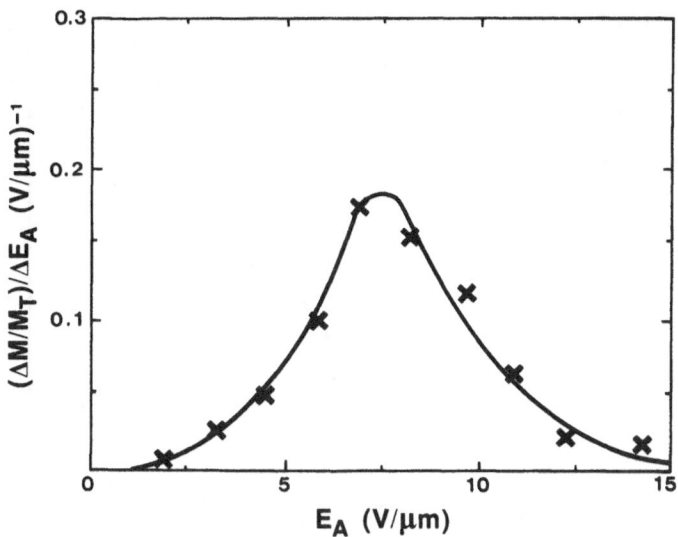

Figure 3. Fractional toner mass transferred per unit electric field as a function of the applied electric field for 21 μm toner loaded on the donor electrode with a M/A = 0.85 mg/cm^2. The residual toner on the donor is 5%.

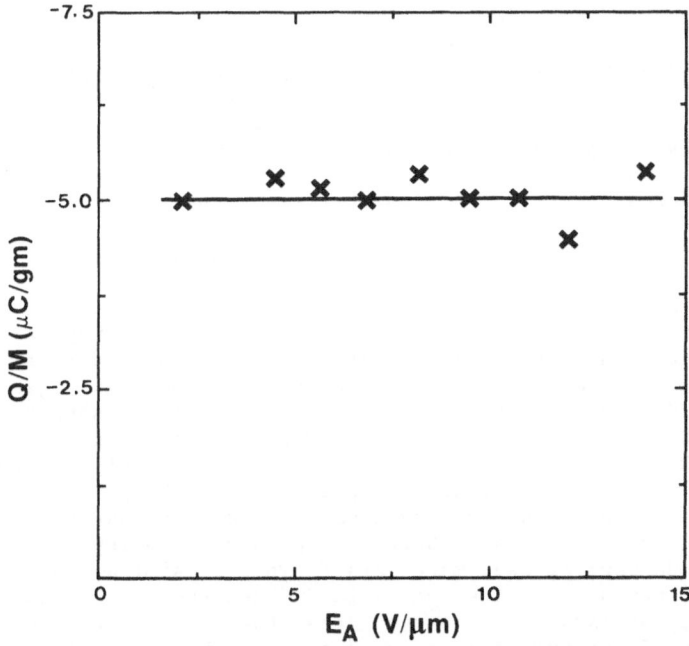

Figure 4. The charge-to-mass ratio as a function of the applied electric field for 21 μm toner loaded on the donor electrode with a M/A = 0.85 mg/cm^2.

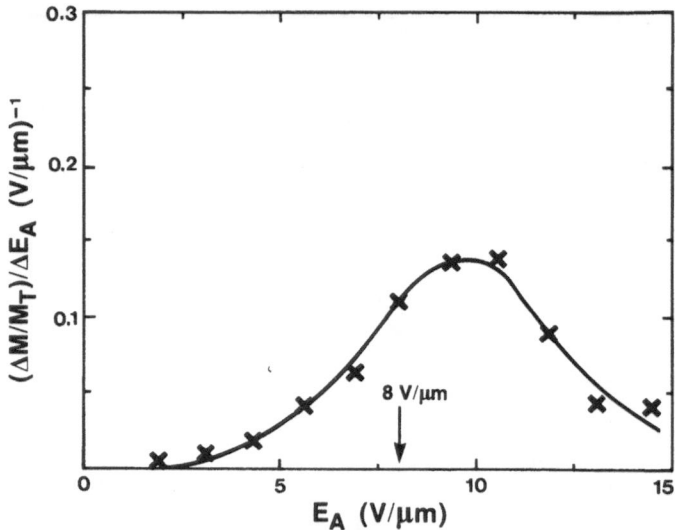

Figure 5. Fractional toner mass transferred per unit electric field as a function of the applied electric field for 21 μm toner transferred to the receiver electrode with an electric field of 8 V/μm. Since 60% of the toner with a coverage of 0.85 mg/cm² was transferred, the receiver electrode coverage is 0.51 mg/cm².

A comparison of Figs. 2 and 3 shows that since toner adhesion does not increase for a higher toner surface coverage, the electrostatic force from the fringe electric field due to neighbors does not contribute appreciably to the toner adhesion.

The data displayed in Figs. 2 and 5 imply that the applied electrostatic force for particle detachment continues to increase for an electric field up to 15 V/μm, in excess of the theoretical limit of 6 V/μm imposed by the polarization force. The nonspherical nature of toner and the surface morphology apparently reduce the influence of the polarization force. This seems plausible since the polarization force being proportional to the square of the radius of curvature would be much less for protrusions in contact with the substrate.

The measurements are interpreted in terms of an electrostatic image force model for a nonuniform distribution of charge on irregularly shaped toner. Since the insulating toner particles are triboelectrically charged by mixing with larger carrier beads, we assume the toner charge will be located only on protrusions from the irregularly shaped particle. The triboelectric charging will only occur at flat areas on the protrusions that are able to contact the larger relatively smoother carrier beads. The area that is triboelectrically charged represents a small fraction of the toner surface area. We assume that toner areas contacted by carrier beads are charged to a charge density that depends on the toner and carrier bead materials. This is similar to a toner adhesion model proposed by Lee and Ayala.[7] When the toner is deposited on a substrate, we assume the charged areas that contact the substrate are a fraction, f, of the total charged areas. The values for f can range between 0 for a uniformly charged sphere and 1 for a single charged area in contact with the substrate. We expect f to be broadly distributed with an average value around 0.2. The average value is based on the assumption that the average number of protrusions on toner is 15, 3 of which are in contact with the substrate.

If the extent of a charged area in contact with the substrate is much larger than the average distance between the charged surface and substrate, the electrostatic adhesion force is equal to

$$F_E = \sigma^2 A_c/(2\varepsilon_0) \tag{7}$$

where σ is the surface charge density and A_c the sum of the charged areas in contact with a conducting substrate. (The validity of Equation (7) depends on the assumption that the average separation between charges on the toner surface is less than the average distance between the charges and substrate. For $\sigma = 50$ nC/cm^2, the charges on the toner surface are separated by 0.02 μm. If the assumption is not valid, the adhesion force is larger.) Since $A_c = fA_t$ where A_t is the sum of all charged areas and $Q = \sigma A_t$, Equation (7) can be rewritten as

$$F_E = f\sigma Q/(2\varepsilon_0). \tag{8}$$

To make a comparison between the model prediction and measurements, we assume the applied electrostatic force for particle detachment is $-QE_A$ since the polarization forces described by Equation (1) are apparently negligible for irregularly shaped particles. Particle detachment will occur when the applied electric field is

$$E_D = -f\sigma/(2\varepsilon_0). \tag{9}$$

The electric field required to detach toner with a distribution in charge is proportional to f and independent of Q. The result displayed in Fig. 4 for the independence of the toner charge on the detachment electric field is consistent with Equation (9). Toner detachment at low fields occurs for all of the charged particles that happen to be oriented with a low value of f. (Particles with a low value of f can have a different value upon redeposition.) If we assume the charge density is -50 nC/cm^2 and f has an average value of 0.2, the average electric field for particle detachment is 5 V/μm which is on the order of the measured values displayed in Figs. 2, 3 and 5. The distribution in detachment fields is attributed to a distribution in f. The results displayed in Fig. 5 are consistent with the model since the f value for a particular toner particle is not unique but varies widely according to the particle orientation during deposition on a substrate.

For a charge density of σ on the toner surface at areas contacted by the carrier bead, one can estimate the fraction of the toner surface area that is charged, f_c, for an average toner charge-to-mass ratio Q/M and average toner diameter D. The fraction of charged areas is

$$f_c = (Q/M)\,\rho D/6\sigma). \tag{10}$$

where ρ is the density of the toner material approximately equal to 1 gm/cm^2. For Q/M = -5 μC/gm, d = 21 μm and $\sigma = -50$ nC/cm^2, $f_c = 0.03$. A 3% fraction of charged areas seems low but considering the toner particle shape and surface morphology, one would expect the true contact area for triboelectric charging to be a small fraction of the total toner surface area.

We assumed the triboelectric charge density at the contact areas was on the order of -50 nC/cm^2. This estimate is based on contact charging measurements reported by Davies[13] for repeated rolling contacts between various metals and polymers. Since toner and carrier bead materials are chosen for good triboelectric charging properties and continually mixed to achieve repeated toner-carrier contacts, an estimated charge density of -50 nC/cm^2 seems reasonable. The electric field associated with the charged area in contact with the substrate is approximately 50 V/μm which according to Equation (3) would be sufficient to initiate air breakdown. But since the electric field is confined to a very small volume, air breakdown should not occur for a charge density of -50 nC/cm^2.

Further adhesion studies on particles for which the uniformity of the charge is either controlled or characterized would be useful. One can anticipate experimental difficulty in accomplishing this since the particles are on the order of 10 μm. But the results would serve to confirm the importance of the electrostatic image force associated with a nonuniform charge distribution on particles.

SUMMARY

We have shown that toner particle adhesion measurements by electric field detachment of toner layers cannot be described by models in which the irregularly shaped particles are approximated as uniformly charged spheres. For 21 μm particles, the measured adhesion force is much larger than the electrostatic image force for a uniformly charged sphere. The van der Waals forces are considered to be negligible since the electric field dependence of the average toner charge is independent of the detachment field. The toner adhesion is attributed to a nonuniform distribution of charge on irregularly shaped particles.

REFERENCES

1. K. L. Mittal, editor, "Surface Contamination: Genesis, Detection and Control", Vol. I and II, Plenum Press, New York, 1979.
2. A. D. Moore, editor, "Electrostatics and Its Applications", John Wiley & Sons, New York, 1973.
3. J. H. Dessauer and H. E. Clark, editors, "Xerography and Related Processes", Focal Press, New York, 1965; R. M. Schaffert, "Electrophotography", Focal Press, New York, 1975.
4. H. Krupp, Adv. Colloid Interface Sci., $\underline{1}$, 111 (1966).
5. D. K. Donald, in "Recent Advances in Adhesion", L. H. Lee, editor, p. 129, Gordon and Breach, New York, 1973.
6. C. J. Mastrangelo, Photogr. Sci. Eng., $\underline{26}$, 194 (1982).
7. M. H. Lee and J. Ayala, J. Imaging Technol., $\underline{11}$, 279 (1985).
8. N. S. Goel and P. R. Spencer, in "Adhesion Science and Technology", L. H. Lee, editor, Part B, p. 763, Plenum Press, New York, 1975.
9. D. A. Hays, Photogr. Sci. Eng., $\underline{22}$, 232 (1978).
10. D. A. Hays, in "Electrostatics 1983," Inst. Phys. Conf. Ser. No. 66, p. 237, The Institute of Physics, London and Bristol, 1983.
11. L. M. Marks, private communication (1977); G. C. Hartmann, L. M. Marks and C. C. Yang, J. Appl. Phys., $\underline{47}$, 5409 (1976). See also M. H. Davis, Am. J. Phys. $\underline{37}$, 26 (1969).
12. J. D. Cobine, "Gaseous Conductors," Dover Publications, New York, 1958.
13. D. K. Davies, J. Phys. D: Appl. Phys., $\underline{2}$, 1533 (1969).

A NEW APPROACH TO THE REMOVAL OF SUB-MICRON PARTICLES FROM SOLID (SILICON) SUBSTRATES

A.F.M. Leenaars

Philips Research Laboratories, P.O. Box 80.000

5600 JA Eindhoven, The Netherlands

The efficiency of all known cleaning methods drastically decreases when particles smaller than about 1 μm are to be removed from a solid substrate. In this paper it is shown both theoretically and experimentally that the passage of a liquid-gas phase boundary along the substrate may result in particle removal. It appears that under properly chosen wetting conditions of the particles and the substrate, the adherence of the particles to the liquid-gas phase boundary is stronger than their adherence to the substrate. Contrary to all other cleaning methods, the efficiency of particle removal by moving phase boundaries is in theory independent of the particle size. Kinetic factors, which are not considered in this theoretical analysis, appear to play an important role.

1. INTRODUCTION

The detrimental influence of particles during the production of Integrated Circuits (I.C.s or "chips") in the electronic industry is twofold: of all I.C.s produced the fraction that is well-functioning decreases (direct influence) and the stability and reliability of this fraction decreases (indirect influence)[1]. For Random Access Memories (RAMs), the minimum particle size that has a direct detrimental influence during production is assumed to be about 10%-20% of the minimum pattern size of the I.C.[2]. For I.C.s to be produced in the near future, i.e. the 1 megabit Static RAM and the 4 megabit Dynamic RAM, the minimum pattern size is 0.7 μm. Particles larger than 0.1 μm will then be critical. There is a general consensus that the minimum pattern size will continue to decrease for many years to come and so will the critical particle size.

Until now, most of the attention has been focussed on the <u>prevention</u> of particle deposition on silicon substrates, by filtering air and processing fluids, by proper clothing for production personnel etc. <u>Removal</u> of particles from the substrate is one of the aims of the cleaning procedures during processing. However, the minimum size of the particles that can be removed efficiently by generally applicable cleaning techniques is about 1 μm. Stowers and Patton[3,4] have compared cleaning methods by measuring the particle removal efficiency on glass and metal surfaces contaminated artificially with alumina particles. For alumina particles larger than 5 μm removal efficiencies were determined, e.g., after spraying with Freon TF (trichlorotrifluoroethane) at 350 kPa and at 6.9 MPa both for 5-30 seconds and after ultrasonic agitation in Freon TF for 1-2 minutes. The values obtained were 97%, 99.7%-99.9% and 24%-92% respectively. The same cleaning procedures tested for alumina particles larger than 1 μm gave removal efficiencies of 3%, 81% and 1% respectively. For megasonic cleaning, effective particle removal down to about 0.3 μm is claimed by the inventors[5], but this has not yet been confirmed by others. It can, therefore, be stated that cleaning techniques which effectively remove particles down to at least 0.1 μm are lacking.

It is worthwhile to elucidate the existence of a lower limit of the particle size, for which only a small fraction of the particles can be removed with a given cleaning method. For particles smaller than about 10 μm, adhesion is primarily caused by Van der Waals forces[6]. For the idealized case of a spherical particle adhering to a flat substrate the following relation can be used[6]:

$$F_A = - \frac{A R}{6 H^2} \tag{1}$$

where F_A is the London-Van der Waals force, A the Hamaker constant, R the particle radius and H the distance of closest separation between the particle and the substrate. Equation (1) shows that at constant separation H, $|F_A|$ decreases linearly with decreasing particle size. A close to linear relationship has also been observed experimentally[6,7]. During cleaning a force is exerted on a cross-section of the particle or on the particle volume and this cleaning-force is therefore proportional to the second or third power of the particle size respectively. A cleaning technique that generates forces which are just large enough to remove particles of 1 μm from a substrate will therefore be ineffective in the removal of similar particles of 0.1 μm under similar conditions.

A fascinating observation concerning the transport of submicron particles adhering on a substrate was made by Visser[8]. He very briefly reported that graphon (i.e. a special type of carbon black) particles of 0.4 μm adhering on a disc rotating in water were removed from the region where an air bubble had touched the glass or cellophane disc surface. No explanation whatsoever was given for this phenomenon. In this paper, I consider the forces which act on particles (adhered on a substrate) during the passage of a phase boundary. Both a theoretical analysis and preliminary experimental results are presented.

2. THEORY

2.1. Analysis of Forces

First the forces which are acting on a particle floating at a liquid-gas interface are analysed. Then a particle adhered on a substrate is considered, during the passage of a liquid-gas phase boundary. It should be noted that the results obtained can be applied equally well to liquid-liquid phase boundaries.

In Fig. 1 a spherical particle is shown in its equilibrium position at the liquid-gas interface. In this case the density of the particle (ρ_p) is assumed to be higher than the density of the liquid (ρ_l). The position of the particle with respect to the interface is given by Ø, i.e., the angle between the vertical and the line going from the centre of the sphere to a point where the liquid-gas interface meets the particle. In this paper the assumption is made that for a given system the contact angle θ (defined in Fig. 1) has a fixed value. The following forces acting on the particle can be distinguished (forces pointing upward are taken positive)[9,10].

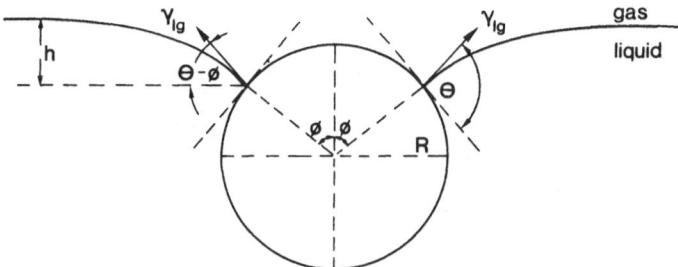

Fig. 1. Spherical particle floating at a liquid-gas interface.

A. The gravity force

$$F_G = - \frac{4}{3}\pi R^3 \rho_p \ g \tag{2}$$

where R is the particle radius and g is the gravitational   acceleration.

B. The buoyant force acting on the immersed part of the sphere

$$F_B = \frac{\pi}{3} R^3 \rho_l \ g \left(-\cos^3\emptyset + 3 \cos \emptyset + 2\right) \tag{3}$$

C. The level of the liquid at the point of contact with the particle differs from the level of the undistorted liquid. As a result a liquid column of height h (see Fig. 1) presses on a surface of $\pi (R \sin \emptyset)^2$. This gives the hydrostatic force F_H as

$$F_H = \pi R^2 \sin^2\emptyset \ h \ \rho_l \ g \tag{4}$$

D. The surface tension of the liquid-gas interface (γ_{lg}) acts on the particle along the line of contact between the particle and the liquidgas interface (i.e. $2\pi R \sin \emptyset$). Only the fraction of this tension that is oriented vertically (i.e. $\sin(\theta-\emptyset)$) is effective. Therefore :

$$F_\gamma = 2\pi R \ \gamma_{lg} \sin \emptyset \ \sin(\theta-\emptyset) \tag{5}$$

It can be shown[10] that for particles smaller than about 10 μm, F_G, F_B and F_H are several orders of magnitude smaller than F_γ, provided that $\emptyset$ is not too close to 0° or θ or 180°. This can be understood by noting that F_G, F_B and F_H decrease much faster with decreasing particle size than F_γ. As only particles smaller than 10 μm are considered in this paper, no more attention will be paid to F_G, F_B and F_H.
    Fig. 2 shows the sequence of situations occuring during immersion of a very small spherical particle in a liquid, when 0° < θ < 180°. In Fig. 2A no contact has yet been established. In Fig. 2B the particle is drawn into the liquid by F_γ, because $\emptyset$ is larger than θ. In Fig. 2C the equilibrium position of the particle is given ($\emptyset$ = 0). When the particle is lower than its equilibrium position (as in Fig. 2D), F_γ is directed upward ($\emptyset$ < θ). In Fig. 2E the particle has passed the phase boundary. From Fig. 2 it becomes evident that a sufficiently small particle for which 0° < θ < 180° has a stable position at the liquid-gas interface.
In Fig. 3, $F_\gamma/2\pi R \ \gamma_{lg}$ is given as a function of $\emptyset$ for different values of θ, according to equation (5). It can easily be proved that for a sphere that is pushed into the liquid, the maximum counter-acting force developed is

$$F_\gamma \ max = 2\pi R \ \gamma_{lg} \sin^2(\frac{\theta}{2}) \tag{6}$$

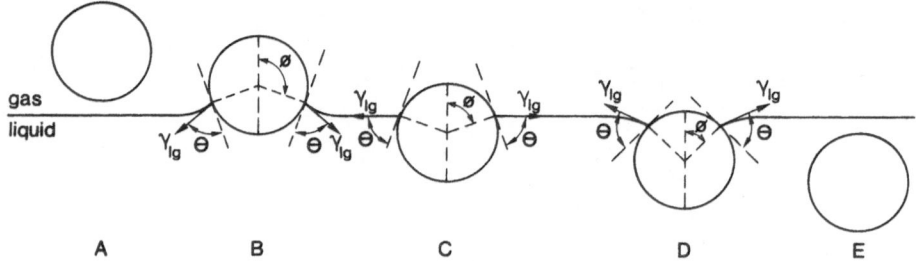

Fig. 2. Different stages of immersion of a particle in a liquid.

The maximum counteracting force that is developed when the particle is pulled out of the liquid is given by :

$$F_{\gamma} \, max = -2\pi \, R \, \gamma_{lg} \, sin^2(90° + \frac{\theta}{2})$$ (7)

For a particle adhering to a substrate, the conditions are somewhat different. The liquid-gas interface close to the substrate is curved, if the contact angle of the liquid with the substrate (α) is unequal to 90°. Then, contrary to the situation given in Fig. 1, the liquid-gas interface close to the particle (but far enough away to be undistorted by the presence of the particle) is no longer horizontal. For the liquid-gas interface a plane is now taken, that makes an angle α with the substrate (see Fig. 4). This planar shape can be justified by noting that the size of the particle is much smaller than the extension of the meniscus, so that the particle will experience only a very weakly curved surface. When α is unequal to 90°, F_{γ} will still be perpendicular to the undistorted liquid-gas interface, but will not be oriented vertically. This gives the opportunity to create situations for which the force of adhesion (F_A)

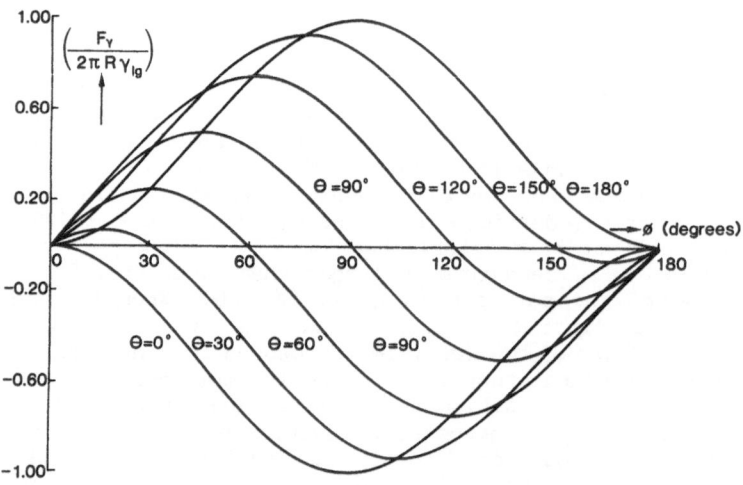

Fig. 3. The vertical component of the liquid-gas surface tension acting on the particle and expressed in units of $2\pi \, R \, \gamma_{lg}$ as a function of $\emptyset$, for different values of θ (according to equation (5)).

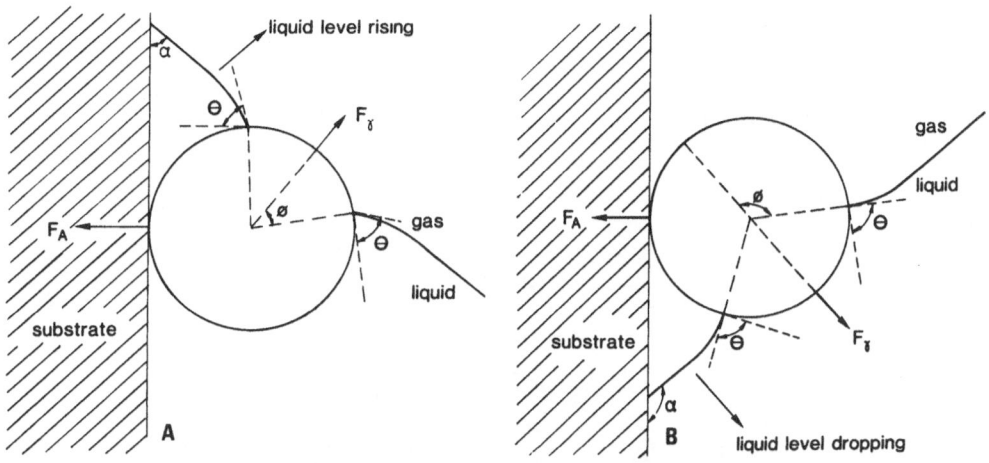

Fig. 4. Spherical particle adhering on a substrate during passage of a
phase boundary. Schematic presentation of conditions in which
F_A is counteracted by a component of F_γ. For Fig. 4A these
conditions are : $\alpha < 90°$ and $\emptyset < \theta$ (liquid level rises) and for
Fig. 4B : $\alpha > 90°$ and $\emptyset > \theta$ (liquid level drops).

is counteracted by the horizontal component of F_γ i.e. F_γ^x. There are
two such situations, which are presented in Fig. 4A and 4B. In Fig. 4A
the following conditions are fulfilled : $\alpha < 90°$ and $\emptyset < \theta$ (the level of
the liquid rises with respect to the adhered particle). The conditions
given in Fig. 4B are : $\alpha > 90°$ and $\emptyset > \theta$ (the level of the liquid drops
with respect to the adhered particle). The component of F_γ that opposes
F_A is $F_\gamma \cos\alpha$ and $-F_\gamma \cos\alpha$ for the cases given in Fig. 4A and Fig. 4B
respectively.
As defined in equation (1) the forces perpendicular to the substrate and
pointing to the liquid or gas phase are taken positive. Using equations
(6) and (7), the maximum force counteracting F_A then becomes

$$F_\gamma^x \text{ max} = 2\pi R \gamma_{1g} \sin^2\left(\frac{\theta}{2}\right) \cos\alpha \qquad (\alpha < 90°) \qquad (8)$$

and

$$F_\gamma^x \text{ max} = -2\pi R \gamma_{1g} \sin^2\left(90°+\frac{\theta}{2}\right) \cos\alpha \qquad (\alpha > 90°) \qquad (9)$$

Fig.5 gives F_γ^x max$/2\pi R \gamma_{1g}$ a function of θ for different values of α.
To determine the conditions for which F_γ^x max is large enough to sur-
mount $|F_A|$, proper values for A and H (see equation (1)) and γ_{1g} are
required. A reasonable value[11] of A for metallic and oxidic particles
adhering to a silicon substrate immersed in water is 15.10^{-20} Joule. For
non-immersed systems, A is somewhat larger. For H the value of 1 nm is
taken, because adhesion forces corresponding to H-values quite close to 1
nm are mostly determined[6,12]. Taking 0.07 Nm^{-1} for γ_{1g} (which is about
the surface tension of water at room temperature), the adhesion force can
be expressed as $F_A/2\pi R \gamma_{1g}$. The dashed line given in Fig. 5 shows
the result. If α and θ are chosen properly F_γ^x max is larger than $|F_A|$.
Under these conditions the particle adheres more strongly to the liquid-
gas interface than to the substrate and the particle will be removed from
its place.

365

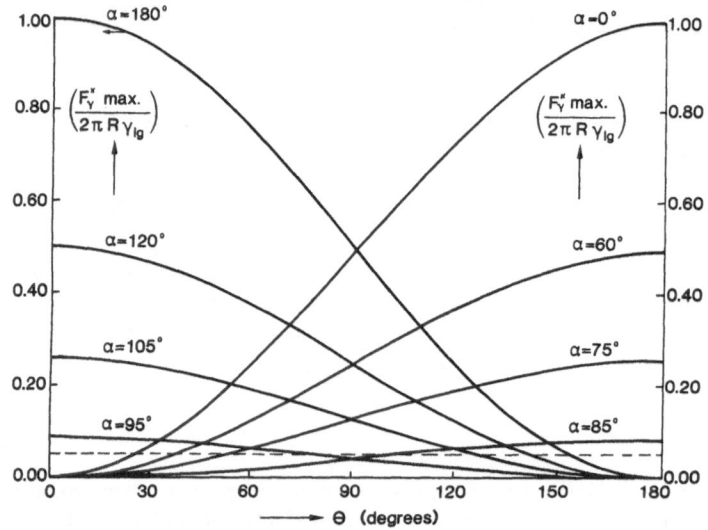

Fig. 5. Plot of equations (8) and (9). The dashed line represents $|F_A|$ divided by $2\pi R \gamma_{lg}$.

2.2. Implications of the Theoretical Analysis

The main feature is that F_A and F_γ^x max are both proportional to R. Therefore, contrary to all other cleaning techniques, any method based upon the principles of section 2.1 is in theory equally effective for micron-sized as for sub-micron particles.

The observation of Visser, concerning the removal of graphon particles from a substrate (see section 1) can now be explained. Here the situation given in Fig. 4A is applicable. The contact angle between graphon and water is known[13] to be 82°, while α will be well below 90°. According to Fig. 5, F_γ^x max is then larger than $|F_A|$.

To test the theory some preliminary experiments were carried out. Special particles and substrates were selected, for which F_γ^x max is expected to be larger than $|F_A|$. In practice however, one can not choose the particles that are on the substrate. Therefore the following test was also performed. Silica particles were deposited onto a silicon substrate. For water α and θ are 0° or close to 0° and removal by moving phase boundaries should not be possible. When this substrate with particles is treated with a silane it is in principle possible to increase both α and θ and bring θ to a proper value. The increase of hydrophobicity is a consequence of the replacement of OH-groups by chemically bonded hydrocarbon groups or fluorinated (hydro)carbon groups.

3. EXPERIMENTAL

The substrates used were monocrystalline silicon wafers of 100 mm with a (1,0,0) crystal orientation, from Wacker Chemitronic Company (Burghausen, West Germany). Pieces of 9x9 mm were prepared with a diamond saw. Sawdust was removed from the substrates as good as possible, with ethanol-saturated lens-tissues. The particles used were TiO_2 (rutile), α-Fe_2O_3 (hematite) and SiO_2 (amorphous silica). The TiO_2 and α-Fe_2O_3 powders were taken because they have suitable wetting properties : when some powder is gently spread on water it remains floating on the surface. This implies that the particles have a contact angle with water larger than 0°. The TiO_2 was obtained from Kronos-Titan, Leverkusen,

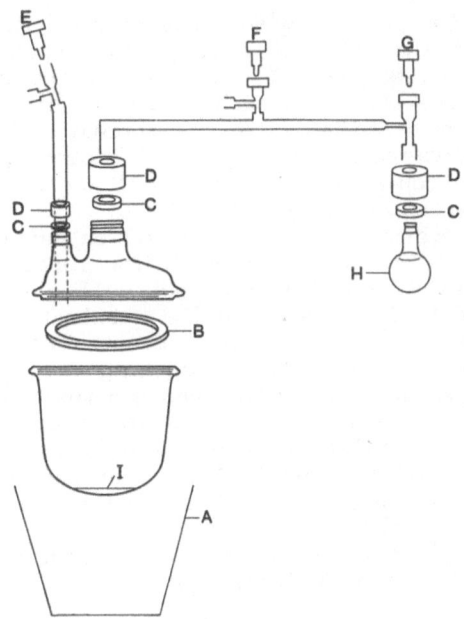

Fig. 6. Reactor used for the silylation experiments : A: Oil bath; B: O-ring; C: connection piece; D: screw; E, F and G: valves, H: Vessel containing silane; I: substrate.

West-Germany (type RLK) and was suspended in water by adding 3.4 mg $Na_4P_2O_7.10H_2O$ (Merck, p.a.) per gram TiO_2. The α-Fe_2O_3 powder was prepared via α-FeOOH (goethite) along the NaOH preparation route[14]. The α-FeOOH powder was then heated at 200°C, giving α-Fe_2O_3. This powder was suspended in water and the suspension was adjusted to pH 3 by adding HCl (Merck, p.a.). The suspension concentration in both cases was about 1% (v/v). To obtain the colloidal fraction of the suspensions, their supernatant was taken after they had settled for a few days at normal gravity. The silica suspension was prepared by the procedure reported by Stöber[15]. Spherical particles of about 0.7 µm were obtained.

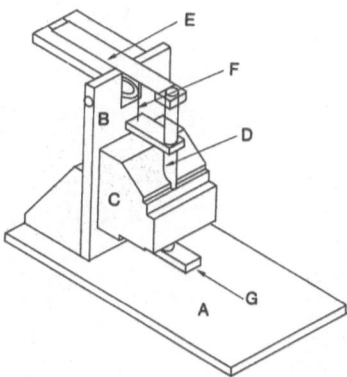

Fig. 7. Apparatus used to move the substrates through the liquid-gas interface at a controlled speed. A: ground-plate; B: mounting-plate; C: motor with friction wheels; D: shaft; E: leaf-spring; F: wire leading to counter weight; G: plate on which substrates are mounted.

The particles were applied to the substrate by placing a 2 µl drop of the suspension on the substrate. To promote spreading of this drop some ethanol was added. No ethanol was added to the drop of the silica suspension because here the dispersion medium consists largely of ethanol. The samples were dried at room conditions. The particles on the silicon substrates were detected by optical microscopy (dark field). A Philips Scanning Electron Microscope (SEM-500) was used to determine the sizes of the particles on the substrate.

To increase the hydrophobicity of the silica particles, the substrates covered with particles were silylated. The reactor used is shown in Fig. 6. Before the reaction was carried out, all glassware was dried in an oven at 125°C for half an hour and the substrates were treated in a UV/ozone reactor (UVP Inc., PR-100) for 5 minutes to remove organic contaminants. The reactor was evacuated by opening valve E, keeping F and G closed. When a pressure of 5 Pa was reached and the oil-bath attained the desired temperature, valve G was opened. The reagent used was (3,3,3 -trifluoropropyl)trichlorosilane from Petrarch Systems Inc. (Bristol, PA, U.S.A.). The reaction with $CF_3(CH_2)_2SiCl_3$ vapour was carried out at 100°C for 2 hours. The contact angles of water with the silylated substrates were measured directly[16] from a sessile drop of 5 µl, using a Ramé-Hart goniometer. The contact angle of water with a single uniform substrate could be measured with a reproducibility of 1°.

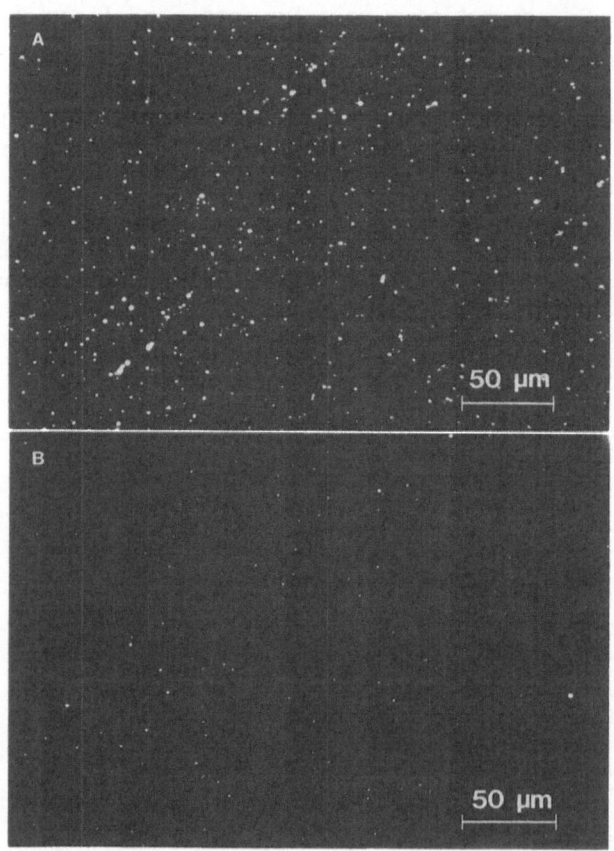

Fig. 8. Dark-field pictures of a silicon substrate covered with TiO_2 particles before (Fig. 8A) and after (Fig. 8B) passage of the water-air phase boundary.

The substrates were mounted on an apparatus (see Fig. 7) to move them vertically through the water-air interface at a controllable speed. Water was deionized and filtered through a 0.2 µm millipore filter before use.

4. RESULTS AND DISCUSSION

The preliminary experiments reported here were carried out with substrates with α-values less than $90°$. According to Fig. 4A, the critical moment then occurs during the process of pushing the substrate into the water. Unless stated otherwise, this process was performed as follows. The substrate was pushed into water at a speed of 3 $µm.s^{-1}$ and when the substrate was entirely submerged it was withdrawn at a speed of 1.5 $mm.s^{-1}$.

Fig. 8 shows the result for TiO_2 particles on silicon. About 85% of the TiO_2 particles were removed. Because the particles are shown in dark field, no information on their size can be obtained. Therefore a sample was investigated with SEM. Clusters of particles with a cluster size between 0.1 µm and 1 µm and more or less spherically shaped were pre-

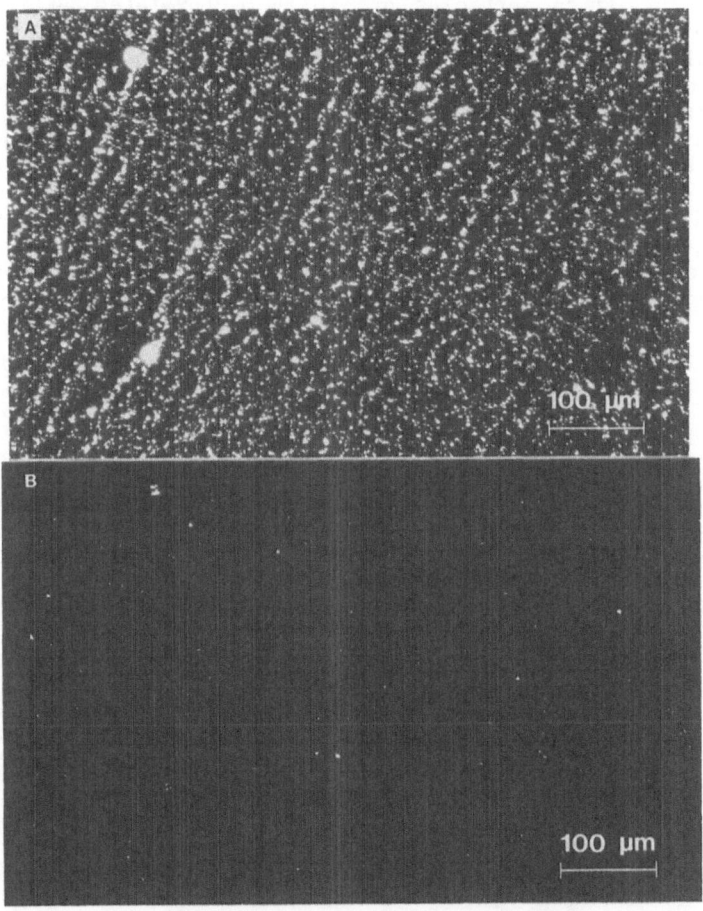

Fig. 9. Dark-field photographs of a silicon substrate covered with α-Fe_2O_3 particles before (Fig. 9A) and after (Fig. 9B) passage of the water-air phase boundary.

sent. For α-Fe$_2$O$_3$ particles the passage of the phase boundary results in removal of about 97% of the particles. Places where particles remain (Fig. 9B) are mostly those where previously (Fig. 9A) large agglomerates were present. Apparently only part of the agglomerate is removed. This might be caused by the particle-particle bonds being weaker than the particle-surface bonds. It is shown by SEM that the α-Fe$_2$O$_3$ crystallites are needle-shaped, have a length of about 0.8 µm and a width of about 0.1 µm. A substantial fraction of the particles is present as individual crystallites and most aggregates consist of a few crystallites only.

The following experiment was carried out to determine the direction of transportation of the particles. A substrate covered with α-Fe$_2$O$_3$ particles was pushed into water slowly. When about half the substrate had passed through the phase boundary, the substrate was quickly withdrawn from the water. Fig. 10 shows the boundary region between the immersed part (lower part) and the non-immersed part (upper part). The boundary line is very sharp and the region immediately above this boundary line does not contain more particles than the average on the upper part. This implies that the particles are really removed from the substrate and not transported along the substrate.

When a silicon substrate covered with SiO$_2$ particles is pushed into water slowly and withdrawn fast, practically no particles are removed. This is expected according to Fig. 5. Here α is about 20° and θ is about 0°, because the silica particles are fully hydroxylated as a result of the wet-chemical preparation method. A substrate treated with CF$_3$(CH$_2$)$_2$SiCl$_3$ gives a contact angle of 76°±5°. For a substrate that is first covered with silica particles and then silylated, the effect of the passage of the phase boundary is shown in Fig. 11. About 70% of the particles are removed. Inspection by SEM shows that the silica aggregates present on the substrate are clusters of a few particles, with a cluster size of about 1-2 µm.

For all cases investigated it was found that almost no particles were removed if the substrate was pushed into water very fast by hand (several cm.s^{-1}) and pulled out manually at a speed of a few mm.s^{-1}. Also

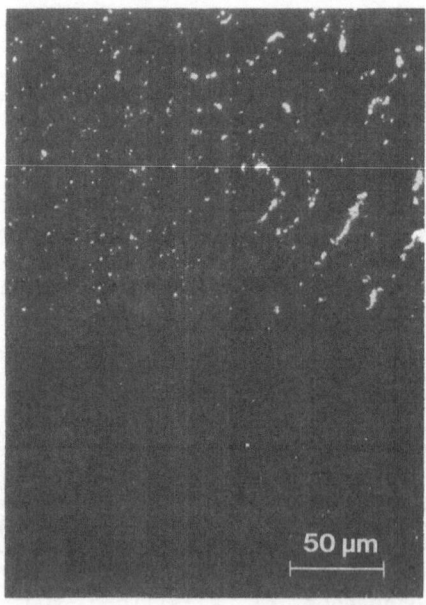

Fig.10. Boundary region of the immersed (lower part) and the non-immersed (upper part) part of the substrate with α-Fe$_2$O$_3$ particles.

no particles were removed if the substrate was immersed into water very fast and left there for an hour. Therefore no spontaneous repeptization (i.e. that the particles move spontaneously into the liquid phase) of the particles occurs.

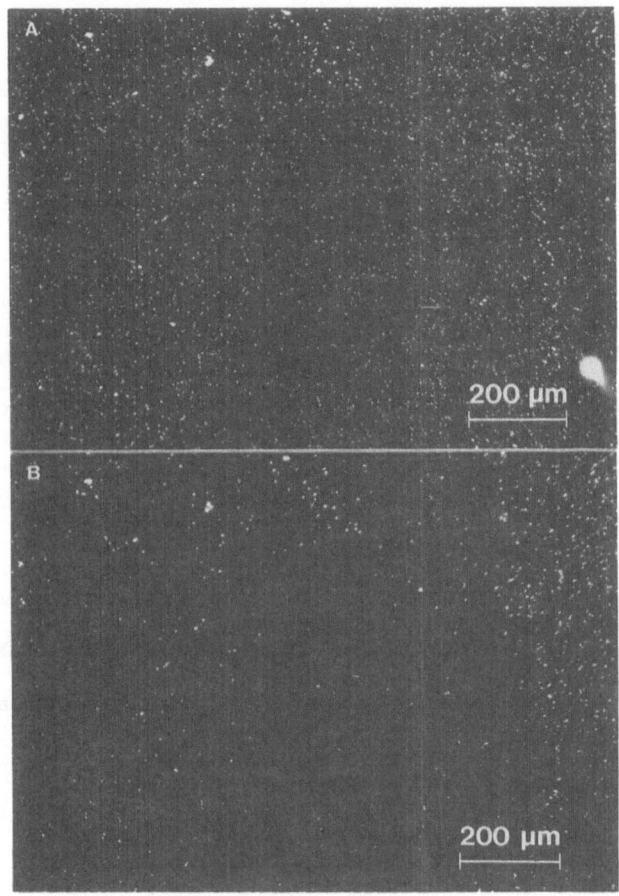

Fig.11. Dark-field pictures of silicon covered with silica particles (and subsequently silylated) before (Fig. 11A) and after (Fig. 11B) passage of the water-air phase boundary.

5. CONCLUSIONS

1. It has been shown both theoretically and experimentally that sub-micron particles can be removed from a (silicon) substrate, by the passage of a liquid-gas phase boundary.
2. Contrary to all other cleaning techniques, the cleaning effect of a method based upon the principles of moving phase boundaries is in theory independent of the particle size.

3. Important parameters governing this process are the wetting properties of the substrate and the particles. For silica particles on a silicon substrate these wetting properties can be brought into the proper range of values by carefully controlled silylation.
4. The kinetics of the process appears to be of prime importance. Almost no cleaning effect is observed when the phase boundary passes at a speed of several $cm.s^{-1}$, whereas efficiencies of 70%-97% were observed at a speed of 3 $\mu m.s^{-1}$.

ACKNOWLEDGEMENT

Thanks are due to Dr. J. Visser (Unilever Research Labs, Vlaardingen, The Netherlands) for a stimulating discussion, to Dr. J. Opitz (Philips Research Labs., Aachen, West Germany) for supplying the silica sample, to J.J. Ponjeé for advice regarding the silylation experiments and to Ing. J.A.M. Huethorst for carrying out the silylation reactions.

REFERENCES

1. A.V. Ferris-Prabhu in "Surface Contamination : Genesis, Detection and Control", K.L. Mittal, Ed., Vol. 2, p.925, Plenum Press, New York, 1979.
2. K. Oshige and T. Kawamata in "Proc. of the 6th Intern. Symp. Contamination Control", Tokyo, 1982, pp. 223-228.
3. I.F. Stowers and H.G. Patton in "Surface Contamination : Genesis, Detection and Control", K.L. Mittal, Ed., Vol. 1, p.341, Plenum Press, New York, 1979.
4. I.F. Stowers, J. Vac. Sci. Technol. 15, 751 (1978).
5. S. Shwartzman, A. Mayer and W. Kern, RCA Review 46, 81 (1985).
6. J. Visser in "Surface and Colloid Science", E. Matijevic, Ed., Vol. 8, p.3, J. Wiley & Sons, New York, 1976.
7. A.D. Zimon, "Adhesion of Dust and Powder", 2nd. edition, Plenum Press, New York, 1982.
8. J. Visser, Ph.D. thesis, Council for National Academic Awards, London, 1973.
9. H.M. Princen in "Surface and Colloid Science", E. Matijevic, Ed., Vol. 2, p.1, J. Wiley & Sons, New York, 1969.
10. H.J. Schulze in "Developments in Mineral Processing", Vol. 4, D.W. Fuerstenau, Ed., Elsevier, Amsterdam, 1984.
11. J.Th.G. Overbeek, Powder Technology, 37, 195 (1984).
12. M. van den Tempel, Adv. Colloid Interface Sci. 3, 137 (1972).
13. A.C. Zettlemoyer, J. Colloid Interface Sci. 28, 343 (1968).
14. C.J. Klomp and G.W. van Oosterhout, U.S. Patent 3,288,563 (1966).
15. W. Stöber, A. Fink and E. Bohn, J. Colloid Interface Sci. 26, 62 (1968).
16. A.W. Adamson, "Physical Chemistry of Surfaces", 3rd edition, J. Wiley & Sons, New York, 1976.

ABOUT THE CONTRIBUTORS

CHARLY D. ALLEMAND is a self-employed optical consultant. He received a
Dr. es Sc. from the University of Neuchatel, Switzerland, in 1954. His
research and development includes pyrometry, inspection instrumentation
for the silicon industry, spectrometric elemental analysis and applied
statistics. He has published several papers in these fields and has
invented or co-invented new instruments in emission -, Raman -, and
fluorescence - spectrometry and in contamination-inspection.

JUAN BARDINA is president of Atcor Corporation, a manufacturer of
equipment and instruments to detect and control contamination, in
Mountain View, CA. He holds Ph.D. MSCE and MSME degrees from Stanford
University and has published a number of articles on research related to
fluid mechanics, thermosciences, and surface contamination.

JOSEF BERGER is a founder and President of VLSI Standards, Inc. He
started his career as a research associate at the Integrated Circuits
Laboratory of Stanford University. He spent eight years in The IC
Operations of Hewlett-Packard Company, and was also Director of
Semiconductor Operations at Trilogy Systems. In 1984 he founded VLSI
Standards, Inc., the first company dedicated to production of measurement
standards and preprocessed test wafers for semiconductor industry.

WILLIAM S. BICKEL is a professor of Physics at the University of Arizona
in Tucson, AZ, where he has been since 1965. He received his Ph.D. in
Physics from Pennsylvania State University in 1965. He has been a
Visiting Research Scientist at the Research Institute for Atomic Physics
in Stockholm, the H.C. Oersted Institute in Copenhagen, the Faculte des
Sciences in Lyon, France, and the Solar Energy Research Institute in
Golden, CO. His present research interests include the experimental areas
of elastic polarized light scattering from particulates, fibers and
surfaces; acoustical measurements, noise pollution, and physics of music;
and adhesion forces between small particles and surfaces.

MARK BLITSHTEYN is Vice-President of engineering at The Simco Company,
Inc., Hatfield, PA, where he has been employed since 1979. He received a
diploma in electrical engineering from Leningrad Instrumentation
Institute in 1973. He is the author of several articles in the field of
applied electrostatics.

R. ALLEN BOWLING is presently a member of the technical staff at Texas
Instruments in Dallas, TX which he joined in 1980 as a process control
engineer in a semiconductor process area. Before joining TI, he spent 18
months in Frankfurt, W. Germany as an Alexander Von Humboldt Research
Fellow at the Institute of Inorganic Chemistry. He received his Ph.D. in
Chemistry from the University of Tennessee in 1979. During 1982-1984, he
supervised the Scanning Auger and Auger/XPS Laboratories at TI and since

that time he has been working on the development of new processes for ULSI.

M.R. COOK is Assistant Professor, Physics and Co-operating Professor, Chemistry at The University of Maine at Orono. He received D. Phil. from the University of York, U.K. (1982) and was IBM Research Fellow at the State University of New York at Stony Brook (1982-1983).

DOUGLAS W. COOPER is currently with the IBM T.J. Watson Research Center in Yorktown Hts., NY which he joined in Sept. 1983 to conduct contamination control research. Before joining IBM, he was on the faculty of Harvard for seven years. He was awarded a Ph.D. in Applied Physics (aerosol science) from Harvard in 1974. He has been involved in aerosol science research and mathematical modeling and analysis since the mid-1960's when he served at the U.S. Army's Biological Laboratories (Ft. Detrick, MD). He is the author or coauthor of over fifty articles published in scientific journals and has been an invited speaker at many meetings.

Hal COWLES is plant manager at Great Western Silicon, Chandler, AZ. Before that he spent one year at the J.C. Schumacher Company, where he carried out characterization of high purity materials for the semiconductor industry. He also spent seven years at the General Electric Company. He holds B.S. degrees on biology and chemistry.

EDWARD F. CUDDIHY is presently a member of the Technical Staff of the Applied Mechanics Technology Section at the Jet Propulsion Laboratory, Pasadena, CA. He received M.S. degree in 1959 in Chemical Engineering from The University of Notre Dame, and has been involved in a Ph.D. program at The University of Southern California in the field of Chemical Physics. He specializes in mechanical and physical properties of polymers, particularly in the analysis of problems and failures of polymeric materials in engineering, biomedical and spacecraft applications. Currently, he in involved in the "Flat-Plate Solar Array Project" (FSA) managed by JPL for DOE. Among his contributions to solar energy technology includes establishing the requirements for low-soiling surface coatings.

VICKIE L. DEBLER is a technician in the Center for Separation Processes Research at Research Triangle Institute, Research Triangle Park, NC. She has conducted adhesion experiments on particles on flat substrates in various media.

R.P. DONOVAN is a research physicist at Research Triangle Institute in Research Triangle Park, North Carolina. He received an M.S. (1959) in Physics from The University of Pennsylvania. He is currently project leader on a program sponsored by the Semiconductor Research Corporation whose purpose is to evaluate the deleterious effects of particles on silicon device manufacturing.

G.L. DYBWAD is currently with AT&T Bell Laboratories in Lee's Summit, MO. He has been with AT&T Bell Laboratories for 19 years. He received his Ph.D. in Physics from Kansas State University, and is the author of 16 papers, 3 patents and 60 internal memoranda.

DAVID S. ENSOR is director of the Center for Aerosol Technology at Research Triangle Institute in Research Triangle Park, North Carolina. He earned his Ph.D. in Engineering at The university of Washington in 1972. His present responsibilities include organizing and supervising research in microcontamination, particulate control devices, air pollution control, and chemical defense.

WILLIAM Y. FOWLKES is Senior Research Scientist at Eastman Kodak Comapny, Copy Products Research and Development Division, Rochester, NY. He received his Ph.D. at New York University in Chemical Physics. His research interests include electrostatics, powder electromechanics, and electronic processes in organic materials. He has published on photoemission from small organic particles and on electrophotography.

LEE K. GALBRAITH is Senior Scientist at Tencor Instruments, Mountain View, CA. He received his Ph.D. in Solid State Physics from Yale University in 1971. Prior to joining Tencor, he designed instrumentation in a number of other fields, including low-temperature physics, nonlinear optics, and high-vacuum surface measurements at Yale University, and meteorology, underground communication and solar energy at Sandia Laboratories in New Mexico and Livermore, California.

M. GRUNZE is Professor, Physics and Cooperating Professor, Chemistry at the University of Maine at Orono and scientist at the Fritz-Haber-Institute in Berlin, W. Germany. He received Dr. rer. nat. from the Free University of Munich in Physical Chemistry in 1974. He was a visiting scientist at IBM Research Laboratories in San Jose (1981) and visiting professor of physics at the University of Osnabruck, West Germany (1982).

DAN A. HAYS is a Principal Scientist in the Xerographic Technology Laboratory,Xerox Webster Research Center, Webster, NY. He received a doctorate in Physics from Rutgers University at New Brunswick in 1966. His areas of research interest include contact electrification, particle adhesion and xerographic development systems.

STUART A. HOENIG is Professor of Electrical and Computer Engineering, University of Arizona, Tucson, AZ. After serving as the Director of the Center for Microcontamination Control (CMC) for the inaugural year, he returned to full-time teaching and research. He received his Ph.D. degree from the University of California, Berkeley. At present he is involved in several programs on contamination detection and control. He has established a research company, Associates in Applied Research, Inc., devoted to the transfer of technology from the laboratory to industrial use.

ANNETTE JAFFE is with IBM Almaden Research Center, San Jose, CA and joined IBM Research in 1974 as a Research Staff Member. She received her Ph.D. in physical organic chemistry from Yale University in 1972 and had a postdoctoral fellowship at the University of Rochester. Currently she is involved in new development systems for electrophotography, the materials aspects of toners, carriers and fusing, contact electrification, and general aspects of non-impact printing.

WALTER JOHN is currently a Research Scientist with the Air and Industrial Hygiene Laboratory, California Department of Health Services, Berkeley, CA. He received his Ph. D. in Physics from the University of California, Berkeley. He is a Fellow of the American Physical Society, a member of a Scientific Advisory Committee of the Electric Power Research Institute and serves on Peer Review Panels for the U.S. Environmental Protection Agency. He is the author of over 70 publications concerned with nuclear physics, x-rays, particulate matter in ambient air, electrical charging of aerosols, filtration, particle-surface interactions and aerosol sampling techniques.

ARVIND KHILNANI is Research Fellow at the Institute For The Future (IFTF) in Menlo Park, CA. Before joining IFTF he was a consultant, performing technology and market assessments. In addition, he has served as Senior Scientist at VLSI Research and has taught in the Department of Engineering-economics at Stanford University where he earned his Ph.D. in the same field. His research centers on the electronics area, with particular emphasis on the computer, telecommunications, and semiconductor industries.

J. KOKOSINSKI is a graduate student in Chemistry at the University of Maine at Orono. She received her M.S. in Chemistry in 1980 from Illinois State University.

GEORGE KREN is Senior Manager for R&D at Tencor Instruments, Mountain View, CA. He received his MSEE at the Czech Institute of Technology in Prague in 1966. Until 1973 he worked for the National Research Institute for Materials, in the Non-destructive Testing Group. In 1974, he joined Perkin Elmer-Ultek where he worked in Sputtering Systems. In 1976, he co-founded Tencor Instruments. He is the author and co-author of more than 20 patents.

MICHAEL H. LEE is with the IBM Almaden Research Center, San Jose, CA. He obtained his Ph.D. in 1975 in Physics from the University of Illinois at Urbana-Champaign working on the light-emitting and lasing properties of III-V semiconductors. Since joining IBM Research Division as a Staff member in 1975, he has been involved in a number of areas including the magnetics of permalloy type materials, metallic thin films, and the physics of smectic liquid crystal displays. His current interest is in the physics of cleaning and development in electrophotographic systems.

AD F.M. LEENAARS has been working since November 1984 at Philips Research Laboratories, Eindhoven, The Netherlands on the subject of sub-micron particle removal from substrates. He studied soil chemistry and colloid chemistry at the Agricultural University in Wageningen, The Netherlands. His Ph.D. thesis dealt with the preparation, structure and separation characteristics of ceramic alumina membrane. This work was carried out at the Twente University of Technology, and he was recipient of the award given by the Shell Company to a Ph.D. student of this university.

ARMAND LEWIS is presently Research Affiliate with MIT Tribology Research Program. Before that he was a Research Associate with the Kendall Company, Special Ventures Group, Walpole, MA. Prior to joining Kendall, he was an Associate Professor of Chemical and Materials Science at Pennsylvania State University, Behrend College Campus, Erie, PA. He has 23 years of industrial experience with Lord Corporation (1971-1982) and American Cyanamid Company (1959-1971). He received his Ph.D. from Lehigh University in 1958 and has over 60 publications and patents in the fields of adhesion and composite materials. His current interests are in the Tribology of Magnetic Media Systems.

BRUCE R. LOCKE is a research assistant for the Department of Chemical Engineering at North Carolina State University, Raleigh. He received an M.S. (1982) in chemical engineering from the University of Houston. From 1982 to 1986 he served as an engineer at Research Triangle Institute where he helped develop an experimental system for the evaluation of particulate contamination in clean rooms. He has written 10 articles on particulate measurement in microelectronics clean rooms.

Mark LOGAN is the CVD Manager at Monkowski-Rhine, Inc., San Diego, CA. Prior to joining MRI, he was the manager of new product development at the J.C. Schumacher Co. working in the areas of particulate contamination and chemical delivery. He received his Ph. D. from U.C. Berkeley in physical chemistry and has authored over 10 scientific publications.

ANGEL M. MARTINEZ is currently with Emde Kinetic Systems, Santa Clara, CA. At the time this paper was presented, he was Corporate Sr. Yield Enhancement Engineer with Advanced Micro Devices in Sunnyvale, CA. He has background in Chemical Engineering from San Jose State University, 1977. He has ten years experience in the semiconductor industry.

VENUGOPAL B. MENON is a Research Chemical Engineer in the Center for Separation Processes Research at Research Triangle Institute, Research Triangle Park, NC, and is working on colloidal and interfacial phenomena associated with emulsions and fine particles, powder preparation, and powder characterization. His current research activities include contamination control in semiconductor processing, dispersion of powders in shock tubes, and measurement of particle/particle interactions in liquid media. He has a Ph.D. degree in Chemical Engineering from Illinois Institute of Technology, Chicago.

ROBERT J. MILLER is currently the manager of the Contamination Technology Group at IBM's T.J. Watson Research Center, Yorktown Hts., NY. He received his Ph.D. degree in Physics from the University of Illinois. He has conducted research and advanced development at IBM since 1976 in the fields of thin film metallurgy, electromigration and contamination control technology.

KASHMIRI LAL MITTAL* is presently employed at the IBM U.S. Technical Education in Thornwood, N.Y. He received his M.Sc. (First Class First) in 1966 from Indian Institute of Technology, New Delhi, and Ph.D. in Colloid Chemistry in 1970 from the University of Southern California. In the last 14 years, he has organized and chaired a number of very successful international symposia and in addition to this volume, he has edited 24 more books as follows: Adsorption at Interfaces, and Colloidal Dispersions and Micellar Behavior (1975); Micellization, Solubilization, and Microemulsions, Volumes 1 & 2 (1977); Adhesion Measurement of Thin Films, Thick Films and Bulk Coatings (1978); Surface Contamination: Genesis, Detection, and Control, Volumes 1 & 2 (1979); Solution Chemistry of Surfactants, Volumes 1 & 2 (1979); Solution Behavior of Surfactants - Theoretical and Applied Aspects, Volumes 1 & 2 (1982); Physicochemical Aspects of Polymer Surface, Volumes 1 & 2 (1983); Adhesion Aspects of Polymeric Coatings, (1983); Surfactants in Solution, Volumes 1, 2 & 3 (1984), Adhesive Joints: Formation, Characteristics, and Testing (1984), Polyimides: Synthesis, Characterization and Applications, Volumes 1 & 2 (1984); Surfactants in Solution, Volumes 4, 5 & 6 (1986); and Surface and Colloid Science in Computer Technology (1987). Also he is Editor of the Series, Treatise on Clean Surface Technology, the premier volume appeared in 1987. In addition to these books he has published about 60 papers in the areas of surface and colloid chemistry, adhesion, polymers, etc. He has given many invited talks on the multifarious facets of surface science, particularly adhesion, on the invitation of various societies and organizations in many countries all over the world, and is always a sought-after speaker. He is a Fellow of the American Institute of Chemists and Indian Chemical Society, is listed in American Men and Women of Science, Who's Who in the East, Men of Achievement and many other reference works. He is or has been a member of the Editorial Boards of a number of scientific and technical journals, and is the Editor of the Journal of Adhesion Science and Technology, which made its debut in 1987.

* As the editor of this volume.

Joseph R. MONKOWSKI is co-founder and president of Monkowski-Rhine, Incorporated, San Diego, CA, a firm which in partnership with Lam Research Corporation, is developing new technology for the chemical vapor deposition of thin films on silicon. Prior to co-founding Monkowski-Rhine, he was vice-president of research and development at the J.C. Schumacher Company, and before that, a professor of electrical engineering at Penn State University. He is advisory editor to Microcontamination and Solid State Technology magazines, and he is the author of over 50 technical and scientific publications.

MICHAEL E. MULLINS is a Research Chemical Engineer in the Center for Separation Processes Research at Research Triangle institute, Research Triangle Park, NC. He has worked extensively in the areas of powder preparation and characterization, particulate adhesion and multiphase dispersions. His current research interests include preparation of novel ceramic powders and dispersion of powders in shock tubes. He has a Ph.D. degree.

ARMAND P. NEUKERMANS is Vice President for R&D at Tencor Instruments, Mountain View, CA. He has a Ph.D. in Applied Physics from Stanford University. He has worked for General Electric, IBM, and Xerox in the area of magnetic recording and xerographic physics. Form 1973 to 1985 he was with Hewlett-Packard, most recently as a manager of the Applied Physics Department, with responsibilities for x-ray lithography and advanced test equipment. He is the author of 23 publications and holds 21 patents.

RICHARD W. NOSKER is a Fellow of the Technical Staff at the David Sarnoff Research Center in Princeton, NJ. He received his Ph.D. from Princeton University in 1970. His areas of interest include contact mechanics, adhesion, surface chemistry, the alignment of liquid crystal materials, triboelectricity, and lubrication.

David L. O'MEARA is a research engineer at the J.C. Schumacher Company, Carlsbad, CA and is involved in new process development and materials characterization. He holds an M.S. in materials science and engineering, from Stanford University.

I.W. OSBORNE-LEE is currently a member of the development staff of the Chemical Technology Division of Martin Marietta Energy Systems, Inc., at Oak Ridge National Laboratory, Oak Ridge, TN. He received his Ph.D. degree in 1985 from the University of Texas at Austin and has four publications in the area of mixed surfactant systems behavior.

JIRI PECEN is Electrical Engineer at Tencor Instruments, Mountain View, CA, which he joined in 1979. He received his MS in electronics from the Czech Institute of Technology, Prague. Until 1979, he worked for the Ore Research Institute in Prague on the development of new instruments for underground geophysical prospecting. He is co-author of several publications and patents.

CHARLIE A. PETERSON is presently Technical Feasibility/Market Analysis Manager at FSI International, Chaska, MN where he has been since 1981. Before his current position, he served in various capacities in the process laboratory in analysis and process and product development. He received his Ph.D. in analytical Chemistry from Iowa State University in 1977.

M.B. (ARUN) RANADE is President of Particle Technology, Inc., Washington D.C. and consults in the area of aerosol and particulate science. Prior to his current position, he was Director, Center for Separation Processes Research at the Research Triangle Institute (RTI) in Research Triangle Park, NC. He was educated in India and in the USA in chemical engineering and researched at IITRI, Chicago for a number of years before joining RTI. He is Adjunct Professor at the University of North Carolina as well as University of Maryland. He has published, lectured and consulted extensively on many aspects of particle technology and is internationally acknowledged in his field.

RANDALL J. ROGERS is presently a Section Head with the Kendall Company, Walpole, MA. Prior to joining Kendall, he was a Scientist with Polaroid Corp. He received his MS degree from MIT in 1973 and has publications concerned with organometallic chemistry and reaction mechanisms. His current interests involve nonwoven fabrics and their applications to data recording.

KELLY S. ROBINSON is currently employed as a research scientist with the Eastman Kodak Company, Rochester, N.Y. He received his Ph.D. degree in 1982 in Electrical Engineering from Colorado State University, Fort Collins. He has several publications in the electromechanics of granular material and is currently interested in electrostatics, the physics of electrophotography and powder electromechanics. He is a member of the Electrostatic Processes Committee of the IEEE Industry Applications Society.

TED ROSS is an Advisory Engineer at IBM in Hopewell Junction, NY, and has been with IBM for 29 years of which 15 years were in contamination control positions. He has received numerous patents and has had technical reports published on contamination. He also received IBM Corporate and Division awards for development of contamination detection systems. He has developed inspection systems using photographic technique, bright field, dark field and laser systems. Also he has initiated new software for interfacing detection systems with host computers to help correlate yield to contamination levels.

CLARENCE L. SMITH is an Advisory Engineer at IBM in Hopewell Junction, NY, and has been with IBM for 28 years. He received his MEE from Syracuse University. He has been in the MLC (Multi Layer Ceramic) Substrate Program for 14 years. For the past five years he has had primary responsibility for contamination control of the MLC Substrate Facility.

GARVIN J. STONE is a Laboratory Specialist in the Technology Assurance Laboratory, IBM Corp. in San Jose, CA. He graduated from U.C. Santa Barbara, and his specialties include multiple particle sampling and identification.

HWA-CHI WANG is currently with the Air and Industrial Hygiene Laboratory, California Department of Health Services, Berkeley, CA, which he joined in 1984. He received his B.S. in Civil Engineering from National Taiwan University and Ph.D. degree in Environmental Engineering from the University of Illinois at Urbana-Champaign. His research activities have concentrated on particle removal mechanisms, dynamic adhesion of particles upon surface impact, and evaluation of aerosol sampling and sizing instruments.

J.J. WEIMER is presently with the Fritz-Haber-Institute in Berlin, W. Germany. Before that he was Research Scientist, Laboratory for Surface Science and Technology, University of Maine at Orono. He received his Ph.D. in Chemical Engineering from MIT in 1983 and was a postdoctorate at the Technical University of Munich in the Physics Department (1982-1984).

THOMAS M. WENTZEL is a Graduate Research Assistant and a doctoral candidate in Physics at the University of Arizona in Tucson, AZ. He graduated cum laude from Bates College in Lewiston, Maine with his B.S. in physics in 1979. While at the University of Arizona working as a research assistant for Dr. John W. Robson he co-authored "The SYM-Timer" a highly versatile data acquisition program now used in the American Association of Physics Teachers Microcomputer Workshop. Currently, he is working with Dr. Bickel on a project involving the time dependence of adhesion forces between small particles.

RICHARD WILLIAMS is a Fellow of the Technical Staff at the David Sarnoff Research Center since 1958. He received his Ph.D. in Physical Chemistry from Harvard University. He is a Fellow of the American Physical Society and has been a Fulbright Lecturer in Brazil and a guest lecturer in Mexico and China. He has worked in the fields of liquid crystals, semiconductor-electrolyte interfaces, internal photoemission, colloidal crystals, electrical properties of insultars, surface chemistry and phosphor materials.

HENRY L. WOLFE is Senior Laboratory Specialist in the Contamination Technology Group at IBM's T.J. Watson Research Center, Yorktown Hts., NY. Before coming to IBM, he assisted in acoustical soundings and sound propagation studies in oceanographic research with Hudson laboratories for a dozen years. Since joining IBM in 1968, he has participated in research on Josephson junctions, magnetic materials, and the behavior of fine particles, and has coauthered several publications in these areas.